Science, Policy and Development in Africa

W0227550

Since gaining political independence in the 1950s, science has rapidly become a prerequisite for national development within many African nations. Supported by international agencies, such as UNESCO, initiatives were taken to direct Africa on the road of scientific development, enabling contributions to world science and significant progress in many specific research areas. However, from a developmental perspective, there remains the question of how science influences national development plans and strategies. How far are science policies integrated into the national development plans? What potential and challenges do science and technology pose for Africa and its prospects for wider development?

Offering a comprehensive historical and empirical study of science in both colonial and postcolonial Africa, R. Sooryamoorthy brings to light the connections between science, policy and development in African nations. Focusing on understanding the widening gap in science and technology between developed and developing regions, and the integration (or lack of) with national development strategies, this study provides important insights into the potential opportunities and challenges facing Africa in the areas of science.

R. SOORYAMOORTHY is Professor of Sociology at the University of KwaZulu-Natal, South Africa. A scientist accredited to the National Research Foundation, he is the co-author of *Science in Participatory Development* (1994) with Mathew Zachariah and author of *Transforming Science in South Africa* (2015), and *Networks of Communication in South Africa* (2017).

Science, Policy and Development in Africa

Challenges and Prospects

R. SOORYAMOORTHY
University of KwaZulu-Natal

CAMBRIDGE
UNIVERSITY PRESS

CAMBRIDGE
UNIVERSITY PRESS

University Printing House, Cambridge CB2 8BS, United Kingdom

One Liberty Plaza, 20th Floor, New York, NY 10006, USA

477 Williamstown Road, Port Melbourne, VIC 3207, Australia

314-321, 3rd Floor, Plot 3, Splendor Forum, Jasola District Centre, New Delhi - 110025, India

103 Penang Road, #05-06/07, Visioncrest Commercial, Singapore 238467

Cambridge University Press is part of the University of Cambridge.

It furthers the University's mission by disseminating knowledge in the pursuit of education, learning and research at the highest international levels of excellence.

www.cambridge.org
Information on this title: www.cambridge.org/9781108816144
DOI: 10.1017/9781108895804

© R. Sooryamoorthy 2020

First published 2020
First paperback edition 2022

A catalogue record for this publication is available from the British Library

ISBN 978-1-108-84203-7 Hardback
ISBN 978-1-108-81614-4 Paperback

To
My Parents

Contents

Figures

Maps

Tables

Preface

Science has been fascinating scientists for a very long time. Interminable possibilities for new discoveries and inventions continue to draw scientists to the world of science. Social scientists, on the other hand, are eager to know about scientists. They want to know what scientists are doing with science and how scientific knowledge is produced. For over two decades, this has been my primary interest. Having had opportunities to study science and scientists in different country contexts, I thought of focusing on a larger canvas. That was the beginning of the study of science in the African continent which is now presented in this book. It is not just about science and scientists in Africa. It is also about the developmental and policy implications of science for Africa. Obviously, it was not an easy task. The canvas was large, covering fifty-four countries on the continent.

No study is without challenges. Apart from the relevant literature, there should be sufficient empirical data for a study like this. Bibliometric records and data from several other sources had to be analysed. Thousands of bibliometric data have to be captured carefully to a data management programme. In order not to compromise the integrity of data, I could not ask any hired assistant to do data capturing. I chose to do it myself, which required a great deal of time. The availability of comparable data for most of the countries in Africa continues to be another handicap for studies on science.

The support I received from my wife was important for me while completing this work. My gratitude goes to my friend Geoff Waters who passed away in 2018. He was the first reader of the manuscript. Geoff was always a source of inspiration and encouragement. I also gratefully acknowledge the support I received from my institution.

It was a pleasure to work with the Cambridge University Press team. The project started with the commissioning editor, Maria Marsh. It was the concerted efforts of Daniel Brown who moved the book to the production line. Dan was a wonderful person to work with who was very prompt with communication. I am so grateful to him. My sincere gratitude also goes to Atifa Jiwa who worked with Dan for the production of this book. Thanks are also due to the excellent content manager, Natasha Whelan, and her production team.

Abbreviations and Acronyms

AAB	Academics Across Borders
AIMS	African Institute of Mathematical Sciences
ANSTI	African Network of Scientific and Technological Institutions
ARC	Agriculture Research Council
ARCT	African Regional Centre for Technology
ARSO	African Regional Organisation for Standardisation
ASEAN	Association of South-East Asian Nations
ASTIPI	African Science, Technology and Innovation Policy Initiative
AU	African Union
COSTECH	Tanzania Commission for Science and Technology
CPA	Consolidated Plan of Action
CSIR	Council for Scientific and Industrial Research
CCTA	Commission for Technical Cooperation in Africa
DHET	Department of Higher Education and Training
DST	Department of Science and Technology
EAC	East African Community
EATPS	Eastern African Technology Policy Studies
ECOWAS	Economic Community of West African States
EFTA	European Free Trade Association
GATT	General Agreement on Tariffs and Trade
GDP	Gross Domestic Product
GERD	Gross Domestic Expenditure on Research and Development
HERANA	Higher Education Research and Advocacy Network in Africa
HSRC	Human Sciences Research Council
ICT	Information and Communication Technologies
IITA	International Institute of Tropical Agriculture
ILCA	International Livestock Centre for Africa

ILRAD	International Laboratory for Research on Animal Diseases
ISHReCA	Initiative to Strengthen Health Research Capacity in Africa
KARI	Kenya Agricultural Research Institute
KEI	Knowledge Economy Index
KEMRI	Kenya Medical Research Institute
MDG	Millennium Development Goals
MOSTIS	Mozambique Science Technology and Innovation Strategy
NAFTA	North American Free Trade Agreement
NCST	National Council for Science and Technology
NEPAD	New Partnership for Africa's Development
NGO	Non-governmental Organisation
NPRSTI	National Programme on Research, Science, Technology and Innovation
NRDS	National Research and Development Strategy
NRF	National Research Foundation
NSI	National System of Innovation
OECD	Organisation for Economic Cooperation and Development
PASET	Partnership for Applied Science, Engineering, and Technology
PFI	Pact for Research and Innovation
PNDES	National Economic and Social Development Plan
SACU	Southern African Customs Union
S&T	Science and Technology
SADC	Southern African Development Community
SCA	Scientific Council for Africa
SCI	Science Citation Index
SKA	Square Kilometer Array
STEI	Science, Technology, Engineering and Innovation
STEM	Science, Technology, Engineering and Mathematics
STI	Science, Technology and Innovation
STISA	Science, Technology and Innovation Strategy for Africa
STISA-24	Science, Technology and Innovation Strategy for Africa 2024
TAI	Technology Achievement Index
TDCA	Trade, Development and Cooperation Agreement

TIMSS	Trends in International Mathematics and Science Study
TOKTEN	Transfer of Knowledge through Expatriate Nationals
TTIP	Transatlantic Trade and Investment Partnership
TYIP	Ten-Year Innovation Plan
UNCTAD	United Nations Conference on Trade and Development
UNESCO	United Nations Educational, Scientific and Cultural Organisation
VOIP	Voice over Internet Protocol
WARDA	West African Rice Development Association
WATPS	West African Technology Policy Studies
WFC	Weighted Fractional Count
WoS	Web of Science
WTO	World Trade Organisation

1 | *Science, Development and Africa*

Africa is the oldest of the continents, the birthplace of civilisations, the storehouse of the earliest remains of humans and possesses the most wonderful works of humans. This is what Gardiner Hubbard wrote about Africa over a century ago in 1889. For ages, Africa has refused to reveal its secrets to the world although explorers had penetrated from every side of the continent (Hubbard, 1889). Africa remains an inadequately explored and studied continent.

A great deal has been written and known about Africa but not much about science from Africa, or what Africa can give to science. Jan Hofmeyr, while serving as the President of the South African Association for the Advancement of Science in 1929, asked questions which are still relevant today: What can Africa give to science? What can science give to Africa (Hofmeyr, 1929a)? One might add, what can science do for the development of the continent? In the words of Mohamed Hassan, who was once the president of the African Academy of Sciences: 'Science alone cannot save Africa, but Africa without science cannot be saved' (Hassan, 2001: 1609).

Africa is a fruitful field for experiments and a place of expert scientific knowledge (Worthington, 1938). The continent once had some of the world's best science departments in its universities.[1] It made its quota of contributions to the solutions of some of the most fundamental problems of science (Worthington, 1938). The great challenge for the world's scientific community, as Nobel laureate David Gross believes, is to connect Africa to the rest of the international scientific world.[2] In a way, Gross was recognising the importance and relevance of African science and its significance for the world of science and for its own development.

---

[1] The universities, among others, include the universities of Lagos (Nigeria), Dar-es-Salaam (Tanzania), Accra (Ghana), and Khartoum (Sudan) (Hassan, 2001).

[2] Quoted in van den Brink and Snyman, 2007, p. 792.

Greg Mills (2016) in his book, *Why Africa Is Poor,* elaborates on why the continent remains poor. In his view, it is not because its people are poor but because their leaders made that choice. It is poor not because its people do not work hard. Rather, their productivity levels are abysmally low because of poor health, poor skills and inefficient land use. It is poor not because it lacks natural resources. Here, science comes to the forefront with a bigger role to play in the development of Africa.

Science is not the monopoly of a few countries and produced in select locations alone. Its presence is felt everywhere and in every field of human life. Dedijer (1968) estimated that between 15 and 30 of the 120 countries which had less than one-third of the world's population had practically all of its science and had spent more than 95 per cent of its R&D (Research and Development) funds. However, this ratio and proportion has changed in the contemporary world. There has been a global shift in the production of science and in the quantity of research publications and their impact (Radosevic and Yoruk, 2014).

The contribution of countries to world science has been changing. Over the years, there has been a shift in core and peripheral countries. Germany was a centre of world science in the 1930s but the United States expropriated that position in the post–Second World War period (Schott, 1998). China, which is a periphery country, has improved its share of the world science. The rise of China on the horizon of science has been striking, overtaking Japan and Europe in its publication output (The Royal Society, 2011).

Science, Growth and Development

The frontier of world science is moving, although not drastically. The production of science and the purpose for which it is done has been evolving. It is a public good. Science is becoming a part of commerce and market goods (Krishna, 2014). In this evolution, the primary purpose of science is changing from the advancement of knowledge more towards wealth (Krishna, 2014). The relevance of science in development comes at this juncture.

Science is a fascinating endeavour, capable of engaging men and women at their best, and enlarging and enriching the human spirit with discoveries (Ziman, 1984). It has become an integral and indispensable component of human history, progress and development.

Science offers a potential source for social change, capable of revolutionising areas of experience (Barnes, 1972) and contributes towards overcoming the conditions of underdevelopment (Sagasti, 1973). Science and its application through technology transforms the world at a rapid pace (IAC, 2004). The new knowledge that is being produced at an unimaginable rate reconstructs society, and influences the way we think about the world today (IAP, 2016) and the way we lead our lives.

While speaking at the International March for Science, Glenda Gray, the president of the South African Medical Council, said: 'Investing in research and development is about investing in the citizens of our country. Science changes lives, shifts paradigms of thought and promotes innovative economic progress' (Pillay, 2017: 3).

With the unprecedented growth in science and technology (S&T), particularly since the Second World War, the relationship between S&T and development has become stronger. In the 1950s and 1960s, the ability of science to deal with human problems generated great levels of optimism (Raina, 1999) and interest in science and development. The growing recognition of the role of S&T has coincided with the Millennium Development Goals as well (Juma and Yee-Cheong, 2005). Science has become a recipe to solve a range of problems of development (Krieger, 2000).

Regardless of the size and wealth of nations, science has evolved into a powerful and influential social institution, sought after by governments, and public and private sectors (Krishna, 2014). As a big global enterprise and a massive human undertaking, science is occurring in more and more locations with millions of researchers and billions of dollars being made available (Moravcsik, 1984; The Royal Society, 2011). There are 8.496 million full-time equivalent (FTE) researchers in the world (UIS, 2016a). The participation of countries in science has been widespread, as science has become a valuable possession to meet their needs (Moravcsik, 1986a). Millions of scientific publications are produced every year, more countries are increasing their participation in the production of science, and the amount of money spent on science is correspondingly increasing across the world. The effects of S&T are more powerful and pervasive than in the past (Rath, 1990). These will continue to increase in future.

Science is part of a whole system. The parts are interdependent. The institutionalisation of science has made it more intimately related to other institutions of society (Merton, 1969). Science maintains its dominant position as an integral tool for development and economic growth. Science is part of progress, for both individuals and nations alike (Schott, 1993). In his studies, Merton (1978) documented the interrelationship between economy and science. He observed the relation between science and economic needs in seventeenth century England. Socially patterned interests, motivations and behaviour in one sphere of life such as the economy are interdependent with the patterned interests, motivations and behaviour in other institutional spheres such as science (Merton, 1973). This shows the interrelatedness of economy, economic growth, development and science (Merton, 1973). In terms of wealth and technology, as Elmslie and Criss (1999) reported, science has advanced to a superior position relative to other countries, and its societal division of labour extended to all fields.

The interplay between socio-economic and scientific development is now well recognised (Merton, 1939, 1978). Scholars have identified the connections between science and several other components of development and growth (Carter, 1968; Enos, 1995; Nour, 2005, 2012). Advances in S&T are crucial for economic success (Enos, 1995), and S&T development is a key to economic growth, industrial competitiveness, social development and improvement of quality of life (Nour, 2005, 2012).

Science is the main driving force behind industrial and post-industrial societies (Krishna, 2014). At the same time, countries that have not adopted the usefulness of science have suffered economically (Debru, 2000). In the globalised economy, S&T are key drivers of economic growth and development (Carty, 2000; Freudenthal, 2014; Gazni and Ghaseminik, 2019; The Royal Society, 2011; Thulstrup et al., 1996; UNESCO, 2000). S&T are also the primary driving force of innovation, increased productivity, wealth creation and social welfare, and play a significant role in the economic performance of countries (ICSU, 2005; OECD, 2000). Evidence suggests that technological change accounts for an overwhelming share of the increase in production output (Rath, 1990).[3] This is obvious from the track record of

---

[3] Nour (2005) quoting *The Second European Report in S&T Indicators 1997* of OECD summarises that the 50 leading S&T countries in the world had long-standing economic growth which is much higher than that of the other 130 countries in the world. The average economic growth of these 50 countries during 1986–94 was three times greater than the rest of the world.

developed countries. Industrialised countries in the past have benefitted from the huge investments they had made in S&T (Nour, 2005, 2012). This is true not only for the developed countries but also for countries such as South Korea, India and Brazil which have all benefitted from their investments in S&T (Wagner et al., 2001).

Economic development requires direct engagement in the generation of knowledge (King, 2004), through which science can be fruitfully exploited to achieve economic growth (Dasgupta and David, 1994). The application and conversion of scientific knowledge into technological products facilitates growth. Indisputably, it is the engine that drives economic growth in developed countries.[4] Science contributes to significant economic growth and productivity (Wagner et al., 2001). The benefits for economic growth of applying science are evident in many areas: agriculture, industry and medicine (Williams, 1968) are but some of them. Any modest improvements in any field – for example, in healthcare, sanitation, drinking water, food and transport – require scientific capabilities in engineering, technology and medicine (King, 2004).

Science continues to be the single most influential force and will remain so for many more years to come (Lane, 2000). A host of factors contributes to the growth and development of science. Dedijer (1968) speaks about three major ones. First is the presence of a scientific community with its own institutions for training and research and its own scientific tradition. A government apparatus with a tradition in dealing with science is the second one. The last one is those institutions (industrial, agricultural, commercial, educational, medical and others) that have learnt the value of research and are able to make reasonable demands on the scientific community and government.

The growing importance of research institutions in the 1950s and 1960s has made people consider science as an essential tool for productivity and economic growth (Hetman, 1979). The productivity of a population is dependent, more than anything else, on advances in

---

[4] In the United States alone, for instance, technology has been responsible for two-thirds of the increase in economic growth (Wandiga, 2000). The per capita growth of the US economy, between 1870 and 1910, was primarily due to improvements in technological and managerial innovations (Colglazier, 1981). At the beginning of the century, the estimated scientific contribution to the Gross Domestic Product (GDP) in developed countries was 10 per cent, which had grown into 40 per cent later in the 1950s, and 60–80 per cent in the 1980s (Yongxiang, 2000).

S&T (Enos, 1995). There is evidence to support the relationship between economic wealth and scientific productivity, although the relationship has not always been positive (Schott, 1991).[5] The relationship between the stock of knowledge and the flow of wealth, and the relationship between the stock of knowledge and the expenditure on R&D have been well documented (Carter, 1968).

From as recently as the 1960s, there has been a general increase in the interest and awareness of science in developing countries. The 1960s were called the decade of awareness-building that occurred through many channels of conferences, workshops, plans and training (Moravcsik, 1975). While the economic benefits that S&T bring to developing countries are considerable, these are not the only reason or justification for them to focus on S&T (Moravcsik, 1982). Other evolving conditions in countries also play a role.

Science is an integral component of national development. Its role in the development of policies for national development has been emphasised. Even modest improvements in science can have remarkable effects on development, especially in poor countries (Kennedy, 2001). Science can provide the knowledge to make informed decisions (Lubchenco, 2000; Odhiambo, 2000) on policies pertaining to national development. Practically, every decision in any field of national endeavour requires scientific knowledge (Dedijer, 1968). Notions of scientific relevance focus less on the language of national growth but more on research towards finding solutions to challenges (Hackmann and Boulton, 2015). Ultimately, science is meant for the living conditions of people, meeting the developmental needs of society. The bond between S&T and the developmental needs of specific societies must be stronger than ever if the former is to serve humankind.

Developments in S&T have become a yardstick of measurement of modernity as well (Godin, 2005). Several dimensions of S&T can be listed. Moravcsik (1987) cites two of these. The first one is that S&T are crucial ingredients in the efforts to find a remedy for a broad range of immediate and pressing problems pertaining to the economy, health and many other aspects of life. The second one is the task of creating an indigenous capability in S&T which is essential for the

---

[5] Schott (1991) provides some specific instances. Japan has relatively little scientific work in relation to its wealth. Arab countries in North Africa and West Asia have the lowest scientific productivity in relation to their economic wealth. But India has the highest scientific productivity in relation to its wealth.

science-technology-production-living style complex, to participate in one of the respected and influential activities of the century, and the exercise of the impact of science for humankind. While S&T alone cannot solve all the complex problems humans face today, those problems cannot be solved without the help of S&T (Lane, 2000).

Increasingly, science leads technology in many areas and is a necessary condition to establishing sustained growth (Lipsey, 2001, 2009). Long-term economic growth and technologies, as Lipsey et al. (2005) argue, transform the entire set of economic, social and political structures of societies. Any technological change requires alterations in the structure of the economy (Lipsey et al., 2005).

One should take into account the way S&T spreads across the world, the levels of progress that have been accomplished in a variegated manner and the benefits those have brought to the well-being of people. But the use and function of science in developing and developed countries are not the same and are debatable (Lomnitz and Salord, 1992). The growing gap between developed and developing countries has also been reflected in a growing distance between countries in terms of S&T (Durant, 2000). Consequently, this gap in the availability and application of S&T is manifested in a range of scenarios, affecting the overall development of economy, society and the countries in general.

Science for African Development

In Africa, science and technological research and development priorities were not always made in line with priority needs (Thisen, 1993). S&T were never thought of as part of the development process in Africa (Forje, 1989). A development option, based on the recognition of scientific knowledge and the potential of Africa, was slow to begin (Worthington, 1938). The national policies of S&T in African countries, by and large, have not been achieved as yet. The integration of S&T in the national development plans of countries is one of the key functions of national S&T policy-making bodies (UNESCO, 1986). The challenge for Africa is to integrate S&T in development (Confraria and Godinho, 2015).

As Edgar Worthington (1952) observed, amidst arguments that there was development for years without any scientific background, the forms of science in Africa have to be closely related to development. Scientists themselves were concerned about the relevance and role of

S&T in society (Rath, 1990). The set of problems evident in Africa requires solutions through research in Africa and by Africans (Dow, 1988). Science has a major role in this, not just for Africa. The 2030 agenda for sustainable development, adopted by the UN in 2015, concedes that the Sustainable Development Goals cannot be achieved without the support of science. Science is critical to meeting the challenge of sustainable development; it lays the foundation for new approaches and technologies that can solve local and global problems (UNESCO, 2015a). Irrespective of the capacity of regions, development in S&T is a challenge for sustainable development (Malcom et al., 2002). Sustainable development can be achieved when there is enough capacity available for the creation and application of S&T in societies.

A principal factor that has been identified as a stumbling block for Africa to achieve economic growth is its shortcomings in S&T (*Current Science*, 2001, cited in Gaillard et al., 2005: 177; Forje, 1989; Gaillard, 2003a; UNESCO, 2000; Wad, 1984). These shortcomings are evident in several forms. There is a deficiency in S&T in African development, which exists at both theoretical and methodological levels (Wad, 1984). In order for Africa to advance, the efficient use of its technical resources is pivotal (World Bank, 1981). The position a country attains in the production of knowledge depends on the degree of technological sophistication it has achieved (Zegeye and Vambe, 2006). They are linked to the scientific capacity of a country. In order for science to advance, both the capacity and the ability of the country to create and absorb science are crucial. Regrettably, the capacity levels for the creation and (or) the absorption of imported S&T are low in underdeveloped countries (Sagasti, 1973). It is also essential to have policies in place that facilitate and encourage science and scientific development. For instance, the cause of underdevelopment in Egypt was inappropriate S&T policies (Zahlan, 1997).

As Wad (1984) points out, the relationship between S&T and socioeconomic development has not been well articulated in Africa. NEPAD (2005), in its report, admitted that science, technology and innovation (STI) have not received any serious attention as engines of development in Africa. One needs to know the history and the contemporary state of science in Africa, and its potential for socio-economic development to make sense of the relationship between S&T and development. The continent does not have a long history of doing science, as compared to about 400 years of history for the developed world. In many areas, science is at its infancy in this part of the world. This study looks into this relationship through the examination of the production of science in Africa.

The idea of science for development was not completely alien to Africa. The vibrant nature of science and science development is becoming more visible in Africa now. In his first presidential speech at the foundation summit of the Organisation of African Unity (OAU) in 1963, Kwame Nkrumah emphasised the possibility of S&T for the development of Africa (AU, 2014a).[6] Prior to the Lagos Plan of Action, the Monrovia Declaration in Liberia in (1979) was announced. The Monrovia Declaration adopted the resolution to put S&T in the service of development by reinforcing autonomous capacity for countries. By the decade 1970–9, several African countries were becoming conscious of the role of S&T for their development (OAU, 1981).

The indispensability of S&T is in laying the foundations for development and in solving fundamental problems (Barré and Papon, 1993). The emphasis of Africa should be primarily on making use of science for its developmental needs. Science affords the opportunity for such development (Merton, 1942), and social and economic development work hand in hand with scientific development (Chatelin et al., 1997).

The series of deliberate efforts under the auspices of UN agencies had an impact on increasing awareness about the role of S&T in development in the postcolonial period (Forje, 1989).[7] These were pursued by a number of missions dispatched to African countries to assist them with the formulation of national science policies (Forje, 1989).[8] Efforts were made to establish institutions for S&T. Despite these, African countries lacked the necessary scientific and technological capacity, resources and skills and were dependent on foreign countries (OAU, 1981). This was not the case at the time of independence. Science development was not a pressing issue or concern for Africa then

---

[6] The OAU became the African Union (AU) in 2002.

[7] Some of these include the UN conference on the application of S&T to the development of less-developed countries (Geneva in 1963), the international conference on the organisation of research and training in Africa (Lagos in 1964), the symposium on science policy and research administration in Africa (Yaoundé, 1967), the UN conference on S&T for development (Vienna in 1979), and the second conference of ministers of African member states responsible for the application of S&T to development (Arusha in 1987) (Forje, 1989).

[8] Forje (1989) records that there were twenty-seven such missions to thirty-two countries and organisations during 1964–87.

(Forje, 1989), and this remained the case for a long time. It was easier to acquire technology from developed countries rather than developing its own research capacity (Forje, 1989). This approach underestimated the part of science in development (Forje, 1989). Perhaps this was the mistake many African countries made on achieving their independence.

There has been a growing interest in S&T for development in Africa. Science and the production of scientific knowledge in Africa have become the interest and concern of scholars and policymakers (Sooryamoorthy, 2018). The role of S&T in economic growth and development for Africa is now being recognised (Wakhungu, 2001). S&T are appreciated as the cornerstones for progress in Africa through economic growth (ESTA, 2006). There is a consensus in African countries that scientific research is a requirement for the creation of long-term sustainable development (Confraria and Godinho, 2015). From this point of view, science is recognised to be crucial for Africa, to address its numerous developmental problems. A heightened interest in the relevance of S&T is obvious, particularly since the Lagos Plan of Action (ESTA, 2006; Odhiambo, 1967; Toivanen and Ponomariov, 2011; Wad, 1984). This was not the case in the early 1930s and the results of scientific research did not influence African development at all (Worthington, 1938).

The general consensus among political leaders, administrators and scientists in Africa is that accelerated science development is useful in dealing with social and economic problems (Odhiambo, 1967). The summit of the OAU had called for its member countries to ensure the development of an adequate S&T base and appropriate application of science in spearheading development in various fields (OAU, 1981). The OAU also developed a programme of action for S&T at its Lagos meeting. The plan emphasised the need for national policies on S&T to be integrated into the national development plan (OAU, 1981). In 2007, Africa declared it to be the year of S&T for Africa and urged countries to set the target of 1 per cent of the GDP (Gross Domestic Product) for R&D. Even earlier, at the first NEPAD ministerial conference on S&T held in November 2003 in Johannesburg, ministers made a commitment to increase funding for R&D to 1 per cent of the GDP. The Science, Technology and Innovation Strategy for Africa 2024 (STISA-24) also encourages the member countries of the African Union to allocate at least 1 per cent of their GDP to R&D. This is yet to be uniformly realised in Africa.

Countries are in the process of increasing the allocation of funds for their S&T portfolios. Efforts are being made to establish universities and scientific research institutions, produce graduates and scientists, and gather funds and resources. Plans are afoot in several African countries that will take them to new levels in science development. For instance, Rwanda has increased its GDP allocation to 1.6 per cent for S&T and is aiming to make it 3 per cent in the near future. Nigeria is investing in creating a national science foundation. Uganda is on the way to fund new research initiatives and Zambia is to offer fellowships to train science and engineering students (Hassan, 2007). At the same time, there are countries in Africa that still do not have a separate ministry for STI (Mouton et al., 2015), or have any proper science policy in place.

Investment in S&T alone is not adequate to produce expected benefits. Ghana, for example, found that its investments in S&T have not yielded the desired improvements in economic growth. Several other conditions should accompany investments in S&T to achieve the intended outcomes of economic growth. Among them are adequate scientific expertise, high levels of science culture in the society, better coordination and management, less reliance on the use of foreign expertise and strong linkages between industry and organisations (ROG, 2010).

The State of Science in Africa

Africa is the second largest continent in the world with fifty-four countries.[9] It is grouped as sub-Saharan countries, Median Africa and Maghreb countries or as north, west, east, central and southern Africa.[10] As per UN statistics, the population in Africa in 2015 was

---

[9] Algeria, Angola, Benin, Botswana, Burkina Faso, Burundi, Cabo Verde, Cameroon, Central African Republic, Chad, Comoros, Congo, Côte d'Ivoire (Ivory Coast), Democratic Republic of the Congo, Djibouti, Egypt, Equatorial Guinea, Eritrea, Ethiopia, Gabon, Gambia, Ghana, Guinea, Guinea-Bissau, Kenya, Lesotho, Liberia, Libya, Madagascar, Malawi, Mali, Mauritania, Mauritius, Morocco, Mozambique, Namibia, Niger, Nigeria, Rwanda, Sao Tome and Principe, Senegal, Seychelles, Sierra Leone, Somalia, South Africa, South Sudan, Sudan, Swaziland, Tanzania, Togo, Tunisia, Uganda, Zambia and Zimbabwe.

[10] Sub-Saharan countries (forty-eight of them) include all countries in Africa except, Algeria, Egypt, Morocco and Tunisia. Median Africa covers sub-

estimated at 1,182,439,000, which was 283,361,000 in 1960 when many African countries gained political independence.[11] This is an increase of four times (416 per cent) within a span of fifty-five years. The population in sub-Saharan Africa has also grown from 220,138,000 in 1960 to 958,577,000 in 2015, and increase of 435 per cent.[12] It is estimated that by 2050 Africa's population will increase to over 2 billion people, about one-fourth of the world population, from 15 per cent today (AfDB et al., 2015). In 2018, the figures were 1,275,921,000 for Africa and 1,038,627,000 for sub-Saharan Africa. The literacy level in sub-Saharan Africa stood at 64.6 per cent in 2016 (UIS, 2016b).

The Knowledge Economy Index (KEI), which incorporates indicators of economy, institutional regime, education, training and ICT, has pertinent figures for Africa.[13] Among African countries, South Africa tops the ranking with an index of 5.64, followed by Mauritius (5.08), Tunisia (4.52), Namibia (3.94), Egypt (3.93), Botswana (3.92), Morocco (3.3) and Algeria (3.07).[14] Rankings of some of these countries have changed between 1995 and 2007. South Africa's ranking declined to 50 in 2007 from 41 in 1995 while some other countries such as Tunisia, Namibia, Morocco and Algeria improved theirs. Compared to leading countries in this index (Sweden, Denmark, Norway, Finland, the Netherlands, Switzerland, Canada, Australia, the United Kingdom and the United States which had the highest indexes of between 9.26 and 8.8 and ranked from 1 to 10), South Africa's position is not low. But there are other African countries such as Eritrea, Rwanda,

---

Saharan Africa without South Africa, and Maghreb countries are Algeria, Libya, Mauritania, Morocco and Tunisia.

[11] United Nation's population statistics, Total population by major area, region and country.

[12] https://population.un.org/wpp/Download/Standard/Population/. Accessed 30 November 2019.

[13] The ranking is based on data from 140 countries and the KEI takes into account four key indices that are the pillars of knowledge economy: economic incentive and institutional regime, education and training, innovation and technological adoption, and information and communications technologies infrastructure (World Bank, 2007).

[14] Other African countries in the declining order of index and ascending order of rankings are Cape Verde, Swaziland, Kenya, Zimbabwe, Senegal, Madagascar, Lesotho, Uganda, Ghana, Nigeria, Tanzania, Mauritania, Benin, Ivory Coast, Angola, Zambia, Cameroon, Sudan, Malawi, Mali, Mozambique, Burkina Faso, Eritrea, Rwanda, Djibouti, Ethiopia and Sierra Leone.

Djibouti, Ethiopia and Sierra Leone which are at the bottom of the index with a score of less than 1 and with low rankings ranging from 133rd to 137th of the 140 countries (World Bank, 2007).

The Technology Achievement Index (TAI) that takes into account national technological capabilities and performance in the diffusion of recent innovations and old innovations, and the development of human skills, presents the technological standing of countries (Nasir et al., 2011). The 2009 TAI index showed lower-ranking African countries. South Africa stands at the 40th position while the other African countries on the list were Tunisia (46), Algeria (50), Ghana (53), Mozambique (55) and Tanzania (56) (Nasir et al., 2011).

The ArCo index developed by Archibugi and Coco (2004, 2005) takes into account the indicators of creation of technology, technological infrastructures and the development of human skills. This plays a comparative role in the making of a country's technological capabilities. In this index, several African countries were in the last group of marginalised countries (ranked between 112 and 162) which did not have wide-spread access to even the oldest technologies such as electricity and telephony (Archibugi and Coco, 2004). However, some had a higher ranking: South Africa (56), Tunisia (92), Algeria (97), and Egypt (99).

The average growth rate for Africa in 2014 was 3.9 per cent, higher than the global average of 3.3 per cent (AfDB et al., 2015). But in 2016, Africa's real GDP growth has been slowed down to 2.2 per cent (AEO, 2017).

Africa was a latecomer in science. It lost much valuable time. Science in colonial Africa did not receive the attention it deserved. During the colonial period, scientific development in Africa was either late or there was no development at all (Gaillard et al., 1997), and it was unevenly distributed (Gruhn, 1984) across the continent.

Science as a social institution suffered under colonialism in Africa more than elsewhere (Waast and Krishna, 2003b). S&T were never featured in official colonial policies to solve problems that were prevailing in the continent (Forje, 1989). The situation under colonial rule was not supportive for people to engage in science either. They were not able to participate in the development of S&T and did not benefit from the scientific research infrastructure (Forje, 1989). There were not enough facilities for scientific training. This led to Africa's isolation from other scientifically progressing countries.

The isolation of science from Africa, as Mbarga (2000) observed, might have been brought about by the colonial education system. Science was largely not a concern or a focus for colonial Africa. Colonial administrators were preoccupied with anything but about development of S&T for Africa, or were not keen on taking advantage of it for the development of Africa. As a consequence, several African countries on the eve of independence did not have any science or technology worth mentioning. Scientific research, research personnel and the capacity to undertake research were rather inconsequential back then. Parts of Africa were an exception. Experimental stations and medical research institutes were operational in some parts of Africa (Waast and Krishna, 2003b). African countries had to start from this background when they gained political freedom in the late 1950s and later.

Under colonial rule, some sort of science policy existed which contained an element of development. It was not meant for the development of Africa but for the remote colonial empire. Colonial science did not adequately prepare the colonies for the greater task of creating and supporting native scientific institutions nor did it foster attitudes conducive to the growth of science (Basalla, 1967). It was an outcome of the realisation that the application of S&T can offset underdevelopment in Africa (Forje, 1992). Since the 1970s, awareness and recognition of S&T in development have caused a great deal of interest in science policy. However, there were hardly any attempts to articulate specific goals or measures to achieve the goals that can be achieved through the application of S&T (Vitta, 1990).

Involvement of Africans in science was absent during the colonial period, which lasted for over 150 years. Shortly after having some strong development of scientific institutions in the 1970s and 1980s, Africa faced a challenging crisis to its science (Arvanitis et al., 2000). The economic crisis in the 1980s was a catastrophe for Africa.[15] The state of science in several African countries began to deteriorate and decline for the support of S&T development was obvious in the 1990s (Gaillard et al., 2005; Mouton et al., 2015). Scientific research in Africa grew more slowly than in the rest of the world (Eisemon and Davis, 1991). According to Waast and Krishna (2003b), the delicate position

---

[15] In most of sub-Saharan Africa, the per capita income deteriorated to the 1960 level and GDP declined significantly between 1980 and 1984 (Vitta, 1990).

of African science in the 1990s, both professionally and institutionally, was due to certain factors. The lack of sustained political and social legitimacy for science, low levels of funding available for R&D in general, the lack of a research climate, the disenchantment with professionalised science, the expansion of market economies and globalisation, and the enduring economic crisis were among them.[16]

Science in African institutions, as Mouton (2008) observed, has been deinstitutionalised and eroded in its capacity. Scientific institutions in several African countries are under-resourced, fragile and susceptible to political instability (Mouton, 2008). Analysis based on publications in the Scopus database showed that the publication output of African Union (AU) member countries in 2005 was small and similar to that of single European countries such as Switzerland and Sweden (AOSTI, 2014). Africa as a whole, during 1999–2008, produced about 27,000 papers a year, which is the same output as the Netherlands (Jonathan et al., 2010). Cuts in government spending on S&T, no recruitment of new scientific personnel, decline in the salaries of scientists and the non-availability of modern laboratory equipment characterised the period (Gaillard et al., 2005; Hudu, 2015). The consequence of low levels of capacity for the production and absorption of S&T is that the countries are incapable of creating and satisfying their own science and technological needs (Sagasti, 1973). These factors contributed substantially to the decline of the state of science on the continent.

Fritz Hahne, director of the African Institute of Mathematical Sciences (AIMS) based at Cape Town in South Africa, concedes that most of the universities in Africa have been in a state of decline and neglect. Therefore, a new generation of faculty and researchers is required to raise the educational standards and to connect Africa to the international scientific community.[17] This is central to Africa being able to deal with its innumerable social, economic, health, environmental and political problems. These problems cannot be tackled effectively without a critical mass of African scientists (Hassan, 2001). The R&D personnel available for countries vary greatly across the continent. African scientists who work and live in other countries

---

[16] Waast and Krishna (2003b) however note that there were some exceptions to this in countries such as Tunisia, Egypt, Morocco and South Africa.

[17] Cited in van den Brink and Snyman, 2007, p. 792.

outweigh those who are currently working on the continent.[18] According to the figures for 1991, sub-Saharan Africa sent the highest percentage of their students overseas for higher education. The rate was 14 per cent for sub-Saharan Africa and 7 per cent for North Africa (Barré and Papon, 1993).

Both the importance of science and the status of the profession decline in such an environment. Migration of graduates ensued, weakening the situation further. Falling oil prices, recession and economic instability caused efforts for scientific growth and advancement to regress. Up until the 1990s, this was the situation. Africa was struggling to stand on its own feet as far as its scientific system was concerned. The decline in the production of scientific publications, as Tijssen (2007) reported, was mainly due to the lack of resources and little willingness to invest in universities, laboratories and research institutions. Added to this is the poor working environment (in terms of pay, benefits, facilities and career advancement) for researchers (Tijssen, 2007).

However, one cannot undermine the modest progress Africa has been making in the later years.

Africa made some notable strides in higher education and in the number of scientists during the 1970s (Gaillard, 1992a). The university systems in the continent experienced rapid transformations in student enrolment and formation of new departments during the 1970s and 1980s (Gaillard, 1992a). With the support of aid from overseas and through fellowships, training, institution building and partnership programmes, there have been tangible benefits for science in Africa (Gaillard et al., 2005). With political independence for many African countries, there has been focused attention on science and scientific institutions. Vigorous efforts were seen from this period up until the mid-1980s. They were evident in the expansion of scientific institutions, development of national research systems, creation of national coordinating bodies, and implementation of national science policies (Gaillard et al., 2005). While there has been progress in institutionalisation and professionalisation of S&T in Africa, they were not enough to create sustainable scientific production and scientific communities (Gaillard, 2003a). A well-developed higher-education system and investment in higher education and research is a *sine qua non* for the

---

[18] It is estimated that about 30,000 PhD holders of African descent, most of them with science degrees, work in other countries (Hassan, 2001).

development of Africa and to address its developmental challenges (Atuahene, 2011).

In scientific manpower, Africa is still at the bottom (Gaillard, 1991). In the matter of scientific research, African universities have not been able to achieve high quality research output (Eisemon and Davis, 1991). Serious constraints, in terms of funding, facilities, resources and the bureaucratic nature of administration of scientific institutions continue to exist. The recent Ibrahim Index of Africa reveals that 40 per cent of African citizens live in a country where the economy has deteriorated and there are chronic power shortages, a major obstacle for growth (Strohecker, 2016).

The findings of a large survey of scientists in Africa are quite revealing. Gaillard and Tullberg (2001) reported on several aspects of science from the perspective of scientists. The survey reported major stumbling blocks that include funding, research equipment (access and maintenance), library facilities that are not adequate, heavy teaching load and administration, and insufficient salaries and benefits. In addition, there are also the issues of African scientists who are highly dependent on foreign funding for their research, and the deteriorating conditions under which scientists perform their work (Gaillard and Tullberg, 2001).

In various measures of global science, technology, innovation, GERD (Gross Domestic Expenditure on Research and Development), number of researchers, and scientific publications and patents, the performance of Africa was not impressive (Zeleza, 2014, cited in Cloete et al., 2015a: 21). The percentage of the GDP spent on GERD is a measure of the importance a country attaches to its scientific research. The rate of expenditure on R&D is a key factor (Williams, 1968) and the growing level of investment and expenditure in R&D is generally correlated with improved growth in GDP (IAC, 2004). Countries now display their S&T performance through the figures of R&D expenditure (GERD) as a percentage of the Gross National Product (GNP) (Godin, 2005).

In 1990, sub-Saharan Africa spent 0.3 per cent of its GDP on GERD. North Africa spent a similar percentage, but in real terms it was much higher than what was spent by sub-Saharan Africa. The percentages for other countries were 0.8 per cent (India and China), 2.8 per cent (the USA), 1.4 per cent (Canada) and 3.1 per cent for Japan (Barré and Papon, 1993). Recent figures, as reported in the *African Outlook of*

Innovation 2014, show that the actual expenditure in Africa is lagging behind the world average (NPCA, 2014). Kenya in 2010 had a GERD of 0.98 per cent, South Africa 0.76 per cent and Ethiopia 0.24 per cent (NCPA, 2014). It was 0.72 for Egypt (in 2015), 0.72 for South Africa (in 2013) and 0.60 for Ethiopia (in 2013) (UIS, 2016c).[19] As to the ratios for scientists and engineers to the population, sub-Saharan Africa in 1990 had 0.1 person per every 1,000 population. It was 0.3 for North African countries, 0.1 for India, 0.4 for China, 3.8 for the USA, 2.3 for Canada and 4.7 for Japan (Barré and Papon, 1993). In Arab countries, some of which are part of North Africa, GERD did not exceed 0.2 per cent of the GNP (UNDP, 2003).

In many countries, the count of scientific publications to GDP is also linked to R&D expenditure, which is GERD (Barré, 1998). King (2004) notes in his analysis of countries that the national citation intensity (in other words, the ratio of citations in all papers to the national GDP) is a function of national wealth intensity or GDP per person. This relationship between GDP and the output of research publications in Africa has been revealed in research. A positive correlation between GERD and the number of publications has been observed by Blom et al. (2016).[20] The analysis of publication data for 1999–2008 for countries in Africa documented the relationship between GDP and output of publications in South Africa, Egypt, Nigeria, Tunisia, Algeria and Kenya (Jonathan et al., 2010). Not only is there a relationship between GERD and the research output in a country but there also exists a linear relationship between funding for R&D and the total impact of research (Pan et al., 2012). This will be examined in detail in the following chapters.

Publications are indicators of the output of a country and citation is a measure to assess the impact of those outputs. Contemporary trends in the publication output from Africa present a growing pattern. Confraria and Godinho (2015) maintain that African science has taken off since 2004. A few countries in the continent had demonstrated a higher level of specialisation and international networks and a higher impact than the world average (Confraria and Godinho, 2015). Schemm's (2013) analysis of publications showed an increase

---

[19] Comparable data are not available for the same year and all countries.
[20] In South Africa, for instance, an increase in the volume of funding (GERD) has given rise to a sharp increase in research output between 1996 and 2008 (Blom et al., 2016).

in publications by Africans during 1996–2012 and Africa's share of world's publications doubled from 1.2 per cent to 2.3 per cent. Studying Scopus data (2003–2012), Blom et al. (2016) found that: all the sub-Saharan regions (West and Central Africa, East Africa and Southern Africa) have doubled their research output from 2003 to 2012; the share of sub-Sahara's global research has increased from 0.44 per cent to 0.72 per cent; and the citations of sub-Saharan publications to global citation grew from 0.06 per cent to 0.28 per cent. This being the case, the research output of sub-Sahara in science, technology, engineering and mathematics lagged behind the outputs that achieved in other subject areas (Blom et al., 2016).[21] The number of publications authored wholly within Africa without any collaborative author from outside the region has doubled since 2000 (Adams et al., 2013). This, according to Adams et al. (2013), is proof of the growth of autonomous research output of Africa, which is a sign of growing self-reliance and independence.

Africa produces science for the world. Recent analyses confirm these trends. The data used by Jonathan et al. (2010) referring to 1999–2008, found Africa's high representation in certain fields. A bibliometric analysis for the period of 2005–2010 revealed that fifty-four member countries in the AU produced 34.3 per cent papers in health sciences, 29.4 per cent in natural sciences and 27.8 per cent in applied sciences (NPCA, 2014). The majority of the publications were in clinical medicine, biology, biomedical research, chemistry, engineering, physics, astronomy and agriculture (NPCA, 2014). The strength of Africa lies in medical research, and it has been the main driver for Africa's growth in science. Blom et al. (2016) noted that sub-Sahara's research output was overwhelmingly driven by its advances in health sciences research that accounted for 45 per cent of all sub-Saharan research with an annual growth rate of 4 per cent. The scientific production of countries in the AU has been central to some specific fields. In the fields of astronomy, meteorology, geology and medical science (tropical diseases) South Africa made significant contributions (Hofmeyr, 1929b).

---

[21] Blom et al.'s (2016) data, drawn from Scopus, showed that research in the physical sciences and science, technology, engineering and mathematics (STEM), made up only 29 per cent of all research in sub-Sahara (excluding South Africa). This share of publications in STEM has declined by 0.2 per cent annually since 2002.

While there is consensus about the dire need for Africa to pursue S&T more intensely, the issue is the direction in which S&T should move (Enos, 1995). The suggestion for this, as Enos (1995) concludes on the basis of his study of a few sub-Saharan countries, is that Africa should choose the direction that best represents its interests and not necessarily what is congruent with the interests of donors. On the basis of an analysis of the state of science in Africa, ideas towards the growth path for Africa can be suggested.

Science Policy and National Development

In underdeveloped countries, policy making in science, from the lowest to the highest level of the social system, remains a hurdle (Dedijer, 1968). More importantly, when science policy comes as an aspect of the national policy, it is the least studied and neglected one in both developed and underdeveloped countries (Dedijer, 1968). For the absence of explicit policies on S&T, as Vitta (1990) found, there was a lack of understanding that development required technology and is the mandate to acquire and use technology. The UNESCO conference in 1974 recommended that the national science policy planning machinery should liaise closely with the national organs responsible for socio-economic development planning (UNESCO, 1974a). The 1979 United Nations Conference on Science and Technology for Development in Vienna called for the governments of all developing countries to formulate their own national S&T policy. The Lagos Plan of Action was adopted in 1980 at the Organisation of African Unity (OAU) extraordinary summit, and urged African countries to formulate national policies on S&T that are integrated into the their national development plans (OAU, 1981).[22] The African agenda for 2063 categorically emphasises that there should be an evolution in education and skills and promotion of STI to build knowledge, necessary human

---

[22] The plan, called the Lagos Plan for Scientific Research and Training in Africa, emerged at the International Conference on the Organisation of Research Training in Africa in Relation to the Study, Conservation and Utilisation of Natural Resources which was held in Lagos, Nigeria, from 28 July to 6 August 1964. The conference made the resolution that governments should devote continued and large-scale efforts for the promotion of science and technical research for development and social progress. The conference also recommended that the governments set up national research organisations and explore the possibilities for regional cooperation (Smith, 1967).

capital and capabilities to drive innovations (AUC, 2015). The agenda at various points in the history of science articulates the role of STI in the development of Africa in the coming years.

The inter-relationship between science policy and the national development plan is very close. The former should be a vital part of the latter, if a country aims to advance on both fronts. This means any national development plan cannot afford to be without a properly formulated national science or STI policy. The components of science policy, again, have to be integrated into the broader national development plan. Linking S&T for development is evident in many partnership programmes with other regions and countries (AOSTI, 2014). NEPAD is one such partnership for African countries to develop their S&T policies that are integrated into the development agenda of Africa. But the experience is that in most countries there is not much harmonisation and integration between science or STI policy and policies meant for national development. The lack of concrete S&T policies in some countries prevented them from setting up structures of management for the conduct of scientific research at centres of research such as universities (Benneh, 2002).

The application of S&T for development is facilitated and supported by the backing of policies that are in place and are implemented in a timeous manner. The relationship between a well-defined national policy on STI and development is to be explored. Considering the relevance of national science policies in building a strong scientific system, policy matters in Africa need to be examined. Many African countries do not have strong STI policies that are well grounded in the existing information on contemporary science in their countries.

This Study

The Context and Rationale

The new knowledge about science is growing (Price, 1965). With the growth of science, there has also been a corresponding need to understand science and its trajectories of development from varied perspectives. Sociologists, anthropologists, historians, economists and political scientists made their contributions towards understanding the role of S&T in society. They are important for providing policy analyses, developing a culture of S&T and promoting S&T (IAC, 2004).

Generally and historically, sociological studies of science were focused on North American and European countries, largely due to their contributions to science (Eisemon, 1981). In many developing countries, S&T have not lived up to expectations while progress was achieved in developed countries (Berlinguet, 1981). The disparities between the developed and developing is not levelling out, but leads to further disparities in the unequal distribution of the benefits of S&T that different countries enjoy (Matsuura, 2000; UNESCO, 2000).

Michael Moravcsik (1975), Hungarian-born American physicist and an authority on science and science development in less developed countries, affirms that our understanding of the science of science or the sociology of science is rather rudimentary. As he says, we are not able to formulate guidelines for building science on the basis of our understanding of science (Moravcsik, 1975). The sociology of science can deal with both the social structure of research institutions and the social conditions of the growth of science (Shils, 1968). In order to understand the widening gap between the developed and developing regions in S&T, studies that map the trajectories of science are unavoidable. Studies that focus on challenges for growth and development of S&T are critical. Research that contributes to knowledge regarding the efficiency of scientists is also essential.

There is no shortage of studies on science in Africa. Reports and documents collected and compiled by agencies and individual scholars are plentiful. Very detailed reports on specific countries are to be found among them, but they are about a few countries and for a particular period of time. They do not provide a comprehensive account of science that is generated in Africa.

Some studies, Arunachalam and Garg, 1986, for instance, have mapped science in the periphery,[23] but not much is known about the science that is generated in the peripheral countries. In contrast, we know a great deal about science in the core countries as we have sources of information to study them. Science in the core countries remains the dominant form of science. We know the context of science and

---

[23] The study examined science in a few ASEAN (Association of South-East Asian Nations) countries, namely, Indonesia, Malaysia, the Philippines, Singapore and Thailand using the publications during 1979–1983.

scientific activity in the scientifically advanced countries that comprise only one-quarter of the world's population, while information about contemporary science in the remaining three quarters of the world's population is extremely fragmentary (Moravcsik, 1985). As for Africa, as Waast and Krishna (2003a) confirm, we know very little about the status of science and the current modes of the production of knowledge. There are no systematic overviews or case studies but rather only general views (Waast and Krishna, 2003a). Given the general lack of reliable and essential data on science, a study that attempts to fill this vacuum, serving as a benchmark for future studies on science in Africa, will be beneficial.

Why study science in the periphery, or in the developing countries? Science in the periphery is characterised by some unique features, patterns, trends and potential.[24] Moravcsik (1985) takes the view that there are basically two kinds of motivation for learning about science in developing countries.

First, developing countries offer a domain of science of science which is different from what is already available to us. This is an opportunity to study the history of science of the unknown world. In many parts of the developing world, science is isolated and therefore variants of science and its development can be observed. The second motivation is pragmatic and functional. Since many developing countries are in the process of building their own scientific systems with the assistance of international agencies like UNESCO, we need factual information on the problems and circumstances in these countries. Again, as the Declaration on Science and the Use of Scientific Knowledge at the World Conference on Science in 1999 at Budapest stressed, all cultures can contribute to scientific knowledge of universal value (UNESCO, 2000). Science produced in both developed and developing countries is knowledge and no knowledge is either inferior or superior. Science from Africa is as valuable as science emanating from any other parts of the world. Information on the state of science in the developing world is incomplete, incorrect and inadequate. Without this knowledge, the building of science or helping build science is hard. This is the main handicap to the countries developing their own sound science

---

[24] Periphery science is known for the absence of a viable scientific community; insular nature reflected in inadequate communication, both within and outside the scientific community; insufficient infrastructure; and dependence on other countries (Arunachalam, 1992).

policies. This can be a further motivation for the study of science in the periphery.

The sociology of science is not well developed in southern Africa and most of the literature is a by-product of other concerns such as political economy (Dubow, 2000). Some periodic analyses of science in Africa are available in the literature. Most of these are based on quantifiable scientometric data, drawn from major databases. Careful attention has not been paid to studying Africa from a scientometric perspective either (Narváez-Berthelemot et al., 2002), and very few such studies have been published on science in Africa (Arvanitis et al., 2000).

What is the need for a study of science that is produced in Africa? Africa contributes to world scientific knowledge in no small measure. Its production has not stagnated, but rather has been growing gradually. Africa has made several groundbreaking discoveries in science and advanced its science in many key fields – for instance, in medical science. However, there are challenges for Africa. Based on a sample study of eighteen countries in Africa, Mouton and Waast (2009) have identified a few areas wherein Africa faces challenges. They observe that Africa needs support for capacity building in policy development in S&T, the quality of data on S&T is to be improved and the visibility of African science needs to be raised. A study of science from Africa can address these and related issues.

The rationale for this study is an interest in the study of science and the evolution and development of science in an understudied region. This study is about the evolution of science in a sizable region of the world and has not yet been adequately documented. Science is an important activity across the world, regardless of whether regions are classified as developed or developing, rich or poor. It is an integral and indispensable component of development. As an instrument for development, growth, progress and advancement, science is unavoidable. The development of science is intimately connected and is in continuous interactions with every other human development (Sarton, 1918). This raises a plethora of issues and concerns regarding the production and utilisation of science. These need to be examined to understand science, its trajectories of growth in different regions of the world, its future and its contribution to development.

The need for the study of science in Africa can therefore be summarised thus: Africa has had a chequered history since the beginning of the independence of African countries in the 1950s; most African countries have gone through an economic crisis in their independence phase (Sooryamoorthy, 2018) which has led to budget cuts, brain drain and the deteriorating situation of science (Arvanitis et al., 2000); and the existing studies of science are mainly country-specific case studies that do not offer a comprehensive and holistic picture of science production in Africa.

Undertaking a study of science and science production in the whole of Africa is not an easy task to accomplish. The issues concerning the availability of reliable data to study science in Africa (Waast, 2010) are not easy to neglect. The poor documentation of data is another real constraint for studies on Africa. Only a handful of countries such as South Africa and Tunisia have systematically collected and produced S&T statistics (AOSTI, 2014). The situation has however changed for the better. Most countries in Africa now have established structures for S&T. Surveys on the status of R&D are becoming more frequent than before. Many countries have drafted policy documents that pertain to S&T. For reasons of development, leaders and planners take S&T more seriously than they did a few decades ago. This is why a study on science from Africa, how it was produced over the past few decades and how it is being done in the recent years became necessary and timely. In this endeavour, the inter-relationships between science, society and development are important.

Objectives and Research Questions

The broad objective of this study is to map science in Africa and its relation to development through the examination of the production of science. Specifically, the study examines the evolution and development of science in both colonial and contemporary Africa. It is concerned with the relationship between the production of science, science in the historical and current periods, national development, research focus areas, scientific collaboration and national scientific policies.

The research questions posed in this study are:

- Is there a relationship between the production of publications and selected years (historical phases) for Africa and African countries?
- What kind of association exists between the production of publications and the publications in recent years, namely, 2011–2015?

- Do countries that had established a stronger scientific system continue to stay at the forefront of scientific production?
- Do the research focus areas in Africa vary across the continent and who leads Africa in specific research areas? Is there any association between the advancement of specific research areas and national development?
- Does association exist between scientific production and respective national policies?
- Is there a relationship between GERD and scientific publications?
- Does the proportion of GERD to GDP determine the production of scientific research?
- More than the percentages of GERD, it is the set of scientific policies that fuel growth in scientific research and publications. Does this mean that policies that encourage and support research, along with supportive structures, are key to scientific growth in Africa?
- Does international collaboration in science exceed continental collaboration in Africa?
- Is the integration of STI policy into the national development policy essential for the success of STI and economic development of Africa?

Theoretical Background

Theoretical development in the world system of knowledge production has not kept pace with the growing interest in it (Collyer, 2014). Scientific development in society is to be examined and explained in terms of its relation to the society. Perceptions of learning and the value of science, its relationship with politics, and the links between science and other spheres of life are relevant (Gaillard et al., 1997).

Scientific research produces and accumulates knowledge. Theoretically, we should be speaking about two types of knowledge, Mode 1 and Mode 2. Mode 1 involves the traditional modes of knowledge production within a single disciplinary and cognitive context, and is governed largely by academic interests. Mode 2 knowledge is generated in much broader transdisciplinary social and economic contexts, application-based and more socially accountable and reflexive (Gibbons et al., 1994). Mode 1 is identical with science. Its cognitive and social norms determine what are significant problems, who is allowed to practice science and what constitutes good science. In Mode 2, knowledge results from a broader range of considerations, intended to be useful for the society, and it is socially distributed knowledge. Knowledge is produced increasingly in the contexts of applications that have become necessary. In this application, the context of the environment in which the problems arise is taken into account, methodologies are developed, outcomes are disseminated and uses are defined (Nowotny et al., 2003).

The new patterns of research in Mode 2 are related to massification of education, involve people located in different locations and require types of funding (more from firms, industries and social lobbies) different from traditional discipline-based research (Gibbons et al., 1994). In other words, knowledge is produced at multi-sites and not only at the conventional and traditional centres of production (universities, research institutions and industry) but also at many new sites. This means research communities now have open frontiers allowing new kinds of knowledge organisations to be part of scientific research (Nowotny et al., 2003).

Compared to Mode 1, Mode 2 production has a closer interaction between scientific, technological and industrial modes of knowledge production. It is characterised by the weakening of disciplinary and institutional boundaries and by the emergence of transient clusters of experts (Gibbons et al., 1994). The transdisciplinary character of Mode 2 allows for the mobilisation of a range of theoretical perspectives to address problems (Nowotny et al., 2003). It relies on both theoretical and empirical information from a variety of disciplinary fields. African science has tended to move more towards Mode 2, which is fragmented, internationally oriented and demand driven (Waast, 2009). More importantly, the relationship between science and society has become stronger in Mode 2. The participation of people in science is increasingly affecting the way science is being viewed by the public. Science is more embedded and integrated in society (ICSU, 2005). Mode 2 is more pertinent to Africa.

In the new form of science, which John Ziman calls post-academic and post-industrial science, science has become more sophisticated, multi- and interdisciplinary, electronically networked, technologically connected and entangled, and economically privatised (Ziman, 2000a, 2000b). Mode 2 is closely aligned to these current patterns in science. No longer are individuals quite free to choose their own research problems or to ignore the putative wealth-creative applications of science (Ziman, 2000b).

In any study of science, its evolution and development are decisive aspects. Krishna et al. (1998) speak about three modes in the evolution of scientific development: colonial, national and private. The colonial mode was characterised by the domination and expansion of colonial powers between the seventeenth and mid-twentieth centuries. For many societies, including Africa, the first encounter with modern S&T was the result of colonisation. The intention of the use of S&T in colonies was profit and colonial expansion. The national mode of scientific development began after the Second World War when countries started building infrastructure and institutions for the development of S&T. At this stage, scientific development was to integrate with the socio-economic interests of countries. In the private mode, countries in the south experienced a new privatised and globalised mode of scientific development. They argue that it has major implications for the south. The existing hierarchy of disciplines, sources of scientific prestige and reputation and research autonomy are being questioned in this emerging private mode of scientific development. This is useful in the understanding of the evolutionary phases of science in Africa.

Coming to the relationships between science and development, there are a few theories that deserve discussion. Goldemberg (1998) elaborates on three models in this regard. The linear model, also called cradle-to-grave approach, is the one the USA has adopted as a blueprint for the establishment of its National Science Foundation which many other countries later copied. But this linear model fails to take into account the interaction between the phases (pure research to technological advancement and to production and marketing of goods). The second model, which has some overlaps between the three phases, is currently the practice in countries like the USA. In the third model all the three phases are completely superimposed, as practised by the Japanese.

A key issue that concerns science development and policy is about the approach. Moravcsik (1986b) proposes two perceptions on this in a developing country. One emphasises the value of science primarily within the context of short-term benefits to economic growth. The aim of science is to help technological activities related to the short-term economic growth of a country. For this, planning is the beginning, and science plans should concentrate on the measurement of inputs and outputs. Science in a country can be evaluated by examining the economic development the country has achieved. If progress has not been achieved, science and scientists are to be blamed. The success of science can also be measured through the success of technology or through the productive sector of the economy as there is a linear relationship between science and technology. Under this perception, a scientist in a developing country is to participate in a research problem that is related to the country's short-term economic problems. Scientific cooperation is to fulfil the short-term economic objectives in the developing country.

Moravcsik's (1986b) second perception contemplates science for its aims and benefits in a broader and far-sighted context in a developing country. It has three inter-related elements for scientific development: science is a necessary condition for technology; science is prevalent aspiration; and science has a powerful influence on human views of the world and their place in it. The perception is that S&T are different human activities and can be separated by their products, which attract people of different talents. People who are differentially talented can result in poor quality and a discouraged workforce. The point of view under this perception is that the most important function of a scientist in a developing country is to act as a funnel of scientific ideas into the country, as more than 99 per cent of science is created outside the country. As planning for science is an uncertain process, evaluation of scientific work should be a process in continuity: evaluation must be carried out by scientists from both within the country and outside. The output of science cannot be measured by the level of economic progress the country has achieved, as there are many intermediate stages between science and production. International cooperation under this perception is aimed at receiving international assistance in enabling the developing country to create strong local scientific infrastructure. Such international cooperation is best achieved through the scientist-to-scientist mode (Moravcsik, 1986b). Moravcsik (1986b) believed that pursuing the short-term perspective is the real cause of poor science development in less-developed countries.

Modernisation theory facilitates an understanding of the relationship between science and progress. It is rooted in the belief that the beneficial effects of scientific rationality should replace traditional beliefs, myths and superstitions (Harding, 2011). Aghion et al. (2009) argue that the intimate connections of technological change and innovation, along with the advances in science, and the set of socio-economic institutions operating in a given context, encourage conceptualisation of science, technology, innovation and growth systems.

Collyer (2014) asserts that the global system of knowledge production operates in a more or less similar way as other aspects of the economic system, and it is hierarchical and characterised by a fundamentally unequal relationship of power between core and periphery. The hierarchy in science is the basis for the core and periphery divisions in the global scientific community. As Schott (1993) details, the centre (core) refers to the sector that attracts more attention: its achievements become exemplars for research in the periphery, its dominance is legitimated, and its authority is conferred by the scientist in the periphery. The centre-periphery division exists for the scientific community. When the scientific community becomes territorially extensive, its peripheries become more extended (Schott, 1991).

According to Schott (1991), two changes can happen at the periphery. One, the previously known periphery countries, by virtue of more scientists and more publications, can become the sites of more scientific activity. Two, countries, for instance those in the sub-Saharan area where there are few scientific activities, can become new peripheral areas as they have enough to make it to the periphery of the world community. If this is the case, possibilities are there for peripheral countries to become core centres of the world scientific community one day and that new peripheral countries can be formed from those countries which did not have any scientific activity worth mentioning. Peripherality is a matter of degree, and there exists a continuum from core to periphery (Schott, 1991), with peripheral countries at various stages of scientific activity. This permits peripheral countries from the extreme entry end of the continuum to move towards sites where scientific activities are greater. They are still on the margin but their position is changed in accordance with their changing scientific activity. The cases of India, Brazil and countries in the Middle East, South East Asia and North Africa are examples that have been developing fast (The Royal Society, 2011), and are on the move from the far periphery towards the other end of the periphery. The emergence of new players and new leaders in science (The Royal Society,

2011) can change the spectrum of core–periphery in the coming years.[25] This situation very well applies to Africa.

Dependency and interdependency theories, on the other hand, emphasise the unequal kind of relationship between the core and the periphery. The practice and production of science in the world is uneven and it is skewed towards some centres, core countries, for that matter. The periphery is increasingly dependent on the core for its scientific advancement (Nagtegaal and de Brun, 1994). The unequal relationship benefits the core more than the periphery while they cooperate with each other. It is mainly because of the advanced capacity of the core countries to disseminate, absorb and use knowledge that this imbalance has been produced (Boshoff, 2009a). This can be explained in terms of scientific cooperation and collaboration among countries.

The inequality dimension of technological advancement and dependence on technology transfer that mainly disadvantages countries in the developing world cannot be ignored. Castells (1993) warns about the serious consequences of this as S&T are development tools of great importance, they are unevenly distributed in the world. The brain drain phenomenon creates structural gaps among countries as the best scientists and engineers are concentrated in a few countries and in a few institutions in these countries (Castells, 1993). Africa remains a region affected seriously by this phenomenon.

The theories discussed here provide useful frameworks for comprehending the evolution, development and connection between S&T, development and the economy in African science. Krishna et al. (1998) outlined the evolution of the development of science appropriate for Africa. The first two modes (colonial and national) apply to Africa. They also connect the development of science in the postcolonial state to the socio-economic interests of the countries. In other words, this theoretical perspective stresses the usefulness of science for development purposes, which is appropriate for this study. The status of science in a country, as envisaged by Moravcsik (1986a), can be examined by its economic development, and there is a linear

---

[25] Paradisia and Dominatia, the terms introduced by Moravcsik and Ziman (1975), are closer to the core–periphery paradigm. These terms refer to the bifurcation to examine the state of science and development and their interactions at various levels. Paradisia refers to medium-sized states such as Ghana, Iran, Korea and Peru and science flows from Dominatia countries that are powerful metropolitan countries like the USA.

relationship between the economy and S&T. Moravcsik's theory acknowledges science as a necessary condition for technology, and science has a powerful influence on society. Modernisation theorists (Aghion et al., 2009, for instance) also recognise the intimate relationship between technological change and system growth. The heightened interest in R&D reflects a shared perception that higher levels of economic growth are attributable to the success of exploiting emerging technologies (Aghion et al., 2009).

The theory postulated by Enos (1995) is grounded in the African context. Concerning development and advancement of S&T, Enos (1995) includes the volume of resources the nation devotes to advancing S&T; the extent to which these resources are made available to economic development in the early years; and the efficiency of the scientific personnel who are employed. These dimensions provide a framework for analysis of the development and advancement of S&T in a given country. The first two issues relate to GERD and percentage of GDP. This can be examined with the help of available GERD data. The third element of this framework can be used, partially at least, to examine the scientific productivity of nations. An examination of science policy and achievements in the domain of S&T and the means (collaboration, for instance) used in the efficient production of science can also contribute to the understanding of the efficiency of both the scientific personnel and the scientific system of a country.

With the support of the these frameworks, namely those of Enos (1995), Krishna et al. (1998) and Moravcsik (1986a) in particular, this study examines the evolution and connection between scientific productivity, economic development and the facilitating conditions for science production (such as STI policies and collaboration) in Africa. The theories we have discussed are beneficial in identifying key factors – namely, the evolution of science and its development, economic indicators, key research areas, scientific collaboration and science policies – for this study.

Methods, Data and Analysis

Science is quite vast and expansive, which led George Sarton, a Belgian historian of science, to wonder: if the growth of science continues at the present rate, will it be possible for any historian to record and understand further progress (Sarton, 1936). He, however, recommends that in order to understand scientific achievements, one must select a relatively few representative documents from among a large mass of others. Any study on science and the scientific community has to depend on certain reliable indicators.[26] Three major indicators are publications, citations and coauthorship (Schott, 1993; Zymelman, 1990). Publication is an index to the production of knowledge, citation indicates impact and influence of science, and coauthorship represents collaboration (Schott, 1993). The research output is yet another indicator in the measurement of the health of scientific systems (Tijssen, 2007). Science produced in Africa, or in any country, can be studied and analysed using these indicators. Although there are reservations about using scientific output as a measure due to the absence of any index of quality, publication counts and citations are still useful tools for estimating scientific output (Moravcsik, 1975). As Moravcsik (1984) confirmed, products of science such as knowledge, human attitude and sense of accomplishment are not quantifiable. It has nevertheless become conventional to measure the scientific output of countries through the analysis of peer-reviewed published papers (The Royal Society, 2011). The study of the publication outputs of scientists can be regarded as a representative set of documents of science that throw light on the way science works today. Scientific papers in learned journals, reports and books therefore represent the end product of science and constitute science (Price, 1965). Since the first scientific publication in the world in 1665,[27] there has been a tremendous surge in publication outlets and scientific publications which marked the production of science. In this study, the production of science is measured in terms of publications.

This study seeks to map science in Africa by drawing on its production and contribution to knowledge as evidenced by scientific publications since 1945. The study focuses on understanding the relationship between science and development in Africa. The difficulty, as Wad (1984) argued, is that the problem of S&T in African development is an inadequate understanding of the problem. Further, discrepancy

---

[26] The list of S&T indicators is very exhaustive. OECD (2016) has compiled 130 indicators, varying from GERD to technology to balance of payments.

[27] The first scientific journal came into being when Henry Oldenberg founded the first issue of *Philosophical Transactions* on 6 March 1665 in London under the auspices of the Royal Society of London for Improving Knowledge (The Royal Society, 2011).

between national development and the orientation towards scientific research in many African countries prevails (Forje, 1989). If this is the case, we need to know how science is going to serve as an effective tool for development in African countries. The attempt in this study is to examine the production of science in Africa and connect it to the issue of African development. In this inquiry, the contribution of Africa to science and vice versa are unavoidable.

Some conceptual clarity on the key concepts used in the study is in order.

Science is a set of characteristic methods by means of which knowledge is certified; a stock of accumulated knowledge stemming from the application of such methods (Merton, 1973). A standard definition of science is that it is an organised attempt by mankind to discover how things work through the objective study of empirical phenomena. Science brings together bodies of knowledge to reconstruct the world (UNESCO, 1974a).

When science is used in the context of development and policy, two other terms always follow: technology and innovation. S&T appear often together as siblings and are interconnected, and one cannot be separated from the other except for definitional purposes. It is the function of science to provide the basis for technology, as became obvious in the nineteenth century (Odhiambo, 2000). Technology is the organic part of knowledge that relates to the production or improvement of goods and services. Technology is applied to satisfy human needs (UNESCO, 1974a). Science produces knowledge while technology makes use of the knowledge to produce tangible products (Moravcsik, 1983). More precisely, science is 'know-why' and technology is 'know-how' (UNESCO, 1974a; Zymelman, 1990).

The bond between science and technology began to strengthen in the period of the industrial revolution, which succeeded on the back of advances made in the fields of S&T and their application in industry. S&T are complex and pervade all aspects of society (Hetman, 1979). S&T are in a continuum of concepts or group of concepts that might overlap and often they are classified in terms of their results (Moravcsik, 1975). The results of science and scientific activity are to produce new knowledge and the result of technology is best materialised in technological products. Both are mutually supportive, and one needs the other for each one's growth and development. It is a two-way crossing, one contributing to the development of the other (Berlinguet, 1981).

S&T are in a symbiotic kind of relationship in which technology adopts and uses the advances in science to produce goods and products, while science advances using equipment and instruments that technology produces (Moravcsik, 1975; Zymelman, 1990). They need each other for their individual existence, rationale, advancement and benefits. Science is a necessary condition for the existence of technology and vice versa (Moravcsik, 1983). The import of new technology into science is to provide new opportunities for new experiments, and to build new instruments (Hoyningen-Huene, 2000). This circle of science to technology and to science proceeds. Technology is required for the future planning of scientific research (Cooper and Zammit, 1964) and the continuous development of science. The rationale for investment in science and support is also the technological benefits science can bring (Madox, 1968).

Innovation is another related concept in S&T studies. It is part of science and technology, and science exists through technology and innovation. S&T are the main sources of innovation (Hetman, 1979). Technology is only one of the fruits of science (Berg, 2000). Innovation is another sibling of science. As Carty (2000) defined it, innovation involves creativity and application and it is a process by which ideas, discoveries, inventions and products are created, developed and finally taken to market. Innovation, like scientific research, can be a long drawn-out process and survives on factors that support and encourage it within a country context and in a competitive world environment. Science or scientists are not enough for marketable innovations.

In this study, science is discussed in its relation to technology and sometimes to innovation, particularly when the discussion revolves around science or science policy. STI form a triad and each one has a mutually serving and contributing purpose and a role to play. The focus of the study is on science, scientific production and the scientific output of scientists rather than technology or innovation.

Employing relevant methods circumvents data inadequacies. Bibliometric methods constitute one such method appropriate to study S&T in Africa (Blom et al., 2016). This method is also useful to understanding the directions in science that are taking place in Africa, with its singularities and experiences in countries within the continent. The heterogeneous character of science in Africa makes the method a feasible one for a meaningful and beneficial inquiry into African science.

Bibliometric studies have served well as an effective tool to map science in developing countries, particularly in Africa (Arvanitis et al., 2000; Chatelin et al., 1997; Confraria and Godinho, 2015; Dahoun, 1999; Gaillard, 1992a, 1992b; Mêgnigbêto, 2013a). The bibliometric method is also used to study the interactions between science and politics and the effects of political barriers in science (Lancaster and Abdullah, 1992). Bibliometric data are valid indicators of publication productivity in countries with lower levels of development (Shrum, 1997). Michael Moravcsik, who did a study of the scientific output of developing countries, urged scholars to use bibliometric measures. He believes that although these are imperfect measures, the greater evil is not practising the evaluation of scientific work and not trying to make progress in understanding the structure of science-making (cited in Garfield, 1990: 10). When the data from relevant databases are considered for longer periods of time it allows us to present the trends and compare the outputs of countries (Waast and Rossi, 2010).

However, very few bibliometric studies have been published on science in Africa, and most of them are rather dated (Davies, 1983).

The primary empirical data used in this study was mined from the Thomson Reuters' Web of Science (WoS) database. Scholars have widely used this database, which is a reliable source for these kinds of studies.[28] The database at the time of this study had 56,990,892 papers published in 12,000 journals that were indexed in the Web of Science Core Collection, including Science Citation Index Expanded (SCI-EXPANDED), Social Sciences Citation Index (SSCI) and Arts & Humanities Citation Index (A&HCI). For the purposes of this study – publications pertaining to science in Africa – only the Science Citation Index Expanded (SCI-EXPANDED) was applicable.

The analysis using bibliometric data was carried out at two levels. Firstly, the data was used to map science in all African countries where

---

[28] Among them were Barré (1998), Confraria and Godinho (2015), Davis (1983), Dahoun (1999), Gaillard (1991, 1997), Mêgnigbêto (2013a), Onyancha and Maluleka (2011), Pouris and Ho (2014), Sancho (1992), Schubert and Braun (1992), Thomas (1992), Toivanen and Ponomariov (2011), Waast and Rossi (2010) and Whitney (1992).

data was available in the WoS,[29] employing all available variables directly drawn from the dataset.[30] Necessary additional variables were created by transforming the basic data drawn from the WoS database. The analysis followed in this first stage referred to two periods. The first was for the entire period from 1945 to 2015, providing a holistic picture of science from Africa and its constituent countries. This was to provide both a single (1945–2015) and a segmented (divided across the years; for example, between 1945 and 1960, between 1960 and 1975, and so on) view of science in the continent.

It is necessary to examine science and its development in recent years, to find where science is going in Africa. The African science ministers resolved that 2011 was to mark the start of an African decade of science, promising increased shares of their budgets and efforts to use S&T for the development of Africa (Nordling, 2010a). This was therefore the second period chosen for the analysis; namely, the last five years, 2011–2015, both inclusive. The same measures as used in the first period (1945–2015), were used in this second period (2011 to 2015). The data for 2011–2015 had to be extracted separately as it could not be filtered from the 1945–2015 dataset for detailed analysis. Following this first step, further analysis was necessary to find answers to the research questions.

Africa cannot be identified and studied as a single entity for reasons of the unequal and uneven state of science in its countries. Science systems in Africa are not homogenous in terms of size, resources, specialisation or the structures of management (Tijssen, 2007). Science is rather unequally distributed, with some countries having far more science than others (Durant, 2000). There are real differences that exist between different parts of Africa in terms of scientific infrastructure, budget, training and publication output (Gaillard et al., 2005). Appropriate distinction should be made (North, sub-Saharan, South and Median) when its science and scientific systems are examined. Keeping the differentiation and heterogeneous nature of the continent in perspective, the analysis

---

[29] There are fifty-four countries in Africa but data was available only for fifty-one countries. The countries not included were the Central African Republic, Democratic Republic of the Congo and Equatorial Guinea.

[30] Scholars such as Pouris and Ho (2014) have used the database for the study of scientific research and collaboration in Africa, selecting African countries whose data are available.

presented in this book follows individual country-specific examination.

African science production has been dominated by a few countries and there are lessons to learn from them. It would add value to the knowledge of science in Africa if a few select countries were chosen for a thorough and detailed analysis. In the second level of analysis in this study, the purpose was to gather lessons from successful countries in sub-Saharan Africa. A few countries, whose total production from 1945 to 2015 was the greatest, were selected, on the basis of the number of scientific publications until 2015 in sub-Saharan Africa. The highest numbers of publications were produced by South Africa (163,112), Nigeria (44,851), Kenya (23,473), Ethiopia (10,581) and Tanzania (10,315). These are the 'Big Five' countries in sub-Saharan Africa. This selection of countries is also based on the findings of several other studies.[31]

The second level of analysis of the selected countries necessitates the capturing of publication records individually. In view of the massive task of capturing and analysing all records for these countries for the period of 1945–2015, another selection criterion was used. Moving backwards from 2015, one year from every five years was therefore selected until there were enough publication records. A sufficient number of publication records prior to 1975 were not available and hence the year 1975 was the initial year chosen for the sample. There were thus nine sampled years of equal interval (2015, 2010, 2005, 2000, 1995, 1990, 1985, 1980 and 1975) and publications for all these years for the five selected countries were captured individually and analysed using a data management programme. There were 55,225 records,

---

[31] Gaillard's (2003a) analysis of scientific productivity in Africa for 1991–7 and 1987–2001 found that S&T capacities in Africa is concentrated in a few countries that include sub-Saharan countries such as South Africa, Nigeria and Kenya. Nigeria, Kenya, Tanzania and Ethiopia were among the major countries in scientific research in Black Africa (Dahoun, 1999; Eisemon, 1979). Bibliometric analyses of publication outputs agree that countries such as South Africa, Nigeria, Kenya, Ethiopia and Tanzania are the most productive countries in sub-Saharan Africa (Lubowa, 1992; Onyancha and Maluleka, 2011). In the Nature Index, the countries that have higher weighted fractional count (WFC) and research efficiency (measured in WFC per 1,000 researchers) were South Africa, Egypt, Kenya, Morocco, Tanzania, Tunisia and Ethiopia (Nature Index, 2014). The AOSTI study (2014), based on Scopus data, found that South Africa, Nigeria, Kenya, Tanzania and Ethiopia are among the top publishing countries in sub-Saharan Africa during 2005–10.

which had to be entered individually into a software programme. On average, each record took three or four minutes to capture, depending on the number of authors and participating countries in each publication record. Some 3,220 hours were required to do this in order to assure the quality of data entry. This focused analysis of each set of publication records is useful to unravel the dynamics involved in scientific production in the selected five countries. The lessons drawn from this will have value for other African countries in their scientific production and development of science.

For the first part of the study of science from Africa (the period extending from 1945 to 2015), there were about 500 variables to individually capture to the data management software. The variables represented publications according to years, collaborating countries, research areas, types and languages of the publications and research areas. This itself formed 25,245 (495 x 51 countries) pieces of information for the period 1945–2015. Another 21,318 (418 x 51 countries) must be added to this for the period 2011–2015. This is a glimpse of the quantity of empirical data used in this pioneering study.

Chapter Scheme

This study is presented in seven chapters. Historical and current trends in science are necessary for our understanding of the context and background of science in Africa. Chapter 2 looks into this aspect of the evolution and development of science, exploring the colonial and postcolonial features of African science. A host of themes are relevant here. How science and scientific institutions were established and administered, what the challenges were, and how science was integrated into the development goals are among them. Chapter 3 is meant for the analysis of scientific production in Africa, measured in terms of indicators such as scientific publications, separately for the sub-Saharan region. The analysis of science and its correspondence with GDP and GERD is carried out. Data was drawn from multiple sources and refers to the period 1945–2015. This chapter relates to the research questions regarding the relationships between the production of publications and historical phases, GERD and GDP.

The nature of science that is produced in Africa can be gleaned from analysis of scientific research areas prominent in African countries. The details of major scientific research areas across countries in Africa are presented in Chapter 4. The analysis makes use of large quantities of scientometric data to discern the trends – past and present. The differentiation within regions such as sub-Saharan Africa is also considered. In this chapter the attempt is also made to examine whether there are any relationships between research areas and the overall development of the country.

In any study of science, scientific collaboration is an integral component. Using the publication data, collaborative features in the production of science and trends of African scholars are demonstrated in Chapter 5. The chapter therefore focuses on the research questions relating to research areas and scientific collaboration. This covers domestic and international collaborations that Africa forged with their continental and international partners. The chapter allows us to see the collaborative patterns across countries, particularly sub-Saharan countries. A few productive countries in sub-Saharan Africa have been considered for detailed analysis.

Since independence, Africa has undertaken the responsibility of formulating science policies for the individual countries. Science policy is fundamental to both scientific and national development. This is one of the major objectives of the study. Chapter 6 reviews science policies in African countries to see how these policies are part of individual national development policies, programmes and strategies. Apart from science policies, the impact of other policy areas (economic, trade and education, for instance) is investigated. The focus of Chapter 7 is on the potential and challenges for African countries in regard to science and development. The themes covered in this chapter include scientific and technological dependency, funding, brain drain/gain, science education and awareness, research capacity, indigenous knowledge, rewards and incentives, and science and development.

2 | Science from Africa
Historical and Current Trends

Before dealing with the production of science in Africa, a consideration of its colonial and recent past is in order. The historical development of science is meaningful for the contemporary development of science. This provides an insight into the current nature of science.

Science Prior to Independence

Africa had a delayed start in scientific research (Gaillard and Waast, 1992). Compared to other countries in the world, research institutes and higher education institutions had a recent birth in Africa.[1] As a natural consequence of this, African countries were generally late in the formation and development of science and scientific capacity, starting only in the first decades of the twentieth century (Tilley, 2011).

The scramble for and partition of Africa also coincided with changes in science. In the colonial period, science in many parts of Africa served the purposes of the colonial administration. These colonial powers – Britain, France, Belgium and Portugal – pursued varying kinds of policies for scientific research. Britain followed its administrative approach of decision-making on scientific policy and administration at the local level while France administered colonies as their overseas departments (Smith, 1967). In the focus of research, there were differences among the colonisers. Britain emphasised more applied research while France went for more long-term research for Africa; the site of laboratories for Britain was primarily Africa but for France they were divided between France and Africa; and scientists were based in Africa for Britain but not in Africa for the French colonies (Smith, 1967).

[1] The first University College in Ibadan in Nigeria was founded in 1948 and the University of Dakar was formed only in 1957 while universities in India and Latin America were established much earlier (Gaillard and Waast, 1992).

In the colonial phase, universities and research institutes were established and supported differently in Anglophone and Francophone Africa. However, many who worked in these universities and institutes could not support research and experienced erosion in their capacity for training in scientific fields (Eisemon and Davis, 1992; Worthington, 1938). Except for training, great universities in Africa were not playing any active role in the scientific development of Africa or African development was influenced by the outcomes of scientific research (Worthington, 1938). Differences in the approaches amongst the colonisers were also evident in the research systems adopted by the British, French and Belgian administrations (Worthington, 1938). British territories, for instance, had permanent research offices and sites while there were fewer such offices in Belgian and French areas.

For Francophone countries, support for the establishment of universities occurred later than in Anglophone countries (ICIPE, 1988). In Anglophone Africa, councils were tasked with the responsibility of defining research priorities, managing budgets and assessing results. These national councils made their objectives consistent with national plans that fit in with the interests of the colonisers. In Francophone Africa, scientific research was directed under a ministry which was in charge of research establishments and coordinated research activities (Waast, 2009).

In the colonised period, scientific research was focused on finding solutions to urgent and immediate problems faced by the colonisers. Such research was mostly related to tropical diseases, health, medicine and agriculture (Lebeau et al., 2000; Worthington, 1952). This explains why certain fields of science, for instance, agriculture, developed rather than others (UNESCO, 1986) in Africa. This is examined in Chapter 4. Some African countries continue to maintain their capacity in these areas. Research was also conducted to support the economic and political policies of the colonial powers (Dow, 1988), to address problems that affected colonial settlers and administrators (Forje, 1992), and to emphasise the relationship between scientific research, economic development and governance (Bonneuil, 2000).

The colonial period under which science had to survive was not beneficial in building science or science capacities for the colonies. Colonial policies were not meant for the creation of appropriate skills and research facilities in Africa nor to address its social and economic needs (Gruhn, 1984). It was not the intention of the colonial policies for the advancement of science in Africa. The colonisers emphasised applied research that was more pertinent to their interests, prosperity and physical survival (Gruhn, 1984). Most often science relied heavily on the scientific institutions of the colonisers (Mouton, 2008). This had implications for the future of science development in Africa. The legacy of the colonisers, their research agendas and policies were not enough to play a facilitative role in Africa to use science and technology (S&T) for its local problems and development.

Scientific research scarcely existed in the colonies of Africa (Ferreira, 1974). The partition of Africa itself was motivated more by economic gains and prestige than anything else (Sanderson, 1985). Scientific growth and development was not part of the main agenda for the colonisers. In colonial Africa, economic development was conceptualised in terms of production of export-oriented goods, and the importation of technology and skilled personnel (Adedeji, 1984). In theory, research was limited to land surveying (cartography, hydrology and hydrography), and agriculture (Ferreira, 1974). Some countries such as Kenya established experimental research stations where scientific research was undertaken mainly in agriculture.

Research undertaken in the colonies was either conducted or administered from colonial locations. Most of the research related to Africa was not done in Africa but in the colonial capitals. Paris, London and Brussels were the headquarters of research institutes for research in tropical agriculture and medicine (Gruhn, 1984). The French colonies in Africa were not the centres of research for Africa-related problems, but Paris served as the main centre for such scientific and research activities. In the case of Anglophone Africa, relatively more African-related research was carried out in British colonies (Gruhn, 1984) than in other colonies. Britain was concerned about tropical diseases in Africa, which prompted the administration to direct research in that specific area (Smith, 1967). For instance, until 1940, agricultural research in Africa was centred around particular products; since 1940, this has gradually changed towards improving production of cash crops and subsistence crops (Smith, 1967).

The early British colonial science policy, called constructive imperialism, adopted before the First World War, was aimed at facilitating development of the vast estates of the colonial empire (Worboys, 1996). By the mid-1920s this science policy was elaborated and expanded so that economic development of the British colonies could be accelerated and their economies integrated into a dependent relationship with Britain (Worboys, 1996). Colonial science was an extension of European nationalism and it was meant to be a science for the empire (Macleod, 1996) and not for the colonies. In the later years, however, a natural tendency in Africa emerged, south of Sahara in particular, for scientific independence from home (colonial) countries facilitating the growth of scientific institutions and organisations (Worthington, 1952). As a cumulative effect of these efforts, tangible changes were to be seen on the horizon of science in Africa. In Worthington's brief (1952), the conditions for science and research in colonial Africa improved in the mid-1930s, development of regional organisations to support research occurred in the 1940s, and the territorial and regional boundaries of science began to dissolve in the 1950s for inter-African cooperation in science. Close to independence, some progress in education in some regions of Africa could be seen. Soon after 1945, there was a rapid expansion in education, both primary and higher levels, in West Africa (Williams, 1984).

Science in Postcolonial Africa

Africa inherited its legacy in science from the colonial past.[2] This legacy is referred to as three contexts from which African science moved to national science (Gaillard et al., 2005; Waast, 2002). Firstly, it is the knowledge in which specific disciplines were developed (agriculture and tropical medicine, for instance). There were detailed inventories and recorded information for knowledge contexts. Secondly, it is organisational models (full-time researchers employed by specialist agencies). Specialised research institutions were created and scientific personnel were employed as civil servants. Thirdly, it is the context of the strategic choices, creating research priorities such as favouring agriculture and health. The legacy was to be carried on later in many African countries.

---

[2] Some of the efforts were evident in organising pan-African conferences, establishing a research agency and its branches in Francophone Africa, and research and experimental stations, the emergence and growth of scientific communities (from 1,000 researchers in 1930 to several thousands by 1950) (Bonneuil, 2000), and inter-territorial services primarily meant for research (Worthington, 1952).

Africa had a dark period of its science. The research climate in Africa remained largely political in the 1960s. The key variable in determining research was the political atmosphere (Rivlin, 1969), which prevailed in these countries but was not essentially similar across Africa. Upon gaining political independence beginning in the late 1950s, African countries worked to strengthen their national systems of S&T. Structures meant for the formulation and implementation of relevant scientific policies have been established since then. For instance, some countries formed ministries of scientific and technical research. Senegal, Ivory Coast, Cameroon and the Central African Republic formed their own ministries of science, technology and development (Gaillard and Waast, 1992).

There were few scientists that Africa could produce in the decade following independence. At the time of the Lagos Conference, the number of research personnel available in Africa was minimal. There were about 3,000 trained researchers and a majority of them were non-Africans (Odhiambo, 1967). But soon after, Africa made progress in its scientific system. During the 1960s and 1970s countries went for the development of universities, Africanisation of research, creation of academic positions and expansion of scientific establishments (Waast, 2002). National policies were introduced through the 1970s and 1980s. Changes in the higher education system were to be seen. An increase in the uptake of students occurred at several universities on the continent. Efforts were intended to build national research systems in several countries (Gaillard and Waast, 1992). International organisations showed interest in establishing research centres and institutions across Africa. In the 1970s, international research institutions in sub-Saharan Africa were in the forefront of science production (Davies, 1983). One-seventh of all the mainstream science authors during this time were based either at bilateral or multilateral research institutions that were supported by international donors (Davies, 1983). International organisations were mostly connected to agriculture research. By 1973, there were more than 300 research institutes in sub-Saharan Africa conducting research in agriculture, and the research produced at many of these institutes was reportedly world class (Dow, 1988).

The recession that began in the 1970s affected Africa and, naturally, its production of science. The fall in oil prices in the 1970s was a blow to many African economies. The recession was accompanied by economic crises. Some countries were more severely affected than others. Due to the economic crises, the growth of GDP in sub-Saharan Africa was compromised between the 1970s and 1990s (Cooper, 1994). There was no real growth in the per capita income. Rather, it declined in many African countries (Cooper, 1994). Scientific research and science development in Africa suffered dearly.

In the event of the economic crisis, there has been a cut in the budget earmarked for scientific research. The crisis in universities and research institutions in Africa depressed the levels of research activities and the production of scientific research (Eisemon and Davis, 1997). Salaries of researchers and university professors were slashed. Research or teaching too, for that matter, as the profession was no longer attractive. The worsening economic and working conditions, and the deteriorating social status of scientists and academics, made the survival of academic and research professions difficult. Institutions were failed for want of maintenance funds. Science lost its glamour among scientists and the public alike. Scientists were frustrated and restricted in their careers.

An extensive survey conducted by Gaillard and Tullberg in 2001 showed this dismal scenario. Scientists were disappointed with job security, salaries and social benefits. Many of them had to supplement their incomes with extra jobs such as consultancies, private businesses or teaching (Gaillard and Tullberg, 2001; Lebeau, 2003). They either changed their professions or migrated to other countries (Gaillard et al., 2005). Similar findings have been reported in an early study of two prominent universities in Nigeria and Kenya. Eisemon (1980) found that many academics in the studied universities (Ibadan and Nairobi) had substantial business and consulting commitments. Budgetary cuts in Nigeria affected scientists losing their social status, purchasing powers and standards of living (Lebeau, 2003). Some of the constraints, as the study of Nigerian scientists showed, were the lack of research equipment, funds and libraries that did not have updated and current material and journals (Ehikhamenor, 1988, 1990). These constraints, caused by cuts in funds to support science, also led to the marginalisation of the research output and the loss of interest in the development of science in Nigeria (Lebeau, 2003). Nigeria experienced a 50 per cent decline in the production of scientific publications during 1991–7. Its scientific establishments were worsened by large budgetary cuts, scientists either changed their professions or migrated to other countries for better pastures, and those who remained in the country sought other work for their living (Gaillard et al., 2005; Hudu, 2015).

Another survey by Court (1988) offered a similar set of findings. The lack or poor maintenance of scientific equipment, degradation of libraries and absence of funds for research and postdoctoral training prevented researchers from staying in Africa (Court, 1988). Many opted to leave their professions in search of other jobs. Some left for greener pastures across the boundaries of the continent. Nigeria is a strong case of this deterioration and decline in science. There had been an exodus of scientists during 1980–95, who found jobs in the USA or in Europe (Waast and Krishna, 2003b).

Countries like Cameroon, which had already built up a strong higher education and research system, also suffered in the economic melt-down. Public funding for research either dried up or stopped in Africa. Many of these countries had to survive with international funding, and researchers had no option but to leave the profession and look for something else (Gaillard et al., n.d.). The drying up of overseas funding sources contributed to declining support for research at universities (Sutton, 1988). Some of the great institutions in Africa lost their standards and credibility. Remember, Africa had excellent universities and research institutions in the world (Makerere and Ibadan, for example). Many of them declined in their standards at this time.[3]

In the 1970s, there was only a slow growth of science produced at universities and public sector institutions (Davies, 1983). Following this, science did not find a climate to grow in Africa in the next decade of 1980s. It was due to the laissez-faire policies of governments, as ineffective and directionless national bodies, that universities failed to meet their research responsibilities, scientific communities were impoverished, and the profession of science became individualised (Gaillard et al., 2005).

Conflicts within the country and wars between ethnic groups and countries have been a regular recurring feature of African history. Science, scientific institutions and scientists suffered under those trying conditions. The efforts that were undertaken to stabilise and develop science in those countries had setbacks and drew them back several

---

[3] The National University of Zaïre, Legon University in Ghana and Makerere University in Uganda were among them (Court, 1988).

years in scientific advancement. For instance, civil wars in Rwanda, Burundi, Mozambique and Angola and the dictatorial regimes in Uganda, Zimbabwe and several other countries had a negative impact on science and institution building (Mouton, 2008). Research institutions and universities were not able to perform their functions under situations of conflicts and political uncertainty. Funding, both internal and external was either cut or stopped and this made scientists unable to pursue their research interests. Political instability, which continued for years in several African countries, prevented these countries from focusing on applying science for national development.

The phase for science after political independence in Africa has been a different one for most parts of sub-Saharan Africa. Incapable and corrupt political leadership was also responsible for the declining state of science in African countries. Dictatorial political leadership did not show much concern for the growth and development of science. The style of leadership in African countries did not augur well for universities and research institutions. There was political interference in the appointments at universities and research institutions, but this was not generally applicable to most of the North African countries.

Political instability caused a decline in science and undermined efforts at scientific growth. In Ghana, for instance, political instability during its independent phase caused the loss of trained professionals, scientists, academics and engineers who left the country (Amankwah-Amoah, 2016). Science did not continue to receive the support it obtained previously. African states began to detach themselves from providing support to science and scientific research and funding from the state became an issue. Researchers and academicians became disillusioned as their status in society began to erode. In a survey of Nigerian university scientists, it was found that a large majority (about three-quarters) felt that they did not enjoy much recognition from society for being scientists (Ehikhamenor, 1988).[4]

The scenario of science in Africa however did not remain like this forever. The scientific landscape of the continent was about to change. Measures were to follow for this change to occur. A series of meetings,

[4] This is partly, in the view of surveyed scientists themselves, because the general public is ignorant about scientific research or what university scientists are doing. Partly it is also due to the fact that the scientists have not distinguished themselves in the work they are doing, which the public cannot appreciate (Ehikhamenor, 1988).

conferences and symposia was held for the advancement of S&T in Africa. In 1964, the International Conference on the Organisation of Research and Training was organised in Lagos (Nigeria). A symposium on Science Policy and Research Administration was conducted in Yaoundé in 1967. Following this, the Conference of Ministers of African Member States Responsible for the Application of Science and Technology to Development (CASTAFRICA) was held in Dakar in 1974.[5] Another major event was the 1979 Vienna Conference on Science and Technology for Development. The Assembly of the Heads of State and Government of the Organisation of African Unity held in Lagos in Nigeria in April 1980 proposed a plan of action for member states. S&T was an integral part of this Lagos Plan of Action. A number of key plans were accepted at this meeting. Among them were the establishment of a centre for S&T with the aim of removing technological dependence and gaining technological self-sufficiency, development of human resources for S&T that entailed advanced training, development of technological entrepreneurship, popularisation of S&T, revision of curriculum that will address the development needs of Africa, prevention of a brain drain, development of infrastructure for an S&T base, development of transfer of technology, institution building, and cooperation (OAU, 1981).

The Kilimanjaro Declaration in 1987 called for the implementation and promotion of S&T. The declaration envisaged that centres of excellence for technological development be established in Africa and educate and train researchers, engineers, technicians and specialists. The declaration called for the member states to encourage the establishment and strengthening of African S&T communities to solve developmental problems that Africa was experiencing. From the policy perspective, the declaration encouraged African countries to define and implement their own national S&T policies with a view to strengthening their capabilities and capacities in S&T. Later in 2005, the African Union Ministerial Conference adopted the Africa's Science and Technology Consolidated Plan of Action as a framework to respond to the socioeconomic challenges of Africa (AU-NEPAD, 2010). The Summit of the Heads of State and

[5] CASTAFRICA I, held in Dakar, Senegal, in 1974, and attended by the majority of the countries in Africa, focused on three main themes: trends of S&T policies, new technologies and possibilities for their development and applications, and scientific and technological cooperation in Africa (Forje, 1992).

Government of the African Union at Addis Ababa in 2007 declared that the ability of Africa to meet the Millennium Development Goals (MDG) depends on the ability to harness S&T for development. The declaration of the summit aimed at advancing development by promoting research in all fields in general and S&T in particular. The plan was to increase funding for S&T, and support the establishment of national and regional centres of excellence in S&T (NPCA, 2014).

In 2000, the Cairo Plan of Action adopted the formulation of programmes with special emphasis on S&T. The Africa's Science and Technology Consolidated Plan of Action (CPA) was devised under the leadership of the New Partnership for Africa's Development (NEPAD) and, extended from 2005 to 2014, became the major consolidated plan for Africa to accelerate its transition towards an innovation-led knowledge economy (Kraemer–Mbula and Scerri, 2015). This plan of action was built on three interrelated conceptual foci: capacity building, knowledge production and technological innovation (NEPAD, 2005). Capacity building, as per this plan, envisaged the creation, improvement and mobilisation of human skills, infrastructure and resources, and the development of necessary scientific policies to solve problems that are specific to Africa. Knowledge production is about the way science is to be conducted in Africa, in order to solve Africa's problems. Specific products, processes and services were covered under technological innovation.

The heads of states and governments of the African Union adopted the Addis Ababa Declaration on Science and Technology and Scientific Research for Development in 2007. The declaration was to implement the summit decision on S&T policy and to support the member countries (UNESCO, 2009). A UNESCO-led initiative in 2007, namely, the African Science, Technology and Innovation Policy Initiative (ASTIPI), was made to develop national science, technology and innovation (STI) policies for all African countries that had not formulated one as yet. ASTIPI was expected to reform the science systems that exist in African countries, focusing on policy analysis, policy formulation and the development of relevant African STI indicators. The World Bank enterprise, namely, the Partnership for Applied Science, Engineering, and Technology (PASET), brought together African ministers to make a joint call at Kigali in Rwanda in 2014. This call was to adopt a strategy that uses investments in S&T to accelerate development in

Africa towards a knowledge-based society within one generation (Blom et al., 2016).

The AU plan, the Science, Technology and Innovation Strategy for Africa (STISA) designed in 2014, aims to prioritise scientific research to drive economic and social development across Africa (AU, 2014a; Nordling, 2014). STISA replaced CPA. The strategy is a long-term plan for Africa, underpinned by STI to achieve developmental goals. STISA, compared to the previous plans, has a sharp focus and emphasis on science for development (Kraemer–Mbula and Scerri, 2015). The strategic plan of the Economic Commission of Africa of the UN recommended that measures should be taken to make scientific research geared towards the social and economic needs of Africa (UN, 1980). This plan was based on the premise that scientific research in Africa is not properly linked to its application for the continent. In 2014, the Science, Technology, Engineering and Mathematics Africa Conference held at the University of Michigan agreed on a charter that reaffirmed the commitment to strengthen and advance STI in Africa.

At the policy level also, there were initiatives to set up ministries in many African countries. The number of countries with separate ministries of S&T increased from 9 to 17 between 1979 and 1984 (Wakhungu, 2001). This was the beginning of a new system of scientific production in Africa. The system brought in some success as well (Waast, 2002). Around this time, scientific publications originating from Africa became internationally known and research establishments gained reputations (Waast, 2002).

Since independence, countries in North Africa were providing considerable support to S&T institutions and higher education (Waast and Krishna, 2003a). The scientific system in Algeria moved at a reasonable pace ever since its independence. At the time of independence in 1962, the new state leadership did not have plans for a national science policy for the country (Kenz and Waast, 1997). The country still had to wait for years before state interest in S&T was born. In 1970, there were attempts to organise its scientific system through numerous research and development programmes. Under its National Office for Scientific Research and through the Ministry of Higher Education and Research, established in 1973, the first National Programme of Priority Research was implemented (Khelfaoui, 2004). New structures were formed, discarding some of the existing ones, which provided directions to scientific research. Algeria introduced its first national programme with hundreds of research projects (Esau and Khelfaoui, n.d.). These efforts had an effect on Algeria gaining its feet in science. In the later period between 1993 and 2001, the country saw the weakening of its system under the pressure of civil war and the loss of government interest in science (Esau and Khelfaoui, n.d.). However, during this period and later, Algeria's contribution to world science did not come to a standstill. Algeria increased its share of publications, as shown in the Web of Science (WoS) database, in fields such as chemistry, physics, astronomy, geosciences, mathematics and engineering (Esau and Khelfaoui, n.d.). Science production in Algeria did not collapse during the civil war (Mouton and Waast, 2009).

In other North African countries, for example, Morocco and Tunisia, S&T was valued and supported. Political leaders and heads of states were aware of the importance of S&T for development. They drew up systematic plans to improve S&T for developmental purposes (Waast and Krishna, 2003a). Institutions for the advancement of S&T were built and nurtured. After independence, Morocco was keen to develop its universities and scientific institutions. Morocco's ministries and industrial firms established research and development (R&D) departments, its university system was decentralised, the status of the teaching staff at universities enhanced, and research turned into a core activity at universities (Waast, 2002).

Recession had its effect on science and scientific research in Morocco as well, but it was not as bad as in the sub-Saharan countries (Waast, 2002). Both teaching and research in Morocco maintained their status and remained attractive (Waast, 2002). Its five-year plan (2000–2004) aligned S&T with socioeconomic developmental goals, choosing the priority areas of agriculture, health, fisheries and ICT, among others (Boshoff and Kleiche, n.d.). In 1976, it created the National Centre for Scientific and Technological Research, responsible for the scientific and technological potential of the country, conducting studies on national priorities to link to the national development plan and providing support to research (Boshoff and Kleiche, n.d.). The research capabilities and output of Morocco have been improving since the beginning of the last decade, with the growing support of the government which is evident in the percentage of the Gross Domestic Expenditure on Research and Development (GERD). Publications of Moroccan scholars that appeared in the best international journals doubled over the fifteen years since 1995 (Mouton and Waast, 2009).

Similar trends were to be seen in Tunisia soon after its independence in 1957. The university system underwent rapid expansion, research and publications were emphasised, schools were opened for professional training and technical services to meet the demands of the country, and research establishments were formed (Wasst, 2002). Tunisian leaders took a supportive stand towards science by harnessing its benefit for nation building. Ever since its independence, science has been regarded as one of the conditions for its survival as a nation (Siino, 2003). As a corollary of these efforts, Tunisia advanced in its production of publications. It presented powerful growth in science publications during the 2000s (Mouton and Waast, 2009).

In independent Egypt, the government did not have much understanding of scientific work, the relevance of science policies, and the importance of innovation, research and development (Zahlan, 1997). The country improved in its science production only in later years.

Burkina Faso made efforts to develop its own scientific system from the days of its independence in 1960. In the 1960s and 1970s, it revamped the available structures for science and created new ones to conduct scientific research (Khelfaoui, n.d.). A completely overhauled national research system gave rise to a strategic plan in 1995 (Khelfaoui, n.d.). Cameroon, which also became an independent country in the same year, began to consolidate its efforts early in the 1970s to develop a national science system. Under its Office National de la Recherché Scientifique et Technique, founded in 1974, Cameroon coordinated research activities taking place at research institutes across the country (Gaillard et al., n.d.). Malawi had a late start in utilising the benefits of S&T for development: a ministry for S&T was only formed in 2004 to support, promote and develop the application of S&T to development (Boshoff, n.d.).

Kenya and Nigeria were open to develop their science soon after their independence (Rabkin et al., 1979). As these countries did not inherit much from the colonial past, they focused on rectifying the deficits of colonial times, which were evident in the policies developed in the 1960s and 1970s (Rabkin et al., 1979). Nigeria had a long tradition of higher education, and was to expand its scientific system at the time of independence (Eisemon, 1981). It established several universities in the first few years of its political freedom (Eisemon, 1981).

The growth and development trajectory of science in Africa passed through a nonlinear path in its history. From a weak starting point in

1960, Africa began to gradually strengthen its science and scientific institutions and universities from the 1970s up until the mid-1980s (Gaillard and Tullberg, 2001; Gaillard et al., 2005). Matters began to improve in the mid-1980s. The scientific output of African scholars appearing in international journals also began to increase (Eisemon and Davis, 1997). This was the beginning phase of growth for Africa. An increase in the number of scientists and academics was s obvious during this period (Gaillard and Tullberg, 2001).

Scholarships to study abroad, offered by the industrialised world, have enhanced the value of science in African countries (Gaillard and Waast, 1992). It was also the intention of the donor countries to hold their influence on these new independent states through assistance (Gaillard and Waast, 1992). The number of African graduates and scientists began to grow faster in the 1970s (Gaillard and Waast, 1992). Except for a few, all countries had institutions for higher learning during this period (Gaillard and Waast, 1992). In some countries, science and scientists were valued. Graduates were attracted to science where a well-organised research system existed and where researchers enjoyed high status and good salaries (Gaillard et al., n.d.).

International research institutes to conduct research in demanding areas of science were formed in Africa. The International Institute of Tropical Agriculture (IITA) was founded in 1967 in Ibadan, Nigeria. In 1970, the International Institute of Insect Physiology and Ecology took shape in Ibandan, Nigeria. The International Laboratory for Research on Animal Diseases (ILRAD), was established in Nairobi in 1974; the International Livestock Centre for Africa (ILCA) in Addis Ababa (Ethiopia) in the same year; and the West African Rice Development Association (WARDA), Monrovia in Liberia in 1976. In 1977, the International Centre for Research in Agroforestry was established in Nairobi. The African Academy of Sciences was formed in Nairobi in 1985 (Gaillard and Waast, 1992; Keay, 1976). Institutions to promote science and research emerged in Africa. The African Academy of Science was created in 1985. Following this, the Pan-African Association for Science and Technology was founded in 1987.

The African Union (AU) launched the New Partnership for Africa's Development (NEPAD) in 2001. NEPAD formed an expert group on science, technology and innovation to promote and strengthen S&T. The Southern African Development Community (SADC), which came into being in 1992, was also engaged in the promotion and development of science, technology and innovation in member countries.[6] The SADC protocol on science, technology and innovation, which was adopted in 2007, worked on collaborative initiatives to support S&T among member countries. This protocol required member countries to develop mechanisms to strengthen regional cooperation in science, technology and innovation. It was also geared to promote public understanding and awareness of S&T.

The research scene in Africa, the mode of production in particular, was transforming but gradually. The mode of production in science, according to Waast and Krishna (2003b), changed to development rather than to investigation; the profession of science was practised on the basis of orders for research work and contracts; international but not national demand determined the programmes and objectives of research; and the system was no longer regulated by peer assessment but by the market. The consequences of these were manifest. Some disciplines of science disappeared in the absence of market demands (Waast and Krishna, 2003b). The structural changes that occurred in some countries had their impact on science. The structural adjustment programmes implemented in Ghana, Kenya, Tanzania and Uganda had their effects on the advancement of S&T (Enos, 1995). There has been a slight upward trend in the domestic expenditures on S&T under the structural adjustment programmes in these countries.

Africa as a whole had 60,900 researchers in 2002, while sub-Saharan Africa had 30,900 researchers. This was 1.1 per cent and 0.6 per cent, respectively, of the total world researchers at that time. For Africa, there were 73.2 researchers per million inhabitants while the figure was only 48 for sub-Saharan Africa. The world average was 894 per million population.[7] Africa needs many more researchers. One estimation for a reasonable number of researchers, as adopted at the Lagos conference, was 200 researchers per million population, which meant about 49,000 researchers for Africa in 1967 (Odhiambo, 1967).

In 2002, the world share of the GDP for R&D was 1.7 per cent which was equivalent to 830 billion US dollars (Tindemans, 2005). These figures also represent the disparities in the allocation of funds for R&D across regions and countries. While there has been an increase in these

[6] SADC consists of countries such as Angola, Botswana, the Democratic Republic of Congo, Lesotho, Madagascar, Malawi, Mauritius, Mozambique, Namibia, Seychelles, South Africa, Swaziland, Tanzania, Zambia and Zimbabwe.
[7] UNESCO statistics for 2004, cited in Tindemans (2005: 6).

funds in certain regions like Asia, the decline in GERD was reported in Africa (2.6 per cent from 3.1 per cent in 1997) (Tindemans, 2005). In 2003, at the first African Ministerial Conference on Science and Technology, African countries had agreed to commit to an increase in spending on R&D to at least 1 per cent of the GDP in the next five years (UNESCO, 2007a). As of 2010, only three countries, namely Malawi, South Africa and Uganda,[8] reached above 1 per cent of GDP while for all other countries it ranged between 0.20 per cent and 0.48 per cent (AU-NEPAD, 2010). The situation has not changed since then. As per UNESCO (2019) statistics, the average figure for sub-Saharan Africa in 2015 was 0.4 per cent as against 1.7 per cent for the world, 1 per cent for Central and Eastern Europe, 2.4 per cent for North America and Western Europe, 2.1 per cent for East Asia and the Pacific, 0.5 per cent for South and West Asia and 0.2 per cent for Central Asia.

The contribution of African scientists to mainstream scientific production was scanty but not less than in other less-developed countries.[9] There was an increase in publications because of the contribution of scientists in some countries (Gaillard and Waast, 1992). In 1991, the share of sub-Saharan Africa to the world scientific production was 0.9 per cent and that of North Africa was 0.4 per cent (Barré and Papon, 1993).[10] Between two decades (1991 and 2001) the production of scientific publications in Africa declined in percentile terms. Its share to world scientific publications was 1.6 per cent in 1991, which was reduced to 1.4 per cent in 2001. The decline was more prominent for sub-Saharan Africa (1.0 per cent and 0.9 per cent), while there was a growth for the Arab countries in Africa (Tindemans, 2005). Presenting the state of science, Waast (2002) showed that Africa lost part of its share in world publications, and the decline was more prominent in some countries. As the African countries contribute

---

[8] Uganda has made substantial progress in the field of S&T in sub-Saharan Africa, by increasing its spending from $73 million in 2005–2006 to $155 million in 2008–2009, which is closer to 1 per cent of its GDP (Irikefe et al., 2011). However it has challenges in terms of fewer researchers per million population (Irikefe et al., 2011).

[9] This is based on the publication productivity of African countries for the 1980s (Gaillard and Waast, 1992).

[10] This can be compared with 2 per cent for India, 1.1 per cent for China, 35.8 per cent for the USA, 4.4 per cent for Canada and 8 per cent for Japan. The figures are based on the analysis of the database of the Science Citation Index (SCI).

only a fraction of scientific knowledge to the world, they also have limited access to the world S&T potential (Barré and Papon, 1993).

Using bibliometric records from sources such as the Web of Science (WoS), publication outputs of countries have been examined. Ondari-Okemwa (2007) made a preliminary analysis of the publications of sub-Saharan Africa for the specific period between 1997 and 2007. The study of Waast (2002) covered fifteen African countries. Covering the production of knowledge in fifteen of the most productive African countries, Narváez-Berthelemot et al. (2002) reported a growing increase in African science during 1991–1997. Arvanitis et al. (2000) did a study on African science for the same period, but using a different database and including all African countries. Shrum (1997) concluded that publication productivity of scientists is a reflection of the national level of development. But a study of the whole of Africa has yet to appear in the literature.

The just-quoted figures apart, scientific production within Africa has shown some recovering statistics in recent years. The AOSTI (2014) study of scientific production in Africa between 2005 and 2010 suggests trends in scientific growth and development in Africa. Publications during this period grew 22 per cent faster than world production. Regionwise, break-up indicated growth of 60 per cent in Arab Maghreb countries and 50 per cent in Sahel-Saharan countries. Scopus data (2003–2012) showed that between 7.5 per cent and 16 per cent of the different sub-Saharan's total publications were included among the world's top 10 per cent of the most highly cited publications (Blom et al., 2016). The trend is repeated in other studies as well. Schemm (2013) reported that scientific research conducted in Africa has seen an increase in recent years. The number of papers published by African scholars has increased from 12,500 in 1996 to 52,000 in 2012 (Schemm, 2013). This means the share of the world's articles with African authors doubled during this period (Schemm, 2013). The propensity of African scholars to publish has risen since 2004 (Confraria and Godinho, 2015). In sub-Saharan Africa, research output doubled between 2003 and 2012, and its share of global research increased from 0.44 per cent 2003 to 0.72 per cent in 2012 (Blom et al., 2016) (Map 2.1 and Figure 2.1).

Some analyses have shown the visibility and impact of the research output of African scholars which, in comparison to the world average,

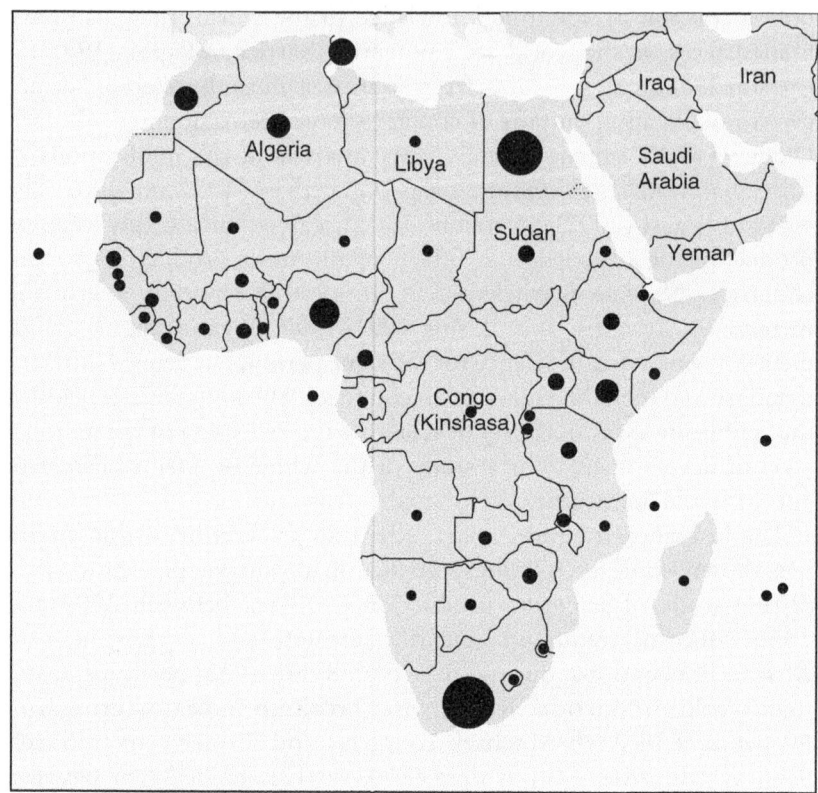

Map 2.1 Africa showing the size of publication output, 1945–2015

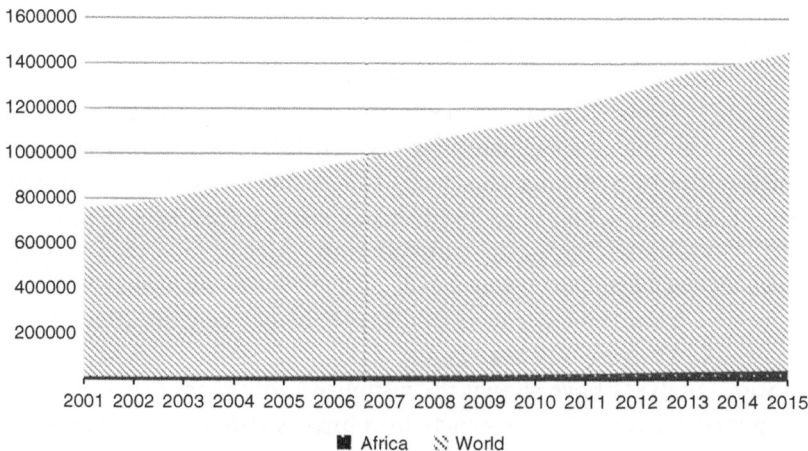

Figure 2.1 Publication output, world and Africa, 2001–15

is not great. Tijssen (2007), who looked at the citation impact for the period between 2001 and 2004, reported poor visibility and impact of African science. In his view, this is not because African science is mediocre or second rate, but there are underlying factors such as outdated or inadequate research facilities, the lack of incentive structures to encourage research, and poor career prospects for those who perform well in research (Tijssen, 2007).

Who are the prolific researching countries in Africa and what characteristics distinguish them from others? According to the Lotka's law of productivity and frequency of publications of scientists, the number of scientists with a lifetime production of publications is proportional to $1/n^2$.[11] The number of persons making n contributions is about $1/n^2$ of those making one contribution and the proportion of all contributors who make a single contribution is about 60 per cent (Coile, 1977; Lotka, 1926; Nicholls, 1989). For Moravcsik (1984), for example, for every scientist who publishes 100 papers in his/her lifetime, there are 100 other scientists who publish only 10 papers and there are 10,000 scientists who will publish only one paper in their lifetime. The findings of the study of Shrum (1997), which also covered African countries such as Ghana and Kenya, refer to the relevance of visible scientists and their production patterns. He found that visible scientists whose works appear in international research are more established, university scientists possessing PhDs and participate actively in professional associations and journals. They have a stronger orientation towards international science and have more influence on problem selection (Shrum, 1997).

The AOSTI (2014) study of Scopus data provides insights into these aspects of science in Africa. By AOSTI definition, leading scientists are the ones who published more than 40 papers during a five-year period (2005–2010) and 505 such scientists were thus selected for the analysis. Half of them had a score of above 1, meaning they have been cited more than the world average, 52 per cent of them had an output growing faster than the world average, more than 50 per cent of them had publications coauthored with authors from other countries, and they had a greater

---

[11] The law is based on the study by Alfred Lotka in 1926 on the frequency distribution of scientific productivity of chemists and physicists (Lotka, 1926).

than average scientific impact. South Africa hosted the largest number of leading scientists (214 of 505), followed by Tunisia (128), Egypt (52), Kenya (25), Algeria (16), Nigeria (11) and Cameroon (10). In regard to scientific disciplines, leading scientists are frequently found in the health sciences (microbiology, virology, tropical medicine, general and internal medicine) and natural sciences (applied physics, inorganic and nuclear chemistry, medicinal and biomolecular chemistry and biotechnology). In many African countries such as Kenya, agriculture research, which was usually conducted at government research institutions, was the centre of development of the scientific community (Eisemon and Davis, 1997).

Africa also has its niche areas in scientific research where it has advanced its capabilities and achievements. A specific study of some flagship universities in Africa indicated that certain scientific fields in those universities have, over the years (2006–12), accumulated critical mass of resources on par with international standards and capabilities (Tijssen, 2015). Some branches of science grew much faster than others in Africa. This growth and decline needs to be viewed in the specific context of Africa. An analysis referring to the 1970s showed that two branches, namely, agriculture and health, jointly accounted for 80 per cent of the total production in Africa (Gaillard and Waast, 1992). Over the years since 1991 there has been an increase in the number of publications in Africa. In 2001 growth had occurred in the fields of biology, biomedical research, chemistry, clinical medicine and in earth and space (Tindemans, 2005). But these increases were not on a par with the increases that occurred simultaneously at the global level.

Scientific collaboration is a decisive factor that causes a positive influence on the production of scientific outputs. Studies have repeatedly confirmed the linkages between collaboration and the productivity of scientists. While Africa recognises the importance of collaboration, this has not fully materialised in scientific production. Scientists in many African countries work in relative isolation and cooperation within Africa, and between Africa and the rest of the world, has not led to any concrete research activities, programmes and institutions (Gaillard, 2003a). Some research fields have however largely benefited from international collaboration, clinical medicine, environmental sciences and technology, chemistry and chemical engineering among them (Tijssen, 2015). Collaboration will be further discussed in a later chapter.

More on the production of science in Africa in the postcolonial period needs to be examined, which is undertaken in the next chapter.

3 | Production of Science in Africa
Decisive Indicators

In comparison to other continents and countries, the national science systems in Africa present a heterogeneous type of orientation (Tijssen, 2007). Within Africa and sub-Saharan Africa, scientific production is no less skewed than in any other region of the world (Mouton and Waast, 2009). Reasons are many. The fractional use of GNP (Gross National Product) for R&D (Research and Development), absence of advanced research in information technology, poor institutional support, inimical social and political contexts for the promotion of science, and the low number of students enrolling in scientific disciplines are seemingly responsible for this heterogeneity (UNDP, 2003).

In S&T studies there are a few relevant key indicators. They include R&D expenditure and the number of people engaged in research, patents obtained, scientific papers and graduates in S&T subjects (UNESCO, 2007b, 2015b). The relationship between the scientific productivity of a country, R&D expenditure and GDP may not be a simple and straightforward one. The relationship is rather complex and therefore it is not easy to interpret the connection (Inönü, 2003). While investment in R&D remains a major input for growth in national scientific capacity, efficient use of investments in R&D cannot be entirely neglected (Hung et al., 2009). Scientific production in a country is not just the function of its economic standing alone. Countries that have an established scientific tradition will have a favourable national and cultural character and will continue to produce well (Inönü, 2003). No less important are factors such as overseas experience and attendance in professional activities (Thomas, 1992).

In this chapter, the production of scientific publications of countries in Africa is analysed. The analysis is carried out in relation to world and other comparable regions and its relationship between R&D, GERD (Gross Domestic Expenditure on Research and Development) and GDP (Gross Domestic Product). The data was drawn from the World Bank, the UNESCO Institute for Statistics, UN world population statistics, and the Web of Science (WoS). Although the dataset of the WoS contained publications since 1945, there were not many publications for the period 1945–70 in the database.

Science Production in Africa

Publication is a measure of productivity, although it is not the only one (Gaillard, 1991). As shown in Table 3.1, a total of 549,427 publications were produced by Africa during 1945–2015. This is an average of 10,806 publications per country, with the lowest number of 27 publications coming from Sao Tome and Principe and the highest number of 163,148 publications from South Africa. The median for these countries is 2,000 publications. In this production, two countries have become the prominent producers of knowledge in Africa: South Africa and Egypt. Being the leader in publications, South Africa's share was about one-third (29.6 per cent). The second major share to African science came from Egypt with a contribution of 20.8 per cent. Both these countries together produced more than half the publications for the whole of Africa. The remaining countries each contributed about 8 per cent or less of the total publications. This is a huge gap between the scientifically large and small countries in Africa. The contributions of the rest of the countries were Nigeria (8.1 per cent), Tunisia (6.6 per cent), Morocco (5 per cent), Algeria (4.6 per cent) and Kenya (4.3 per cent). About 2 per cent each were made by Ethiopia, Tanzania, Cameroon, Ghana, Senegal and Uganda, Sudan and Zimbabwe. Others were in the region of less than 1 per cent of the total publications. Fourteen countries produced fewer than 500 papers: Angola, Burundi, Cape Verde, Chad, Comoros, Djibouti, Eritrea, Guinea-Bissau, Lesotho, Liberia, Mauritania, Sao Tome and Principe, Seychelles and Somalia. Some of them had more than 10,000 publications to their credit: Algeria, Egypt, Ethiopia, Kenya, Morocco, Nigeria, South Africa, Tanzania and Tunisia.

The data was further grouped into five-year periods from 1970 to 2015, except for the last group of 2010–15 to accommodate the latest year of 2015. This is to examine the historical trends in production. To start with, the major players – South Africa and Egypt – have to be considered. Half the publications that originated from Africa during 1970–4 were from South African researchers. In the following periods

Table 3.1 *Production of scientific publications in Africa, 1945–2015*

Country	1945–2015 No.	%	2010–2015 No.	%	2005–2009 No.	%	2000–2004 No.	%	1995–1999 No.	%
Algeria	25408	4.60	12684	6.10	5907	6.01	2700	4.25	1600	3.15
Angola	490	0.09	244	0.12	91	0.09	50	0.08	0	0.00
Benin	2979	0.54	1523	0.73	683	0.70	354	0.56	230	0.45
Botswana	2875	0.52	1071	0.52	719	0.73	518	0.82	282	0.55
Burkina Faso	3600	0.65	1731	0.83	830	0.84	493	0.78	292	0.57
Burundi	459	0.08	125	0.06	43	0.04	32	0.05	85	0.17
Cameroon	9215	1.67	4002	1.92	2112	2.15	1177	1.85	773	1.52
Cape Verde	139	0.03	102	0.05	21	0.02	5	0.01	7	0.01
Chad	338	0.06	107	0.05	90	0.09	47	0.07	25	0.05
Comoros	56	0.01	21	0.01	14	0.01	8	0.01	0	0.00
Congo	1918	0.35	817	0.39	366	0.37	173	0.27	135	0.27
Djibouti	103	0.02	54	0.03	15	0.02	6	0.01	0	0.00
Egypt	114746	20.78	46725	22.47	18812	19.14	13178	20.75	10560	20.77
Eritrea	335	0.06	92	0.04	118	0.12	91	0.14	0	0.00
Ethiopia	10599	1.92	4763	2.29	1844	1.88	1212	1.91	1003	1.97
Gabon	2000	0.36	716	0.34	398	0.41	278	0.44	229	0.45
Gambia	1845	0.33	669	0.32	414	0.42	330	0.52	275	0.54
Ghana	7595	1.38	3466	1.67	1337	1.36	838	1.32	607	1.19
Guinea	4240	0.77	1096	0.53	543	0.55	488	0.77	440	0.87
Guinea-Bissau	459	0.08	184	0.09	102	0.10	83	0.13	57	0.11
Ivory Coast	1331	0.24	0	0.00	0	0.00	0	0.00	0	0.00
Kenya	23477	4.25	7814	3.76	3761	3.83	2747	4.33	2445	4.81
Lesotho	353	0.06	123	0.06	64	0.07	24	0.04	22	0.04
Liberia	308	0.06	103	0.05	0	0.00	5	0.01	0	0.00
Libya	2979	0.54	1014	0.49	512	0.52	269	0.42	227	0.45
Madagascar	2134	0.39	1196	0.58	0	0.00	234	0.37	0	0.00
Malawi	4239	0.77	1906	0.92	841	0.86	538	0.85	368	0.72
Mali	2100	0.38	927	0.45	458	0.47	263	0.41	174	0.34
Mauritius	1268	0.23	541	0.26	235	0.24	217	0.34	115	0.23
Mauritania	389	0.07	134	0.06	100	0.10	77	0.12	41	0.08
Morocco	27515	4.98	9594	4.61	5548	5.65	5269	8.30	3782	7.44
Mozambique	1802	0.33	943	0.45	376	0.38	172	0.27	132	0.26
Namibia	1711	0.31	708	0.34	364	0.37	228	0.36	168	0.33
Niger	1659	0.30	596	0.29	293	0.30	174	0.27	245	0.48
Nigeria	44851	8.12	12437	5.98	7839	7.98	3731	5.88	3599	7.08
Rwanda	1221	0.22	714	0.34	165	0.17	40	0.06	75	0.15
Reunion	2074	0.38	1036	0.50	462	0.47	274	0.43	177	0.35
Sao Tome and Principe	27	0.00	15	0.01	2	0.00	10	0.02	0	0.00
Senegal	5671	1.03	2129	1.02	1111	1.13	871	1.37	786	1.55
Seychelles	454	0.08	206	0.10	97	0.10	50	0.08	50	0.10
Sierra Leone	702	0.13	245	0.12	46	0.05	33	0.05	56	0.11
Somalia	221	0.04	28	0.01	6	0.01	3	0.00	0	0.00
South Africa	163148	29.55	52841	25.41	25960	26.42	18260	28.76	16312	32.09
Sudan	5793	1.05	1862	0.90	763	0.78	445	0.70	454	0.89
Swaziland	542	0.10	242	0.12	99	0.10	56	0.09	50	0.10
Tanzania	10315	1.87	4110	1.98	2077	2.11	1180	1.86	988	1.94
Togo	1067	0.19	376	0.18	192	0.20	143	0.23	122	0.24
Tunisia	36385	6.59	18490	8.89	9028	9.19	3851	6.07	2004	3.94
Uganda	8631	1.56	4375	2.10	1825	1.86	896	1.41	536	1.05
Zambia	3735	0.68	1400	0.67	607	0.62	372	0.59	319	0.63
Zimbabwe	6669	1.21	1630	0.78	972	0.99	1001	1.58	992	1.95
Total	552170	100	207927	100	98262	100	63494	100	50839	100

this percentage has declined. For the period of 2010–15, South Africa produced only 25 per cent of all publications in Africa. Although the percentile contribution was declining, the actual number of publications have increased substantially by about 20 times from 2,623 (1970–4) to 52,841 (2010–15). In the current period, 2010–15, South Africa generated one-fourth of all the science publications in Africa. Egypt, on the other hand, had a small beginning in the first period (1970–4). It gradually increased the percentage to the total publications in Africa and actual number of publications. In the second period of 1975–9, Egypt grew its percentage to 17. Since then the country consistently produced in the region of 20 per cent in all the selected periods. From 34 publications, Egypt produced 46,725 publications in recent years (Figure 3.1).

Table 3.1 *(cont.)*

Country	1990–1994 No.	%	1985–1989 No.	%	1980–1984 No.	%	1975–1979 No.	%	1970–1974 No.	%
Algeria	990	2.44	593	1.65	510	1.79	291	1.41	131	2.53
Angola	20	0.05	12	0.03	0	0.00	10	0.05	11	0.21
Benin	117	0.29	50	0.14	13	0.05	8	0.04	0	0.00
Botswana	109	0.27	81	0.23	59	0.21	36	0.17	0	0.00
Burkina Faso	176	0.43	0	0.00	0	0.00	0	0.00	0	0.00
Burundi	65	0.16	74	0.21	27	0.09	7	0.03	1	0.02
Cameroon	575	1.42	234	0.65	157	0.55	154	0.75	31	0.60
Cape Verde	3	0.01	1	0.00	0	0.00	0	0.00	0	0.00
Chad	12	0.03	4	0.01	2	0.01	27	0.13	34	0.66
Comoros	6	0.01	1	0.00	0	0.00	0	0.00	0	0.00
Congo	159	0.39	127	0.35	76	0.27	51	0.25	14	0.27
Djibouti	19	0.05	1	0.00	0	0.00	0	0.00	0	0.00
Egypt	8851	21.86	7171	19.97	5961	20.95	3439	16.71	34	0.66
Eritrea	2	0.00	0	0.00	0	0.00	0	0.00	0	0.00
Ethiopia	734	1.81	455	1.27	317	1.11	177	0.86	94	1.82
Gabon	163	0.40	116	0.32	65	0.23	25	0.12	10	0.19
Gambia	50	0.12	8	0.02	40	0.14	39	0.19	11	0.21
Ghana	362	0.89	238	0.66	266	0.93	309	1.50	164	3.17
Guinea	511	1.26	539	1.50	418	1.47	201	0.98	2	0.04
Guinea-Bissau	29	0.07	4	0.01	0	0.00	0	0.00	0	0.00
Ivory Coast	8	0.02	374	1.04	511	1.80	354	1.72	84	1.62
Kenya	2171	5.36	1696	4.72	1513	5.32	1019	4.95	304	5.88
Lesotho	37	0.09	36	0.10	22	0.08	21	0.10	4	0.08
Liberia	40	0.10	48	0.13	59	0.21	15	0.07	7	0.14
Libya	237	0.59	296	0.82	252	0.89	151	0.73	21	0.41
Madagascar	0	0.00	0	0.00	0	0.00	0	0.00	0	0.00
Malawi	202	0.50	141	0.39	112	0.39	101	0.49	30	0.58
Mali	115	0.28	75	0.21	63	0.22	21	0.10	0	0.00
Mauritius	70	0.17	20	0.06	30	0.11	30	0.15	10	0.19
Mauritania	17	0.04	12	0.03	5	0.02	3	0.01	0	0.00
Morocco	1656	4.09	853	2.38	521	1.83	240	1.17	50	0.97
Mozambique	67	0.17	44	0.12	27	0.09	15	0.07	26	0.50
Namibia	116	0.29	73	0.20	38	0.13	16	0.08	0	0.00
Niger	198	0.49	90	0.25	45	0.16	16	0.08	2	0.04
Nigeria	3929	9.70	4957	13.80	4411	15.50	3216	15.63	720	13.92
Rwanda	93	0.23	78	0.22	31	0.11	18	0.09	7	0.14
Reunion	53	0.13	32	0.09	31	0.11	8	0.04	3	0.06
Sao Tome and Principe	0	0.00	0	0.00	0	0.00	0	0.00	0	0.00
Senegal	162	0.40	13	0.04	172	0.60	334	1.62	87	1.68
Seychelles	16	0.04	6	0.02	15	0.05	14	0.07	0	0.00
Sierra Leone	78	0.19	64	0.18	78	0.27	68	0.33	34	0.66
Somalia	58	0.14	73	0.20	31	0.11	12	0.06	2	0.04
South Africa	14364	35.47	14132	39.36	10249	36.01	8376	40.70	2623	50.72
Sudan	488	1.21	594	1.65	549	1.93	486	2.36	139	2.69
Swaziland	28	0.07	26	0.07	25	0.09	13	0.06	3	0.06
Tanzania	685	1.69	471	1.31	382	1.34	328	1.59	93	1.80
Togo	86	0.21	52	0.14	43	0.15	41	0.20	12	0.23
Tunisia	1257	3.10	899	2.50	514	1.81	276	1.34	66	1.28
Uganda	229	0.57	135	0.38	154	0.54	245	1.19	226	4.37
Zambia	234	0.58	230	0.64	182	0.64	309	1.50	82	1.59
Zimbabwe	849	2.10	680	1.89	484	1.70	61	0.30	0	0.00
Total	40496	100	35909	100	28460	100	20581	100	5172	100

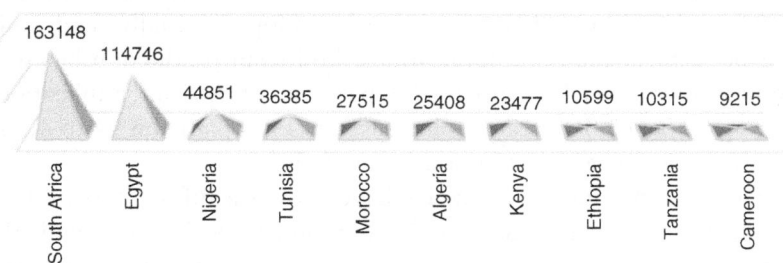

163148   114746   44851   36385   27515   25408   23477   10599   10315   9215

South Africa | Egypt | Nigeria | Tunisia | Morocco | Algeria | Kenya | Ethiopia | Tanzania | Cameroon

Figure 3.1 Scientific publications in top ten African countries, 1945–2015

Nigeria, the third largest producer in Africa, had more publications than Egypt in 1970–4. In the beginning, Nigeria produced 14 per cent and continued to produce more or less the same measure up until 1985–9. Since then the Nigerian share in African science deteriorated from 10 to 6 per cent during 2010–15. In number of publications, Nigeria grew from 720 to 12,437 publications, increasing by 17-fold.

Turning now to the medium countries, close to 7 per cent of papers emanated from Tunisia. Since 1984, Tunisia made a significant mark in its science production by increasing its percentage from 2. The period of 2000–4 and after proved to be a very productive time for Tunisia. Its production percentages touched the region of 6–9 per cent in 2000 and later. Noticeably, Tunisia doubled its production between 2005–9 and 2010–15. Half its total production during 1945–2015 was made during 2010–15. Algeria produced about 5 per cent of the publications in Africa for the period 1945–2015. It started with 3 per cent during 1970–4 and doubled its percentage to 6 per cent during 2010–15. Morocco made a contribution of 5 per cent to the total publications in Africa between 1945 and 2015. It started with less than 1 per cent (1970–4) and steadily increased to 8 per cent during 2000–4. Since then, production declined in Morocco. The percentage went down to 5.7 per cent during 2005–9 and further to 4.6 per cent during 2010–15. But during 2010–15 it produced almost double the number of publications over the previous years.

Kenya is another medium country in science production with a share of 4.3 per cent of the total. Beginning from 1970–4 to 2000–4 Kenya had a contribution of about 4–5 per cent which declined later to less than 4 per cent after 2000. It doubled its number of papers during 2010–15 over the previous period. Tanzania, which produced over 10,000 publications, has been consistent in its share of the total African publications at around 2 per cent. Increase in the number of papers was more obvious recently. Ethiopia initially had only 1.8 per cent of the total publications. Ethiopian production has seen some growth, reaching up to 2.3 per cent of African publications during 2010–15. Publications increased in Ethiopia from 94 to 4,763 between the first and last periods of analysis.

The least productive countries have shown a different pattern in production. They include Angola, Burundi, Cape Verde, Chad, the Comoros, Djibouti, Eritrea, Guinea-Bissau, Lesotho, Liberia, Mauritania, Sao Tome and Principe, Seychelles and Somalia. The smallest in number of publications were the Comoros, Djibouti, Sao

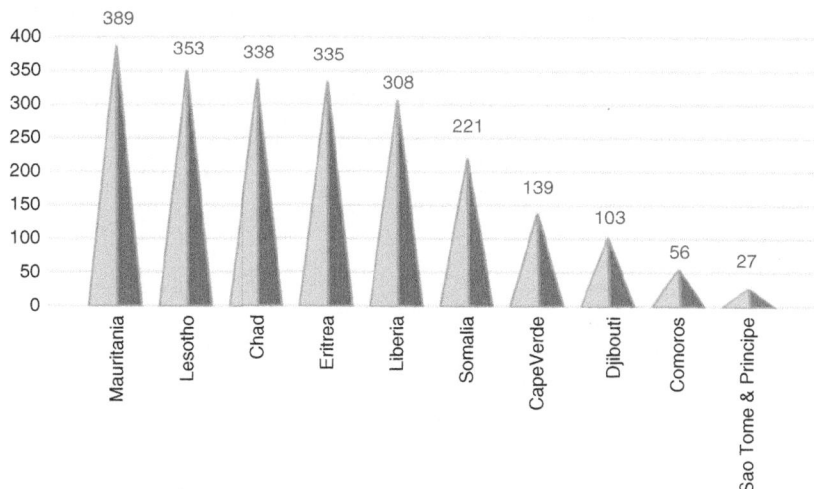

Figure 3.2 Lowest publications in ten African countries, 1945–2015

Tome and Principe. Generally most of these countries have appeared on the map of science very late, largely since 2000 (Figure 3.2).

Concerning the production of publications within individual countries, in most cases nearly half or more of the total publications for all years (1945–2015) were produced only in the recent past (i.e., 2010 or later). For instance, Algeria, Angola, Benin, Burkina Faso, Cameroon, Cape Verde, the Comoros, Djibouti, Egypt, Ethiopia, Ghana, Madagascar, Malawi, Mali, Mozambique, Reunion, Sao Tome and Principe, Seychelles, Swaziland, Tunisia and Uganda. There can be at least two reasons for this trend. Firstly, it is plausible that these countries recognised the growing importance of science and made efforts to increase their production. Secondly, some of these countries did not have the human and infrastructure resources to produce scientific publications in the past, between 1970 to 2010. They could enter the scene only in recent years for many economic reasons. As seen in the review of the literature, Africa had a late entry into science. This however does not seem to be the case for Egypt, Ghana, Kenya, Nigeria and South Africa, which had a head start as early as 1970. North African countries such as Algeria, Morocco and Tunisia have taken science seriously from the 1980s. Egypt embarked on science production more vigorously from 1975.

Table 3.2 *Production of scientific publications in sub-Saharan Africa, 1945–2015*

Country	1945–2015		2010–2015		2005–2009		2000–2004		1995–1999	
	No.	%	No.	%	No.	%	No.	%	No.	%
Angola	490	0.14	244	0.20	91	0.15	50	0.13	0	0.00
Benin	2979	0.86	1523	1.26	683	1.16	354	0.92	230	0.70
Botswana	2875	0.83	1071	0.89	719	1.22	518	1.35	282	0.86
Burkina Faso	3600	1.03	1731	1.44	830	1.41	493	1.28	292	0.89
Burundi	459	0.13	125	0.10	43	0.07	32	0.08	85	0.26
Cameroon	9215	2.65	4002	3.32	2112	3.58	1177	3.06	773	2.35
Cape Verde	139	0.04	102	0.08	21	0.04	5	0.01	7	0.02
Chad	338	0.10	107	0.09	90	0.15	47	0.12	25	0.08
Comoros	56	0.02	21	0.02	14	0.02	8	0.02	0	0.00
Congo	1918	0.55	817	0.68	366	0.62	173	0.45	135	0.41
Djibouti	103	0.03	54	0.04	15	0.03	6	0.02	0	0.00
Eritrea	335	0.10	92	0.08	118	0.20	91	0.24	0	0.00
Ethiopia	10599	3.04	4763	3.95	1844	3.13	1212	3.15	1003	3.05
Gabon	2000	0.57	716	0.59	398	0.67	278	0.72	229	0.70
Gambia	1845	0.53	669	0.56	414	0.70	330	0.86	275	0.84
Ghana	7595	2.18	3466	2.88	1337	2.27	838	2.18	607	1.85
Guinea	4240	1.22	1096	0.91	543	0.92	488	1.27	440	1.34
Guinea-Bissau	459	0.13	184	0.15	102	0.17	83	0.22	57	0.17
Ivory Coast	1331	0.38	0	0.00	0	0.00	0	0.00	0	0.00
Kenya	23477	6.74	7814	6.49	3761	6.38	2747	7.14	2445	7.43
Lesotho	353	0.10	123	0.10	64	0.11	24	0.06	22	0.07
Liberia	308	0.09	103	0.09	0	0.00	5	0.01	0	0.00
Libya	2979	0.86	1014	0.84	512	0.87	269	0.70	227	0.69
Madagascar	2134	0.61	1196	0.99	0	0.00	234	0.61	0	0.00
Malawi	4239	1.22	1906	1.58	841	1.43	538	1.40	368	1.12
Mali	2100	0.60	927	0.77	458	0.78	263	0.68	174	0.53
Mauritius	1268	0.36	541	0.45	235	0.40	217	0.56	115	0.35
Mauritania	389	0.11	134	0.11	100	0.17	77	0.20	41	0.12
Mozambique	1802	0.52	943	0.78	376	0.64	172	0.45	132	0.40
Namibia	1711	0.49	708	0.59	364	0.62	228	0.59	168	0.51
Niger	1659	0.48	596	0.49	293	0.50	174	0.45	245	0.74
Nigeria	44851	12.88	12437	10.33	7839	13.29	3731	9.69	3599	10.94
Rwanda	1221	0.35	714	0.59	165	0.28	40	0.10	75	0.23
Reunion	2074	0.60	1036	0.86	462	0.78	274	0.71	177	0.54
Sao Tome and Principe	27	0.01	15	0.01	2	0.00	10	0.03	0	0.00
Senegal	5671	1.63	2129	1.77	1111	1.88	871	2.26	786	2.39
Seychelles	454	0.13	206	0.17	97	0.16	50	0.13	50	0.15
Sierra Leone	702	0.20	245	0.20	46	0.08	33	0.09	56	0.17
Somalia	221	0.06	28	0.02	6	0.01	3	0.01	0	0.00
South Africa	163148	46.87	52841	43.88	25960	44.02	18260	47.43	16312	49.59
Sudan	5793	1.66	1862	1.55	763	1.29	445	1.16	454	1.38
Swaziland	542	0.16	242	0.20	99	0.17	56	0.15	50	0.15
Tanzania	10315	2.96	4110	3.41	2077	3.52	1180	3.07	988	3.00
Togo	1067	0.31	376	0.31	192	0.33	143	0.37	122	0.37
Uganda	8631	2.48	4375	3.63	1825	3.09	896	2.33	536	1.63
Zambia	3735	1.07	1400	1.16	607	1.03	372	0.97	319	0.97
Zimbabwe	6669	1.92	1630	1.35	972	1.65	1001	2.60	992	3.02
Total	345373	100.00	120434	100.00	58967	100.00	38496	100.00	32893	100.00

Nigeria, South Africa and to a certain extent Kenya, had a good beginning in 1970, which they continued over the subsequent years. These countries either maintained or improved on their production figures in the later years. Each of three countries began to grow their production of scientific publications in absolute numbers particularly after 2005. All of them doubled their number of publications during 2005–9. There has been a wide gap between the first two major countries and the remaining major countries in producing their share of African science.

Table 3.2 *(cont.)*

Country	1990–1994		1985–1989		1980–1984		1975–1979		1970–1974	
	No.	%	No.	%	No.	%	No.	%	No.	%
Angola	20	0.07	12	0.05	0	0.00	10	0.06	11	0.22
Benin	117	0.42	50	0.19	13	0.06	8	0.05	0	0.00
Botswana	109	0.39	81	0.31	59	0.28	36	0.22	0	0.00
Burkina Faso	176	0.63	0	0.00	0	0.00	0	0.00	0	0.00
Burundi	65	0.23	74	0.28	27	0.13	7	0.04	1	0.02
Cameroon	575	2.07	234	0.89	157	0.75	154	0.94	31	0.63
Cape Verde	3	0.01	1	0.00	0	0.00	0	0.00	0	0.00
Chad	12	0.04	4	0.02	2	0.01	27	0.17	34	0.70
Comoros	6	0.02	1	0.00	0	0.00	0	0.00	0	0.00
Congo	159	0.57	127	0.48	76	0.36	51	0.31	14	0.29
Djibouti	19	0.07	1	0.00	0	0.00	0	0.00	0	0.00
Eritrea	2	0.01	0	0.00	0	0.00	0	0.00	0	0.00
Ethiopia	734	2.65	455	1.72	317	1.51	177	1.08	94	1.92
Gabon	163	0.59	116	0.44	65	0.31	25	0.15	10	0.20
Gambia	50	0.18	8	0.03	40	0.19	39	0.24	11	0.22
Ghana	362	1.30	238	0.90	266	1.27	309	1.89	164	3.35
Guinea	511	1.84	539	2.04	418	1.99	201	1.23	2	0.04
Guinea-Bissau	29	0.10	4	0.02	0	0.00	0	0.00	0	0.00
Ivory Coast	8	0.03	374	1.42	511	2.44	354	2.17	84	1.72
Kenya	2171	7.83	1696	6.43	1513	7.22	1019	6.24	304	6.22
Lesotho	37	0.13	36	0.14	22	0.10	21	0.13	4	0.08
Liberia	40	0.14	48	0.18	59	0.28	15	0.09	7	0.14
Libya	237	0.85	296	1.12	252	1.20	151	0.92	21	0.43
Madagascar	0	0.00	0	0.00	0	0.00	0	0.00	0	0.00
Malawi	202	0.73	141	0.53	112	0.53	101	0.62	30	0.61
Mali	115	0.41	75	0.28	63	0.30	21	0.13	0	0.00
Mauritius	70	0.25	20	0.08	30	0.14	30	0.18	10	0.20
Mauritania	17	0.06	12	0.05	5	0.02	3	0.02	0	0.00
Mozambique	67	0.24	44	0.17	27	0.13	15	0.09	26	0.53
Namibia	116	0.42	73	0.28	38	0.18	16	0.10	0	0.00
Niger	198	0.71	90	0.34	45	0.21	16	0.10	2	0.04
Nigeria	3929	14.16	4957	18.78	4411	21.05	3216	19.69	720	14.72
Rwanda	93	0.34	78	0.30	31	0.15	18	0.11	7	0.14
Reunion	53	0.19	32	0.12	31	0.15	8	0.05	3	0.06
Sao Tome and Principe	0	0.00	0	0.00	0	0.00	0	0.00	0	0.00
Senegal	162	0.58	13	0.05	172	0.82	334	2.04	87	1.78
Seychelles	16	0.06	6	0.02	15	0.07	14	0.09	0	0.00
Sierra Leone	78	0.28	64	0.24	78	0.37	68	0.42	34	0.70
Somalia	58	0.21	73	0.28	31	0.15	12	0.07	2	0.04
South Africa	14364	51.78	14132	53.54	10249	48.91	8376	51.28	2623	53.63
Sudan	488	1.76	594	2.25	549	2.62	486	2.98	139	2.84
Swaziland	28	0.10	26	0.10	25	0.12	13	0.08	3	0.06
Tanzania	685	2.47	471	1.78	382	1.82	328	2.01	93	1.90
Togo	86	0.31	52	0.20	43	0.21	41	0.25	12	0.25
Uganda	229	0.83	135	0.51	154	0.73	245	1.50	226	4.62
Zambia	234	0.84	230	0.87	182	0.87	309	1.89	82	1.68
Zimbabwe	849	3.06	680	2.58	484	2.31	61	0.37	0	0.00
Total	27742	100.00	26393	100.00	20954	100.00	16335	100.00	4891	100.00

Science Production in Sub-Saharan Africa

The figures for sub-Saharan Africa are presented in Table 3.2. Compared to the total publications of 552,170 for Africa, there were 345,373 publications for sub-Saharan Africa. This is 62.6 per cent of the total publications in Africa. Four countries in North Africa (Algeria, Egypt, Morocco and Tunisia) jointly produced as much as one-third of total publications in Africa for the period 1945–2015. The average number of publications for sub-Saharan countries was 7,406, which is 3,400 short of the average figure for the whole of Africa. The median number of publications was 1,845, against 2,000 for Africa.

In sub-Saharan Africa, the major player in science production is South Africa, with 47 per cent of the total publications between 1945 and 2015. Nigeria follows South Africa as the second major contributor with 13 per cent of the total sub-Saharan production. The other significant players were Kenya (6.7 per cent), Ethiopia (3 per cent), Tanzania (3 per cent), Cameroon (2.7 per cent), Uganda (2.5 per cent) and Ghana (2.2 per cent). Many of these major players had consistently produced more or less the same share of publications of sub-Saharan Africa over the years. Figures for countries that have not performed well are the same as seen in the analysis for Africa.

Researchers and Publications in Science

For the purpose of understanding science that comes out of Africa, and in relation to the world, more information is required. Compiling and merging data from sources such as the World Bank, the UNESCO Institute for Statistics, UN world population statistics and the Web of Science variables that are relevant have been examined. Tables 3.3 and 3.4 indicate the publications, percentage of world GERD, GERD percentages of GDP, researchers (percentage of world researchers and per million inhabitants), and GERD per researchers for the world, Africa, sub-Saharan Africa and for Asia. These figures present Africa in perspective and in relation to world and other regions. The data drawn from multiple sources allows for the period between 1996 and 2013. The division of this period according to three decades provides insights into trends and patterns over the years (Figure 3.3).

One important measure in science production is the number of researchers available to conduct scientific research. Africa had an average of 2.4 per cent of the total of world researchers during 1996–9. For sub-Saharan Africa, it was only 0.85 per cent for the same period. This has not changed for Africa (2.4 per cent) but changed for sub-Saharan Africa (0.95 per cent) during the following decade of 2000–9 (actual numbers are in Table 3.3). A growth in the number of researchers was evident for 2001–13 but only for sub-Saharan Africa. The average percentage was 1.05 for sub-Saharan Africa and 2.4 for Africa (Figure 3.4).

Researchers per million inhabitants are also shown in Tables 3.3 and 3.4. This measure shows that Africa had an average of 150.5 researchers per every million (per year) of its population during

Table 3.3 *Publications, GERD, GDP and researchers, world, Africa and sub-Saharan Africa, 1996–2013*

Year	Publications					% of world GERD		GERD % of GDP		
	Africa	% share to world	Sub-Sahara	% share to world	World	Africa	Sub-Sahara	Africa	Sub-Sahara	World
1996	9665	1.41	6645	0.97	686106	1.1	0.7	0.33	0.37	1.42
1997	10279	1.47	7125	1.02	699217	1.0	0.6	0.33	0.37	1.43
1998	10725	1.50	7360	1.03	716527	1.0	0.6	0.33	0.37	1.45
1999	11474	1.57	7841	1.07	729739	1.0	0.6	0.34	0.38	1.49
2000	11414	1.53	7602	1.02	745009	1.0	0.6	0.35	0.39	1.53
2001	11924	1.60	7922	1.06	747396	1.1	0.6	0.36	0.40	1.55
2002	12612	1.64	8389	1.09	767347	1.1	0.7	0.38	0.41	1.54
2003	13490	1.67	8976	1.11	805840	1.1	0.7	0.37	0.41	1.54
2004	14054	1.66	9458	1.12	846846	1.2	0.7	0.37	0.41	1.52
2005	15143	1.70	10427	1.17	890370	1.1	0.7	0.36	0.41	1.54
2006	17233	1.84	12032	1.28	938756	1.2	0.8	0.37	0.42	1.55
2007	19671	2.00	13772	1.40	981848	1.1	0.7	0.36	0.42	1.57
2008	22036	2.11	15329	1.47	1042781	1.2	0.8	0.37	0.42	1.61
2009	24729	2.28	16985	1.57	1084803	1.3	0.8	0.40	0.42	1.65
2010	27190	2.43	18620	1.66	1120597	1.3	0.7	0.40	0.41	1.63
2011	30773	2.58	20803	1.74	1192615	1.3	0.7	0.42	0.41	1.65
2012	32534	2.59	21649	1.72	1256290	1.3	0.7	0.42	0.41	1.68
2013	35328	2.66	23354	1.76	1328031	1.3	0.8	0.45	0.41	1.70

Year	Researchers			% of world researchers		Researcher/million inhabitants		
	Africa	Sub-Saha	World	Africa	Sub-Saha	Africa	Sub-Sahara	World
1996	110821	37737	4567682	2.4	0.8	150.9	65.8	784.7
1997	112934	38884	4582310	2.5	0.8	150.1	65.7	776.8
1998	116248	40847	4583681	2.5	0.9	150.9	67.6	767.1
1999	118441	43057	4719263	2.5	0.9	150.1	69.4	779.9
2000	121702	45354	4923702	2.5	0.9	150.6	71.2	803.5
2001	125678	48338	5099984	2.5	0.9	151.8	74.0	822.0
2002	129011	48901	5225028	2.5	0.9	152.2	72.9	831.9
2003	131283	50851	5478570	2.4	0.9	151.2	73.9	861.7
2004	139580	56195	5603507	2.5	1.0	156.9	79.5	870.7
2005	144770	58299	5908603	2.5	1.0	158.8	80.4	907.0
2006	148592	58755	6133699	2.4	1.0	159.1	78.9	930.3
2007	150139	58849	6400939	2.3	0.9	156.8	77.0	959.2
2008	143874	64798	6647311	2.2	1.0	146.6	82.5	984.3
2009	152704	69375	6901881	2.2	1.0	151.8	86.0	1009.8
2010	164837	74159	7074185	2.3	1.0	159.9	89.5	1022.8
2011	173409	77072	7350365	2.4	1.0	164.1	90.6	1050.4
2012	181790	80111	7572578	2.4	1.1	167.8	91.7	1069.6
2013	187488	81988	7758852	2.4	1.1	168.8	91.4	1083.3

Sources: World Bank Statistics, UNESCO Institute for Statistics, UN world population statistics and Web of Science data

1996–9 and sub-Saharan Africa had 67.1 researchers per million. This is against 777.1 researchers per million inhabitants for the world as a whole. That is, the number for Africa was only 20 per cent or one-fifth of the number of researchers available for the world. Sub-Saharan Africa had only 8.6 per cent of the average world number of researchers per million population during 1996–9. The difference between Africa and sub-Saharan Africa is conspicuous.

Table 3.4 *GERD per capita and per researchers, world, Asia, Africa and sub-Saharan Africa, 1996–2013*

	GERD per capita, in PPP$ current prices				GERD per researcher, in '000 PPP $ current prices			
Year	Africa	Sub-Sahara	Asia	World	Africa	Sub-Sahara	Asia	World
1996	7.9	6.3	40.8	94.1	52.2	95.2	92.4	119.9
1997	8	6.4	44.1	99.5	53.5	97.6	99.5	128.1
1998	8.3	6.6	45.3	103.2	54.9	97.3	107.3	134.6
1999	8.7	6.8	48.2	109.8	58	98.3	111.7	140.8
2000	9.3	7.2	53.6	119.4	61.7	101.3	116.8	148.6
2001	10.1	7.7	57.4	125.6	66.3	104	120.5	152.8
2002	10.9	8.1	61.7	129.3	71.3	110.5	128.2	155.4
2003	11.1	8.4	66.1	134.9	73.4	113.7	132.1	156.5
2004	12.1	9.4	72.9	141.9	77.4	118.6	142.2	163
2005	12.4	10.1	82.7	153.3	77.8	126	147.3	169
2006	13.5	11.1	92.4	167.2	84.9	140.2	156.9	179.7
2007	14.3	11.7	102.5	180.8	91.2	151.8	162.6	188.5
2008	15.3	12.4	110.2	193.4	104.1	150.2	168.1	196.5
2009	16.8	12.4	117.5	198.4	110.7	145.4	171.6	196.5
2010	17.2	12.5	127.8	205.6	107.6	139.9	180.4	201
2011	18.2	13.2	141.2	218.8	110.7	145.4	190.6	208.3
2012	19.1	13.7	154.9	231.9	113.7	149.7	202.2	216.8
2013	20.8	14.4	168.7	242.5	123.2	157.3	214.7	223.8

Sources: Extracted and compiled from World Bank Statistics, UNESCO Institute for Statistics and UN world population statistics.

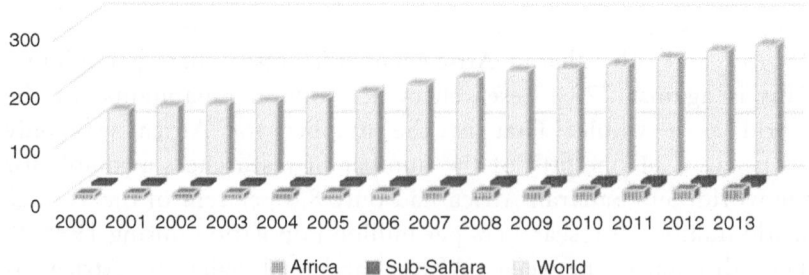

Figure 3.3 GERD per capita in PPP$ current prices, Africa, sub-Saharan Africa and world, 2000–13

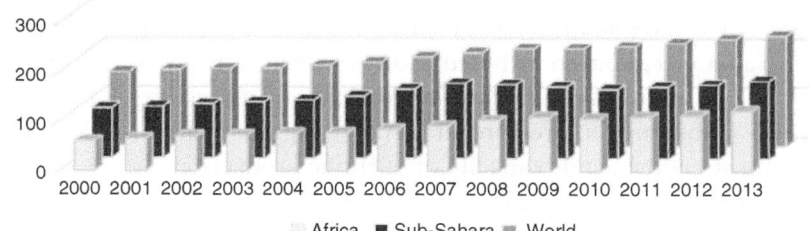

Figure 3.4 GERD per researcher, in '000 PPP$ current prices, Africa, sub-Saharan Africa and world, 2000–13

In the following decade of 2000–9, the situation improved by an increase of researchers. This improvement was not uniform for all the three regions of world, Africa and sub-Saharan Africa. The average number of researchers per million was 898 for the world, 153.6 for Africa and 77.6 for sub-Saharan Africa. There was an increase of 15 per cent for the world, 2 per cent for Africa and 15 per cent for sub-Saharan Africa from the previous period of 1996–9. Sub-Saharan Africa gained during the period, reaching 50.5 per cent of the number of Africa. However, sub-Sahara is far below the average number for Africa by about half the number (77.6 and 153.6, respectively).

For the 2010–13 period there was an average of 1,056.5 researchers per million in the world, 165.1 in Africa and 90.8 in the sub-Saharan region.[1] The growth rate from the previous decade was 17.6 per cent (world), 7.5 per cent (Africa) and 17 per cent (sub-Sahara). This data for the recent years indicates that Africa has only 15 per cent of the researchers (per million inhabitants) of the world. For sub-Saharan Africa it was only 9 per cent of the number of the world and 55 per cent of Africa.

As against the figures previously discussed, the production of scientific publications can be compared for Africa and sub-Saharan Africa with world (Table 3.3). There was an average of 707,897 science publications per year in the world during 1996–9. The average publication count for Africa for the same period was 10,535 and for sub-Sahara it was 7,242. Africa thus produced 1.48 per cent of the total world output. For the same period, Africa had 2.4 per cent of the total

---

[1] The world had 1,478 researchers in 2015 for which comparable African and sub-Saharan figures are not available.

world researchers. However, the contribution was not equivalent to the percentage of researchers Africa had at this time. The researchers in the sub-Saharan region contributed 1.02 per cent of the world output, which had more than the percentage (0.85) of its researchers to the world community of researchers. Africa lagged behind while sub-Sahara strengthened its capacity of scientific production.

During the next ten-year period, 2000–9, an average of 885,099 publications per year was produced in the world. Africa's contribution (average per year again) was 16,230 publications, while it was 11,089 for sub-Sahara. Africa thus produced 1.83 per cent of the world's scientific publications. Sub-Sahara, on the other hand, produced 1.25 per cent of the total world production with its share of 0.95 per cent of world researchers. Africa's productivity increased from the previous period of 1996–9 but not on a par with its share of researchers to the world. Sub-Saharan researchers continued to strengthen their capacity in scientific publications during 2000–9.

Referring to 2010–13, there was a substantial increase in the production of publications in science in all three regions of world, Africa and sub-Sahara. The average production per year during 2000–13 was 1,224,383 publications for the world, 31,456 for Africa and 21,106 for sub-Sahara. Production in Africa was 2.57 per cent of the total world production with 2.4 per cent of total world researchers. Sub-Sahara expanded its production capacity by another 0.47 per cent from the previous years. The sub-Saharan region contributed to 1.72 per cent of the world production with a share of 1.05 per cent of world researchers. Africa, for the first time, contributed more than its share of researchers while sub-Sahara moved ahead with improved production figures. While the percentage of researchers in relation to world researchers has not increased in Africa in the recent years (remained at 2.4 per cent since 1996), its contribution to world science is now growing. More significantly, sub-Sahara made greater progress in scientific production than Africa, often exceeding its share of researchers to world researchers.

GERD and Scientific Production

A large gap in GERD exists in different parts of the world, as the percentage to the GDP varies from less than 1 per cent to 4 per cent. In the 1980s, the developed world spent about 0.3 per cent of the GDP on R&D, which has increased to about 3 per cent while the developing world was spending between 0.5 per cent to 1.5 per cent (Moravcsik, 1984). Current figures, as shown in previous chapter, have been changed.

The countries that spent the greatest amount of money on R&D in 2014, in order, were the USA, China, Japan, Germany, Republic of Korea, France, India, the Russian Federation, the UK and Brazil (UIS, 2017). Germany, a leader in science and innovation, spent 2.8 per cent of its GDP on R&D in 2013 which has been increased to 3 per cent in 2017. In 2015, Israel spent 4.3 per cent of the GDP on GERD, Sweden 3.3 per cent, Japan 3.3 per cent, Denmark 3.1 per cent, Finland 2.9 per cent, the US 2.7 per cent, France 2.3 per cent, China 2.1 per cent, Canada and the UK 1.7 per cent each.[2] These increases have been translated into increases in the production of scientific publications in the respective countries.

Countries have been classified according to the variables of publications per million population and per capita GDP (Inönü, 2003). This classification is useful in assessing the relationship between the GDP, GERD and production of science. The GDP of a country has a relationship with its productivity of scientific publications. The relationship between the amount of money spent on science (GERD) and the production of science is quite illuminating from world experience. Singapore, in 1996, had 1.37 per cent of its GDP allocated for R&D, which was increased to 2.61 per cent later in 2007. During this period, its scientific publications have grown from 2,620 to 8,506 (The Royal Society, 2011). Higher outputs in publications in South Africa, Egypt, Nigeria, Tunisia, Algeria and Kenya were related to their GDP (Adams et al., 2010). A growth in the production of scientific publications in Nigeria during 1975–9 occurred when there was a growth in federal government funding of universities and the country's GDP (Adamson, 1992). This was however reversed when there was a decline and erratic funding in the country. GERD remained at low levels for a long time in Africa. In the 1980s, GERD was less than 1 per cent of the GNP in sub-Sahara (Eisemon and Davis, 1997). In 2016, the figures were not better at 0.42 per cent for sub-Saharan Africa and 0.61 per cent for Northern Africa.[3]

---

[2] http://data.uis.unesco.org/. Accessed 1 December 2019.   [3] Ibid.

Reliable data on GERD in African countries is not readily available. Official data on S&T is particularly weak in sub-Saharan countries (Cooper, 1994). What kind of relationship can be expected between GERD and the production of scientific publication in the world in general and in Africa in particular? The details need to be considered and related to these two variables. Some comparable figures for other regions are also relevant.

The percentage of GERD for Africa was in the region of 1.02 per cent of the GDP during 1996–9 (Table 3.3). For the same period, it was 0.63 per cent for sub-Saharan Africa. Between 2000 and 2009 the average percentage of GERD was 1.14 per cent for Africa and 0.71 per cent for sub-Saharan Africa (Figures 3.5 and 3.6). The percentage grew slightly during 2010–13 to 1.29 (an increase of 13 per cent) for Africa and 0.75 per cent (increased by 5 per cent from the previous period) for sub-Saharan Africa. These figures can be compared with the GERD percentage of the GDP. The average world percentage of GDP for GERD during 1996–9 was 1.45 per cent against 0.33 per cent for Africa and 0.37 per cent for sub-Saharan Africa. During 2000–9, the average percentage was 1.56 per cent for world, 0.37 per cent for Africa and 0.41 per cent for sub-Saharan Africa. This has improved for both world and Africa but not for sub-Saharan Africa during 2010–13: 1.67 per cent for world, 0.42 per cent for Africa and 0.41 per cent for sub-Saharan Africa.

In 1980, African governments agreed to increase their R&D expenditure to 1 per cent of GDP but this goal is still elusive. In 2007, it was an average of 0.5 per cent for sub-Saharan African countries.

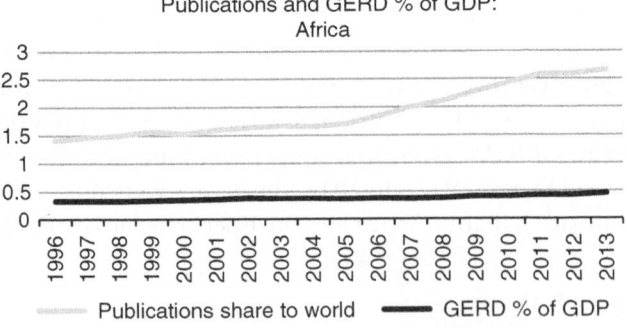

Figure 3.5 Publication share to world and GERD percentage of GDP: Africa

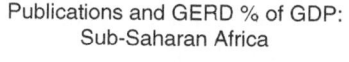

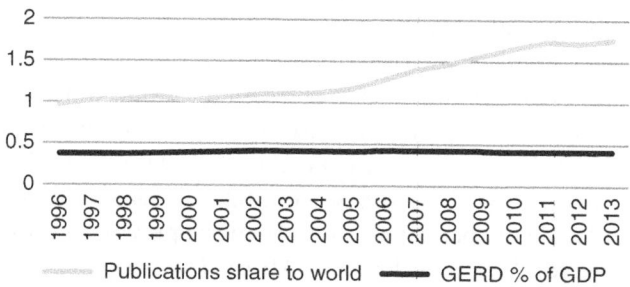

Figure 3.6 Publication share to world and GERD percentage of GDP: sub-Saharan Africa

Compared to the world average of 1.7 per cent, Africa spends between 0.1 per cent (Kenya) and 0.98 per cent (Angola) (Nature Index, 2014). Since the adoption of structural adjustment programmes in some African countries such as Ghana, Kenya, Uganda and Tanzania, there has been a degree of increase in the expenditure on advancing S&T (Enos, 1995).

The percentage allocation of funds for R&D is not comparable. The size of allocation varies according to the GDP. A 1 per cent of GDP of a poor country is not comparable in real terms to that of a developed country. The difference lies in actual fund increases when a developed and an industrialised country are compared. The implications of this for science and its growth are considerable. The more underdeveloped a country, the smaller the proportion of its income invested in scientific research. This condemns these countries to falling further and further behind in science (Dedijer, 1968). An allocation in the region of 1 per cent for a small and developing country is therefore not even adequate for it to make any step towards becoming a strong scientific society.

Further analysis using correlation is appropriate to show the relationship between GERD, GDP and publications. This also reveals additional dimensions of scientific production in Africa.

GERD per capita and per researcher provides more information. For comparison, Asia is also included along with world, Africa and sub-Sahara (Table 3.4). During the first decade of analysis, 1996–9, the average per capita GERD at current prices was $8.23 for Africa, $6.53 for sub-Sahara and $101.65 for the world. This can be compared with

Asia, which had $44.6 at this time. In the following decade (2000–9), the averages were $12.58 (Africa), $9.85 (sub-Sahara), $154.42 (world) and $81.47 (Asia). These were obviously an improvement from the previous period. For Africa, sub-Sahara and world, the increase was uniform at about 150 per cent. At the same time, the increase was 183 per cent for Asian countries. The trend continued generally during 2010–13. The average per capita GERD for the period was $18.83 for Africa, $13.45 for sub-Sahara, $224.7 for world and $148.15 for Asia. While the per capita share from the previous period of 2000–9 improved for Asia, it declined for Africa (149 per cent), sub-Saharan Africa (137 per cent) and for world (146 per cent).

Data for GERD per researcher is available for 1996–2013 (Table 3.4). These figures are at current prices (purchasing power parity or PPP). Africa spent an average of $54,650 per researcher a year during 1996–9. But for sub-Sahara, the expenditure was almost double at $97,000 per researcher. The countries in the world as a whole allocated an average of $130,850 per researcher per year. This is 240 per cent of what Africa and 135 per cent of what sub-Sahara spent during the period. Asian countries spent an amount of $102,275 per researcher. This means the world spent 112.7 per cent of what Asian countries spent. In relation to GERD per capita, sub-Sahara had more money to spend on their researchers than Africa had. Sub-Sahara's allocation was 178 per cent higher than that of Africa. As seen in the data, Asia's allocation was above the levels of both Africa as a whole and sub-Sahara.

In the following decade, 2000–9, the pattern was repeated. In ascending order of expenditure were Africa, sub-Sahara, Asia and world. The average expenditures per researcher per year were $81,800 for Africa, $126,170 for sub-Sahara, $144,530 for Asia and $170,650 for the world. The world allocation in relation to the three regions was 208 per cent of Africa, 135 per cent of sub-Sahara and 118 per cent of Asia. Sub-Saharan researchers received 154 per cent more money than an average African researcher received during this decade. Asian researchers, in comparison to their sub-Saharan counterparts, had 114 per cent more money for their research. Between the two periods of 1996–9 and 2000–9, the gap in the expenditure by the world and Africa had narrowed down (240 per cent to 208 per cent of world expenditure) and between world and Asia (178 per cent to 118 per cent of world expenditure). It remained unchanged for sub-Sahara (135 per cent of world expenditure) during 2000–9.

For 2010–13, the average expenditures per researcher per year were $113,800 (Africa), $148,075 (sub-Sahara), $196,975 (Asia) and $212,475 (world), maintaining the pattern that existed in the previous decade. The expenditure for world countries against Africa was 187 per cent, sub-Sahara 143 per cent and Asia 108 per cent. The gap between the expenditure by world countries and Africa has been narrowed down from the previous decade (from 208 per cent to 187 per cent), but increased between world and sub-Sahara (135 per cent to 143 per cent). Asian countries are striving to close the gap with world. Still, the allocation for sub-Saharan researchers is more than that of an average African researcher, who received 135 per cent more money during 2010–13.

A high positive correlation existed between the number of publications in the world and in Africa ($r = 0.99$), and between the publications in the world and in sub-Sahara Africa ($r = 0.99$) during 1996–2013. A similar level of correlation ($r = 0.99$) prevailed between the publications in Africa and in sub-Sahara. Clearly, there was a positive correlation between the publications in the world and the GERD percentage to GDP ($r = 0.94$). There was also a relationship between the percentage of world GERD for Africa and world GERD for sub-Saharan Africa ($r = 0.74$). As for the publications and GERD percentage of GDP, there was a high positive relationship for world ($r = 0.94$) and for Africa ($r = 0.92$), and a moderate relationship for sub-Saharan Africa ($r = 0.58$). The correlation coefficient for sub-Sahara was not as strong as for Africa. What this finding reveals is that any increase in the GERD to the percentage of GDP can stimulate a corresponding increase in the production of scientific publications. The moderate level of correlation that was found in sub-Saharan Africa, however suggests that GERD is not the only factor that determines the production count of scientific publications. As seen earlier, the allocation of funds per researcher was higher for sub-Saharan researchers than for African researchers. The increase was between 130 per cent and 177 per cent during the three periods of analysis (1996–2013). This explains why the contribution of sub-Sahara to publication output was above the level of the contribution of Africa. Obviously, a more or less similar level of positive correlation existed between the number of publications and the numbers of researchers available to all three regions (world, Africa and sub-Saharan Africa).

There have been attempts to map the relationship between the amount spent on R&D per researcher and the impact of that research in countries (Pan et al., 2012, for instance). Setting the minimum target of $100,000 per researcher to have any impact of their research, Pan et al. present their estimations. If the spending is less than $100,000 per researcher per year, an increase in this amount is followed by an increase in the number of citations. But when the amount exceeds this threshold no increase was found and citations are independent of research funding. They also found some exceptions to this in the case of some countries. Kuwait and Luxemburg, for example, have high investment per researcher per year, which is not translated into a high number of citations, as is the case with the USA, Switzerland, Costa Rica, Germany, Austria and the Netherlands. India and Brazil also have made high investments per researcher but received low citation averages. Needless to say, investments in S&T yield better returns when they are properly planned and monitored.

As proposed in the first chapter, the two research questions related to GERD and publications have answers here. The relationship between GERD and number of publications (for Africa and sub-Saharan Africa especially) is strong in the data. Given the data, we should also note that the proportion of GERD to GDP alone is a determinant factor in the production of scientific publications. This can be explored further. Africa, like other developing regions and countries in the world, has challenges in the practice and development of science. The growth in the production of scientific publications is the cumulative effect of many things, including increased funding, S&T policies that have been developed in many African countries, better infrastructure for research, and measures to improve research capacity (Schemm, 2013). An increased share of GERD in GDP is an essential prerequisite for all these to occur. This is to yet to materialise in several African countries. As Mouton (2008) argued, the symbolic commitment to increased investment in R&D by African governments has to be put into practice as many of them still consider research and knowledge production a luxury.

In the production of science, GERD is not the sole determining factor. The ability of a country to produce science is dependent on the levels of literacy, education and higher education. Literacy is linked to economic and technical development in societies (Ntiri, 1993), and education is a decisive condition not only for development but also for the advancement of science and technology. But scientific achievement is not simply a product of teaching science in schools (McConney et al., 2014). Science, technology and engineering are undergoing a process of being distanced from compartmental and unidimensional enterprises and becoming multidimensional and cross-boundary enterprises. This poses new challenges to the science, technology, engineering and society nexus (Zoller, 2013). Higher levels of education are therefore a prerequisite for Africa (Oanda and Sall, 2016). In order to achieve scientific and technological progress, countries need to have a strong foundation of education from school to tertiary levels, making them competent to face such challenges. Africa has not made sufficient progress in its literacy levels (Omolewa, 2008) and it has numerous factors inhibiting it from attaining universal literacy (Ntiri, 1993). Although access to basic education in sub-Saharan Africa has improved in the last two decades, many children are still not in school (Lewin, 2009). Some key indicators, like education, that have significance in science production in Africa are therefore to be examined here.

In a pattern similar to GERD data, relevant education variables are shown in Table 3.5 for the same period of 1996–2013. The literacy, gross enrolment (pre-primary, primary, secondary and tertiary) and dropout rates for sub-Sahara are compared with world rates.

In the decade of the 1990s, sub-Sahara had an adult (15 years of age or older) literacy rate of about 57 per cent compared to the world average of 82 per cent (Table 3.5). Sub-Saharan Africa lagged behind by 22 per cent in the adult literacy rate. In the following decade (2000–9), sub-Sahara increased its position to an average of 61 per cent, which is a gain of 4 per cent over the previous decade. The world average continued to improve to 85 per cent during 2000–9. In the current decade, beginning from 2010, the situation has not changed for both sub-Saharan Africa and the world. The difference between sub-Sahara and the world thus continues to be significant.

The gross enrolment data on preprimary, primary, secondary and tertiary education points to the strength of an education system. In the 1990s, sub-Saharan Africa had an enrolment ratio of about 11 per cent. This can be compared to 31 per cent for the world. By the end of the decade (2000–9), the ratio improved to 17 per cent for sub-Sahara and 38 per cent for the world. The ratio for sub-Sahara in 2009 was not

Table 3.5 *Literacy, enrolment and dropout rates, sub-Sahara and world, 1996–2013*

Year	Literacy rate (%)*		Gross enrolment rate (%)			
			Pre-primary		Primary**	
	Sub-Sahara	World	Sub-Sahara	World	Sub-Sahara	World
1996	56.98	81.90	10.48	29.91	73.42	97.78
1997	56.98	81.90	10.62	32.77	76.93	--
1998	56.98	81.90	10.78	30.47	78.59	97.81
1999	56.98	81.90	10.77	30.73	79.14	97.96
2000	56.98	81.90	11.61	30.95	81.63	98.48
2001	56.98	81.90	12.07	31.20	83.78	98.96
2002	56.98	81.90	13.27	31.28	86.03	100.17
2003	56.98	81.90	14.41	32.00	88.63	102.57
2004	56.98	81.90	15.64	32.70	90.69	102.48
2005	60.88	85.31	16.51	34.02	92.99	102.21
2006	60.88	85.31	17.91	35.92	94.71	103.13
2007	60.88	85.31	16.33	36.55	95.47	104.64
2008	60.88	85.31	16.14	37.79	96.58	105.49
2009	60.88	85.31	17.10	38.31	96.81	105.58
2010	60.88	85.31	17.42	38.81	96.61	105.29
2011	60.88	85.31	17.81	41.02	97.10	105.52
2012	60.88	85.31	19.64	43.78	97.73	106.02
2013	60.88	85.31	20.35	46.02	97.84	105.74

even half that of the world average. As of 2013, the figures were 23 per cent for sub-Sahara and 46 per cent for the world. The gap has therefore seemingly not closed.

In the enrolment ratio at the primary education level, sub-Sahara had 79 per cent as against 98 per cent for the world. However, in the subsequent decade and in the current decade the differences between sub-Sahara and the world have been reduced. By 2013, sub-Sahara had an enrolment ratio of 98 per cent. In secondary education, the gross enrolment ratio improved from 25 per cent in the 1990s to 38 per cent in the 2000s for sub-Saharan Africa. In the current decade, the ratio is 42 per cent. This is in comparison with world ratios, namely 59 per cent, 69 per cent and 75 per cent for the three decades.

Dropout rate is a significant indicator of the school system. For this purpose, the cumulative dropout rates of the last grade of primary education were extracted and are presented in Table 3.5. In the 1990s, more than 40 per cent of primary students in their last grade left school. The average rate for the world at this time was lower, at

Table 3.5 (cont.)

| Year | Gross enrolment rate (%) | | | | Dropout rate(%)*** | |
| | Secondary | | Tertiary | | | |
	Sub-Sahara	World	Sub-Sahara	World	Sub-Sahara	World
1996	24.60	58.25	3.87	16.12	47.43	23.61
1997	24.93	59.08	3.95	16.89	45.18	23.43
1998	25.42	59.28	3.99	17.27	41.64	23.14
1999	25.76	59.40	4.05	18.36	41.26	23.15
2000	26.33	60.06	4.41	19.04	38.59	23.65
2001	27.59	60.44	4.77	20.10	36.28	22.42
2002	28.71	61.23	5.12	21.63	37.39	23.56
2003	29.83	62.16	5.50	22.78	39.28	24.48
2004	31.36	62.96	5.75	23.59	38.00	24.88
2005	32.31	63.78	6.07	24.28	37.69	25.18
2006	33.37	64.73	6.31	25.03	41.82	24.70
2007	34.39	66.33	6.54	25.89	43.64	24.09
2008	36.44	67.95	6.94	26.92	40.78	22.28
2009	38.28	68.89	7.39	27.97	42.49	22.59
2010	40.21	70.81	7.86	29.34	42.78	22.96
2011	41.06	72.38	8.05	31.05	41.43	22.77
2012	41.80	73.71	8.31	32.20	43.17	24.20
2013	42.19	75.38	8.53	32.78	43.05	24.32

Source: Extracted from World Bank statistics, https://datacatalog.worldbank.org/dat
aset/education-statistics
Notes: * Adult literacy of 15 + years. ** Cumulative to the last grade of primary
education. *** Gross enrollment includes students of all ages. If there is late
enrollment, early enrollment, or repetition, the total enrollment can exceed the
population of the age group that officially corresponds to the level of education leading
to ratios greater than 100 per cent (World Bank, https://datahelpdesk.worldbank.org
/knowledgebase/articles/114955-how-can-gross-school-enrollment-ratios-be-over
-100)

23 per cent. In the decade of the 2000s, the rates for sub-Saharan Africa
ranged between 36 per cent and 42 per cent, with an average rate of
40 per cent. The world average for the same period was 24 per cent,
ranging between 22 per cent and 25 per cent. Since 2001, the dropout
rate for sub-Sahara deteriorated to 43 per cent while it remained at
24 per cent for the world.

A significant difference between sub-Sahara and the world in tertiary
education enrolment is obvious. This is an important area for the
advancement of S&T as the capacity of the population depends largely
on the number of university graduates who can take part in the

production of S&T. The low enrolment ratios, varying from 3 per cent to 9 per cent between the three decades, do not augur well for Africa. In contrast, the world average varied between 16 per cent and 33 per cent in the last three decades. Over the last three decades, sub-Saharan Africa showed some improvement. In the 1990s, the mean tertiary enrolment rate was 4 per cent (against 17 per cent for the world), which rose to 6 per cent during 2000–9 (24 per cent for the world), and to 8 per cent during 2010–13 (31 per cent for the world). The decadal increase for sub-Saharan Africa was 2 per cent against 7 per cent for the world.

While discussing the above indicators it is relevant to consider some related facts and figures. Some African countries in recent years (2011–15), as the UNESCO statistics reveal, have recorded better tertiary enrolment rates.[4] Mauritius (39 per cent), Tunisia (35 per cent), Algeria (33 per cent), Egypt (29 per cent), Botswana (22 per cent), South Africa (19 per cent), Morocco (19 per cent) and Ghana (14 per cent) are among the highest.

The percentage of students enrolled in science programmes in tertiary education demonstrates the importance of science in countries. The data for 2011–15 shows that some African countries have as high as 27 per cent of student enrolment being in science programmes. Gambia had 27 per cent, Tunisia 24 per cent, Botswana 17 per cent, South Africa 11 per cent, Ghana 10 per cent, Zimbabwe 8 per cent, Egypt 5 per cent and Tanzania 5 per cent. In engineering, manufacturing and construction programmes, Tunisia had an enrolment rate of 19 per cent, Algeria 17 per cent, Angola 14 per cent, Zimbabwe 13 per cent, Egypt 9 per cent, South Africa 9 per cent, Botswana 9 per cent, Mozambique 9 per cent, Ghana 8 per cent, Gambia 5 per cent and Tanzania 4 per cent. That countries are paying attention to science and engineering is evident from the enrolment figures in higher education institutions. Science enrolment, as shown in the percentages, leads engineering subjects.

As regards the percentage of graduates from science programmes (average between 2011 and 2015), it ranged between 2 per cent (Lesotho) to 48 per cent (Gambia). Other major countries with

---

[4] Calculated from UNESCO Institute for Statistics (http://data.un.org/Data.aspx? q=education&d=UNESCO&f=series%3aGER_56). Information for all African countries for all years (2001 to 2015) was not available.

their percentages of science graduates were Tunisia (26 per cent), Algeria (12 per cent), South Africa (11 per cent). Egypt had only 4 per cent. Countries that are not prominent on the map of science production also had some good percentage of graduates: Burundi (12 per cent), Burkina Faso (12 per cent), Zimbabwe (10 per cent) and Eritrea (9 per cent).

In tertiary graduates from engineering, manufacturing and construction programmes, the percentages varied between 1 per cent (Mauritania) and 26 per cent (Eritrea). Major countries are Tunisia (17 per cent), Algeria (14 per cent), South Africa (8 per cent), Egypt (7 per cent), Ghana (7 per cent) and Gambia (6 per cent).

The above percentages should be viewed in relation to the respective populations and the student populations. A smaller percentage of science graduates for a country like Egypt (85.9 million population in 2011) may be a higher number than for a small country like Eritrea with a population of 4.4 million (in 2011). Similarly, large countries will have more tertiary institutions with bigger student populations and a greater number of graduates coming out of these institutions.

Africa at this point needs to maintain steady growth in the number of students enrolled in university education, specifically in science and engineering courses, and the production of qualified graduates who can lead the country in the production of science and technology. If we examine the trends in these areas during 2011–15, some countries are more prominent. For instance, the enrolment ratio in tertiary education has increased in Algeria, Benin, Botswana, Burundi, Chad, Côte d'Ivoire (Ivory Coast), Egypt, Ethiopia, Ghana, Mauritius, Morocco, Mozambique, Niger, Seychelles, South Africa and Sudan. Conversely, there were a few countries, such as the Democratic Republic of the Congo and Tunisia, in which the percentage declined. There has been an increase in the enrolment rates of students in science programmes in a number of countries; namely, Benin, Burkina Faso, the Democratic Republic of Congo, Egypt, Lesotho, Madagascar, Mauritius and Mozambique. But the rates were declining in Botswana, Burundi, Côte d'Ivoire (Ivory Coast), South Africa and Zimbabwe. In the enrolment of students in engineering programmes, countries such as Cabo Verde, Botswana, Burkina Faso, Burundi, the Democratic Republic of the Congo, Ghana, Lesotho,

Mozambique and Tunisia have registered an increase. In contrast, the ratio has decreased in Côte d'Ivoire (Ivory Coast), Benin, Egypt, Madagascar, Mauritius and Seychelles.

In the production of graduates in science programmes, an increase was observed in Algeria, Burundi, Ghana, Lesotho, Mozambique, Sudan and Seychelles during 2011–15. A decline was evident in Burkina Faso, Egypt and South Africa. In engineering, manufacturing and construction, the number of graduates produced between 2011 and 2015 grew in Algeria, Burkina Faso, Tunisia and Zimbabwe. In some other countries the trend was the reverse: Botswana, Egypt, Lesotho, Mozambique, Seychelles and Sudan, for instance.

From the above analysis the potential relationship between population, education and society is obvious. Science production in small African countries has not been comparable to that of other larger countries. Countries need people in numbers who can and want to work in S&T. Population is the human resource for the production of science. Although unbridled population growth has undesirable implications for unemployment, poverty and economic growth, it is not so adverse for science. A large population will have a greater number of inventive minds than a small population (Merton, 1937). As Merton (1937) observed, there is a relationship between the density of population and advancement in S&T, which is facilitated in two ways. Firstly, density of population evokes new needs which have the effect of directing inventive interests. Secondly, the higher density of population induces a higher estimation of inventive activity due to the greater economic value of inventions among large populations.

In discussing the less-developed state of 'providing for science' in less-developed countries, Moravcsik (1975) referred to the planning, organisation and management of science. He elaborated further on these three elements in science. In less-developed countries there are fewer scientists who can engage in scientific activities, the interest in matters relating to planning, organisation and management is less pronounced, and there is a basic lack of knowledge in these in terms of experience and familiarity with the body of literature (Moravcsik, 1975). As has been discussed, Africa, and sub-Saharan Africa in particular, has fewer researchers per million inhabitants of the population. In comparable terms, the figures are only one-eleventh of the world average for sub-Saharan Africa and one-sixth of the world average for Africa. This ratio has to change drastically if Africa is to have a critical mass of scientists. One positive outcome of this might be that there will be scientists who will tend to show interest in the planning, organisation and management of science in their respective countries, as all of these affect them as scientists and researchers. Managing science will become as important as conducting research. The third aspect of experience and familiarity with the body of knowledge that Moravcsik mentions will also be improved in due course, when the size of the scientific community is enlarged and the interest in the efficient management of science is addressed.

The chapter which follows closely examines the scientific research areas in African countries, the strengths of African countries in specific research areas and the recent research foci of Africa.

4 | *Scientific Research Areas*

Countries are able to specialise in certain scientific areas while other areas are still being formed and developed with interest and expertise available within the country. Africa as a whole can be seen as a place where most known scientific areas have a presence at varying levels of development.

This chapter examines the scientific publications of Africa according to research areas, following the classification of the Web of Science (WoS). The WoS categorises research areas under broad subject areas of Arts and Humanities, Life Sciences, Physical Sciences, Social Sciences and Technology.[1] Since only the subset of the Science

---

[1] The research areas are:

Life Sciences and Biomedicine (agriculture, allergy, anatomy and morphology, anesthesiology, anthropology, behavioural sciences, biochemistry and molecular biology, biodiversity and conservation, biophysics, biotechnology and applied microbiology, cardiovascular system and cardiology, cell biology, critical care medicine, dentistry, oral surgery and medicine, dermatology, developmental biology, emergency medicine, endocrinology and metabolism, entomology, environmental sciences and ecology, evolutionary biology, fisheries, food science and technology, forestry, gastroenterology and hepatology, general and internal medicine, genetics and heredity, geriatrics and gerontology, health care sciences and services, hematology, immunology, infectious diseases, integrative and complementary medicine, legal medicine, life sciences, biomedicine and other topics, marine and freshwater biology, mathematical and computational biology, medical ethics, medical informatics, medical laboratory technology, microbiology, mycology, neurosciences and neurology, nursing, nutrition and dietetics, obstetrics and gynaecology, oncology, ophthalmology, orthopedics, otorhinolaryngology, paediatrics, palaeontology, parasitology, pathology, pharmacology and pharmacy, physiology, plant sciences, psychiatry, public, environmental and occupational health, radiology, nuclear medicine and medical imaging, rehabilitation, reproductive biology, research and experimental medicine, respiratory system, rheumatology, sport sciences, substance abuse, toxicology, transplantation, tropical medicine, urology and nephrology, veterinary sciences, virology, and zoology)

Physical Sciences (astronomy and astrophysics, chemistry, crystallography, electrochemistry, geochemistry and geophysics, geology, mathematics,

88

Citation Index was used, the research areas applicable in the analysis are those under Life Sciences, Physical Sciences and Technology. There were, however, a few publications that overlapped into Arts and Humanities and Social Sciences as well.

The analysis of research areas is carried out over two broad periods. The first period refers to 1945–2015 and the second to the more recent five-year period 2011–15. The purpose of the first period is to find out the prominent research areas that Africa had been pursuing over all these years. The analysis of research areas in the second period is meant to capture recent research trends and foci in contemporary Africa and in specific African countries.

Predominant Research Areas in Africa, 1945–2015

There were 141 research areas in which publications were produced by African countries during 1945–2015. All the research areas were reflected in varying numbers in a total of 837,034 publications. These figures are more than the actual number of publications that Africa had produced during this period, due to the classification system adopted by the WoS in which a paper is sometimes grouped under more than one research area. For the purposes of this analysis, the total count of research areas is acceptable.

meteorology and atmospheric sciences, mineralogy, mining and mineral processing, oceanography, optics, physical geography, physics, polymer science, thermodynamics, and water resources)

Technology (acoustics, automation and control systems, computer science, construction and building technology, energy and fuels, engineering, imaging science and photographic technology, information science and library science, instruments and instrumentation, materials science, mechanics, metallurgy and metallurgical engineering, microscopy, nuclear science and technology, operations research and management science, remote sensing, robotics, science, technology and other topics, spectroscopy, telecommunications and transportation)

Arts and Humanities (architecture, art, arts, humanities and other topics, Asian studies, classics, dance, film, radio and television, history, history and philosophy of science, literature, music, philosophy, religion, and theater)

Social Sciences (archaeology, area studies, biomedical social sciences, business and economics, communication, criminology and penology, cultural studies, demography, education and educational research, ethnic studies, family studies, geography, government and law, international relations, linguistics, mathematical methods in social sciences, psychology, public administration, social issues, social sciences other topics, social work, sociology, urban studies, and women's studies)

The number of research areas ranged between 238 (audiology)[2] and 62,500 (chemistry) showing the differing levels of prominence of scientific research areas in African science. These two figures translate into 0.02 per cent of the total publications (according to research areas) for audiology (with the lowest number of publications) and 7.5 per cent for chemistry (with the highest number of publications). There was an average of 5,936 publications per research area in Africa. A few research areas had 1 per cent or above the total count: agriculture, biochemistry and molecular biology, biotechnology and applied microbiology, computer science, engineering, entomology, environmental sciences and ecology, food science and technology, general and internal medicine, geology, immunology, infectious diseases, marine and freshwater biology, materials science, mathematics, microbiology, parasitology, pharmacology, physics, plant sciences, public, environmental and occupational health, science, technology and other topics, surgery, tropical medicine, veterinary sciences, water resources, and zoology.

For more details further analysis is required. As the average number of publications according to research areas was 5,936, yet another level of analysis was done for publications with a count of more than 5,000. The focus is on prominent and active research areas in Africa. The data is presented in Table 4.1. Of the 141 research areas, there were 41 research areas with over 5,000 publications: agriculture, biochemistry and molecular biology, biotechnology and applied microbiology, chemistry, computer science, energy and fuels, engineering, entomology, environmental sciences and ecology, food science and technology, genetics and heredity, general and internal medicine, geology, immunology, infectious diseases, marine and fresh water biology, materials science, mathematics, mechanics, metallurgy, microbiology, neurosciences and neurology, nuclear sciences and technology, nutrition and dietetics, obstetrics and gynaecology, oncology, parasitology, paediatrics, pharmacology, physics, plant sciences, polymer science, public, environmental and occupational health, science, technology and others, surgery, thermodynamics, tropical medicine, veterinary sciences, virology, water resources, and zoology.

Chemistry, the research area with the highest number of publications, accounted for 7.5 per cent of total publications in Africa.

---

[2] This is not, however, shown in the current WoS list which might have been incorporated into the research area of acoustics.

Following chemistry were the research areas of engineering, physics (4.8 per cent each), agriculture (4 per cent), environmental sciences and ecology (3.6 per cent), general and internal medicine (3.3 per cent), materials science (3.1 per cent), public, environmental and occupational health (2.8 per cent), infectious diseases, plant sciences, mathematics, science, technology and others (2.6 per cent each), pharmacology (2.5 per cent), tropical medicine (2.3 per cent), biochemistry and molecular biology (2.1 per cent), immunology (1.9 per cent), veterinary sciences (1.8 per cent), geology (1.7 per cent), food science and technology (1.6 per cent), microbiology (1.6 per cent), biotechnology and applied microbiology (1.5 per cent), parasitology (1.5 per cent), zoology (1.5 per cent), entomology (1.2 per cent), water resources (1.2 per cent), marine and fresh water biology (1.1 per cent), surgery (1.1 per cent), astronomy and astrophysics (1 per cent), computer science (1 per cent), energy and fuels, (0.9 per cent), paediatrics (0.9 per cent), polymer science (0.9 per cent), genetics and heredity (0.8 per cent), mechanics (0.8 per cent), metallurgy (0.8 per cent), neurosciences and neurology (0.8 per cent), nuclear sciences and technology (0.7 per cent), nutrition and dietetics (0.7 per cent), obstetrics and gynaecology (0.7 per cent), thermodynamics (0.7 per cent), virology (0.7 per cent), and oncology (0.6 per cent). Publications with at least 1 per cent can be considered as the core and active research areas of Africa.

As seen above, the first ten research areas that dominated scientific research in Africa were chemistry; engineering; physics; agriculture, environmental sciences and ecology; general and internal medicine; materials science; public, environmental and occupational health; infectious diseases; plant sciences; and mathematics. Who are the leaders in Africa in the production of scientific publications in these ten areas? This query takes us to the scientific strengths of individual African countries. A country-by-country analysis can reveal this.

Major Contributors to Research Areas in Africa, 1945–2015

To commence, the first ten major research areas in Africa are considered.

In the area of chemistry, Egypt made the largest single contribution with 42 per cent of all African publications. South Africa made

Scientific Research Areas

Table 4.1 *Prominent scientific research areas of Africa, 1945–2015*

Country	Agriculture		Bioch & Mol Bio		Biote Appl Micro		Chemistry		Comp Sci		Ener & Fuels	
	No.	%	No.	%	No.	%	No.	%	No.	%	NO.	%
Algeria	500	1.52	669	3.84	330	2.67	4532	7.25	1195	14.11	867	13.33
Angola	10	0.03	9	0.05	6	0.05	23	0.04	6	0.07	10	0.15
Benin	498	1.52	53	0.30	139	1.12	98	0.16	13	0.15	8	0.12
Botswana	217	0.66	93	0.53	42	0.34	311	0.50	39	0.46	45	0.69
Burkina Faso	451	1.37	86	0.49	110	0.89	117	0.19	20	0.24	50	0.77
Burundi	49	0.15	3	0.02	8	0.06	24	0.04	0	0.00	4	0.06
Cameroon	614	1.87	404	2.32	205	1.66	758	1.21	90	1.06	85	1.31
Cape Verde	5	0.02	2	0.01	1	0.01	2	0.00	2	0.02	1	0.02
Chad	18	0.05	4	0.02	5	0.04	7	0.01	3	0.04	0	0.00
Comoros	2	0.01	1	0.01	1	0.01	2	0.00	0	0.00	0	0.00
Congo	88	0.27	31	0.18	35	0.28	60	0.10	0	0.00	5	0.08
Djibouti	1	0.00	4	0.02	0	0.00	6	0.01	0	0.00	3	0.05
Egypt	3786	11.53	4397	25.21	2395	19.35	25788	41.26	2162	25.53	2066	31.77
Eritrea	36	0.11	6	0.03	11	0.09	7	0.01	1	0.01	3	0.05
Ethiopia	1908	5.81	178	1.02	188	1.52	482	0.77	36	0.43	54	0.83
Gabon	28	0.09	100	0.57	31	0.25	37	0.06	5	0.06	7	0.11
Gambia	41	0.12	42	0.24	23	0.19	4	0.01	3	0.04	3	0.05
Ghana	866	2.64	159	0.91	148	1.20	342	0.55	34	0.40	78	1.20
Guinea	231	0.70	104	0.60	28	0.23	96	0.15	12	0.14	24	0.37
Guinea-Bissau	2	0.01	4	0.02	2	0.02	0	0.00	0	0.00	0	0.00
Ivory Coast	337	1.03	41	0.24	8	0.06	100	0.16	1	0.01	1	0.02
Kenya	2832	8.62	730	4.18	441	3.56	462	0.74	73	0.86	150	2.31
Lesotho	30	0.09	3	0.02	5	0.04	38	0.06	2	0.02	23	0.35
Liberia	7	0.02	4	0.02	1	0.01	1	0.00	0	0.00	4	0.06
Libya	65	0.20	68	0.39	29	0.23	458	0.73	48	0.57	170	2.61
Madagascar	120	0.37	114	0.65	31	0.25	96	0.15	6	0.07	17	0.26
Malawi	415	1.26	51	0.29	55	0.44	73	0.12	15	0.18	30	0.46
Mali	339	1.03	56	0.32	33	0.27	38	0.06	5	0.06	10	0.15
Mauritius	146	0.44	49	0.28	46	0.37	148	0.24	34	0.40	29	0.45
Mauritania	33	0.10	7	0.04	3	0.02	14	0.02	3	0.04	4	0.06
Morocco	1022	3.11	747	4.28	416	3.36	4464	7.14	580	6.85	296	4.55
Mozambique	113	0.34	25	0.14	23	0.19	0	0.00	6	0.07	18	0.28
Namibia	57	0.17	22	0.13	21	0.17	33	0.05	10	0.12	6	0.09
Niger	406	1.24	35	0.20	35	0.28	44	0.07	8	0.09	12	0.18
Nigeria	5820	17.72	1001	5.74	2307	18.64	3395	5.43	276	3.26	932	14.33
Rwanda	106	0.32	19	0.11	21	0.17	25	0.04	6	0.07	8	0.12
Reunion	179	0.55	121	0.69	45	0.36	68	0.11	53	0.63	28	0.43
Sao Tome and Principe	0	0.00	0	0.00	0	0.00	0	0.00	0	0.00	0	0.00
Senegal	490	1.49	139	0.80	144	1.16	270	0.43	44	0.52	37	0.57
Seychelles	7	0.02	13	0.07	1	0.01	1	0.00	1	0.01	7	0.11
Sierra Leone	110	0.33	15	0.09	11	0.09	48	0.08	1	0.01	6	0.09
Somalia	32	0.10	16	0.09	2	0.02	17	0.03	0	0.00	1	0.02
South Africa	5685	17.31	5623	32.23	3170	25.61	13937	22.30	2303	27.19	1541	23.69
Sudan	672	2.05	161	0.92	116	0.94	463	0.74	39	0.46	56	0.86
Swaziland	66	0.20	7	0.04	6	0.05	32	0.05	8	0.09	4	0.06
Tanzania	855	2.60	135	0.77	144	1.16	261	0.42	28	0.33	110	1.69
Togo	155	0.47	13	0.07	16	0.13	27	0.04	0	0.00	8	0.12
Tunisia	1485	4.52	1612	9.24	1256	10.15	4948	7.92	1231	14.53	633	9.73
Uganda	656	2.00	117	0.67	151	1.22	115	0.18	26	0.31	46	0.71
Zambia	278	0.85	32	0.18	47	0.38	58	0.09	12	0.14	22	0.34
Zimbabwe	968	2.95	119	0.68	87	0.70	170	0.27	30	0.35	52	0.80
Total	32837	100.00	17444	100.00	12379	100.00	62500	100.00	8470	100.00	6504	100.00

the second largest contribution with 22 per cent. Tunisia, Morocco and Algeria produced about 7 per cent each in the area. With 32 per cent of all research publications in engineering, Egypt led Africa. Another quarter of the publications in engineering came from South Africa, 13 per cent from Algeria and 10 per cent from Tunisia.

Four countries are known for their prominence in physics: Egypt (31 per cent), South Africa (25 per cent), Algeria (15 per cent) and Morocco (11 per cent). Nigeria became the largest contributor in agriculture in Africa with 18 per cent. South Africa followed closely with 17 per cent of publications in agriculture. Research in environmental sciences and ecology for Africa was mostly done in South

Table 4.1 *(cont.)*

Country	Engg		Entmlgy		Envmtl Sci & Ecolgy		Food Sci Tec		Gene & Here	
	No.	%	No.	%	No.	%	No.	%	No.	%
Algeria	5387	13.43	143	1.47	790	2.60	501	3.65	159	2.37
Angola	44	0.11	6	0.06	14	0.05	1	0.01	2	0.03
Benin	41	0.10	322	3.32	178	0.59	164	1.20	46	0.68
Botswana	147	0.37	0	0.00	367	1.21	103	0.75	12	0.18
Burkina Faso	98	0.24	121	1.25	232	0.76	72	0.53	70	1.04
Burundi	3	0.01	10	0.10	33	0.11	5	0.04	2	0.03
Cameroon	360	0.90	202	2.08	531	1.75	405	2.95	87	1.29
Cape Verde	8	0.02	4	0.04	35	0.12	3	0.02	3	0.04
Chad	4	0.01	1	0.01	11	0.04	7	0.05	2	0.03
Comoros	0	0.00	2	0.02	7	0.02	2	0.01	2	0.03
Congo	32	0.08	47	0.48	140	0.46	42	0.31	32	0.48
Djibouti	4	0.01	0	0.00	7	0.02	0	0.00	0	0.00
Egypt	12915	32.21	1629	16.79	3645	12.00	3054	22.28	1030	15.33
Eritrea	12	0.03	9	0.09	26	0.09	1	0.01	6	0.09
Ethiopia	173	0.43	132	1.36	713	2.35	259	1.89	142	2.11
Gabon	14	0.03	17	0.18	142	0.47	30	0.22	49	0.73
Gambia	8	0.02	17	0.18	30	0.10	4	0.03	85	1.26
Ghana	287	0.72	185	1.91	693	2.28	371	2.71	121	1.80
Guinea	116	0.29	112	1.15	204	0.67	46	0.34	73	1.09
Guinea-Bissau	2	0.00	2	0.02	11	0.04	1	0.01	9	0.13
Ivory Coast	10	0.02	29	0.30	48	0.16	0	0.00	10	0.15
Kenya	332	0.83	1495	15.41	2186	7.20	495	3.61	447	6.65
Lesotho	17	0.04	10	0.10	13	0.04	4	0.03	2	0.03
Liberia	13	0.03	10	0.10	8	0.03	0	0.00	1	0.01
Libya	455	1.13	17	0.18	100	0.33	32	0.23	23	0.34
Madagascar	30	0.07	26	0.27	287	0.94	52	0.38	65	0.97
Malawi	38	0.09	43	0.44	146	0.48	96	0.70	38	0.57
Mali	25	0.06	58	0.60	93	0.31	0	0.00	65	0.97
Mauritius	81	0.20	35	0.36	156	0.51	66	0.48	34	0.51
Mauritania	13	0.03	16	0.16	30	0.10	6	0.04	1	0.01
Morocco	1819	4.54	124	1.28	932	3.07	515	3.76	261	3.88
Mozambique	45	0.11	35	0.36	121	0.40	41	0.30	22	0.33
Namibia	36	0.09	31	0.32	339	1.12	12	0.09	21	0.31
Niger	55	0.14	55	0.57	135	0.44	39	0.28	47	0.70
Nigeria	2331	5.81	695	7.16	2352	7.74	2810	20.50	320	4.76
Rwanda	16	0.04	14	0.14	73	0.24	21	0.15	11	0.16
Reunion	60	0.15	55	0.57	213	0.70	97	0.71	93	1.38
Sao Tome and Principe	0	0.00	0	0.00	6	0.02	0	0.00	2	0.03
Senegal	112	0.28	86	0.89	351	1.16	97	0.71	65	0.97
Seychelles	17	0.04	7	0.07	104	0.34	2	0.01	13	0.19
Sierra Leone	17	0.04	14	0.14	35	0.12	17	0.12	3	0.04
Somalia	0	0.00	3	0.03	9	0.03	5	0.04	4	0.06
South Africa	10124	25.25	2835	29.22	10555	34.74	2204	16.08	2158	32.11
Sudan	182	0.45	103	1.06	205	0.67	266	1.94	96	1.43
Swaziland	27	0.07	15	0.15	57	0.19	14	0.10	0	0.00
Tanzania	322	0.80	228	2.35	982	3.23	211	1.54	118	1.76
Togo	21	0.05	46	0.47	43	0.14	13	0.09	11	0.16
Tunisia	3926	9.79	189	1.95	1515	4.99	1189	8.67	681	10.13
Uganda	105	0.26	193	1.99	700	2.30	131	0.96	78	1.16
Zambia	62	0.15	89	0.92	201	0.66	51	0.37	26	0.39
Zimbabwe	156	0.39	184	1.90	576	1.90	152	1.11	72	1.07
Total	40102	100.00	9701	100.00	30380	100.00	13709	100.00	6720	100.00

Africa which had 35 per cent of all publications in Africa. Most of the remaining publications with a wide gap originated either in Egypt (12 per cent), Nigeria (8 per cent) or in Kenya (7 per cent). Scientific research in the field of general and internal medicine was concentrated in South Africa. It produced 46 per cent of all publications for the continent. Nigeria had another 14 per cent while Kenya was the third largest producer with 8 per cent of publications.

Materials science was among the top ten research areas on the continent. This branch of science developed most in four countries, namely, Egypt, South Africa, Algeria and Tunisia. Egypt made up the highest percentage of publications (39 per cent). The other three

Table 4.1 *(cont.)*

Country	Gen Int Med No.	%	Geology No.	%	Immulgy No.	%	Infec Dis No.	%	Mari & Fres No.	%
Algeria	103	0.38	592	4.21	143	0.92	193	0.90	141	1.61
Angola	17	0.06	38	0.27	27	0.17	69	0.32	23	0.26
Benin	28	0.10	39	0.28	96	0.62	256	1.20	37	0.42
Botswana	96	0.35	269	1.91	135	0.87	214	1.00	29	0.33
Burkina Faso	47	0.17	69	0.49	268	1.73	517	2.42	21	0.24
Burundi	9	0.03	18	0.13	17	0.11	41	0.19	39	0.45
Cameroon	172	0.63	331	2.35	338	2.18	679	3.17	64	0.73
Cape Verde	2	0.01	14	0.10	0	0.00	3	0.01	12	0.14
Chad	4	0.01	41	0.29	13	0.08	27	0.13	25	0.29
Comoros	1	0.00	0	0.00	3	0.02	7	0.03	0	0.00
Congo	62	0.23	70	0.50	159	1.03	284	1.33	25	0.29
Djibouti	6	0.02	12	0.09	2	0.01	8	0.04	1	0.01
Egypt	937	3.43	2250	15.99	1174	7.58	722	3.38	591	6.75
Eritrea	6	0.02	56	0.40	3	0.02	13	0.06	18	0.21
Ethiopia	1018	3.73	396	2.81	357	2.31	769	3.59	114	1.30
Gabon	82	0.30	42	0.30	254	1.64	394	1.84	6	0.07
Gambia	121	0.44	1	0.01	430	2.78	395	1.85	3	0.03
Ghana	297	1.09	187	1.33	349	2.25	583	2.73	95	1.09
Guinea	648	2.37	93	0.66	358	2.31	494	2.31	113	1.29
Guinea-Bissau	40	0.15	0	0.00	133	0.86	138	0.65	6	0.07
Ivory Coast	79	0.29	17	0.12	17	0.11	55	0.26	46	0.53
Kenya	2135	7.82	316	2.25	1783	11.51	2210	10.33	447	5.11
Lesotho	17	0.06	8	0.06	11	0.07	25	0.12	0	0.00
Liberia	20	0.07	4	0.03	26	0.17	33	0.15	3	0.03
Libya	167	0.61	133	0.95	32	0.21	39	0.18	24	0.27
Madagascar	16	0.06	59	0.42	60	0.39	198	0.93	39	0.45
Malawi	256	0.94	77	0.55	491	3.17	783	3.66	67	0.77
Mali	49	0.18	33	0.23	197	1.27	363	1.70	11	0.13
Mauritius	16	0.06	19	0.14	13	0.08	24	0.11	33	0.38
Mauritania	2	0.01	18	0.13	11	0.07	31	0.14	27	0.31
Morocco	512	1.88	1050	7.46	215	1.39	279	1.30	262	2.99
Mozambique	86	0.32	72	0.51	154	0.99	266	1.24	80	0.91
Namibia	32	0.12	154	1.09	33	0.21	49	0.23	127	1.45
Niger	43	0.16	100	0.71	43	0.28	145	0.68	10	0.11
Nigeria	3663	13.42	1025	7.29	436	2.82	816	3.81	357	4.08
Rwanda	81	0.30	13	0.09	144	0.93	214	1.00	10	0.11
Reunion	37	0.14	121	0.86	45	0.29	102	0.48	138	1.58
Sao Tome and Principe	0	0.00	0	0.00	0	0.00	7	0.03	0	0.00
Senegal	153	0.56	119	0.85	380	2.45	740	3.46	152	1.74
Seychelles	10	0.04	7	0.05	2	0.01	5	0.02	60	0.69
Sierra Leone	31	0.11	14	0.10	25	0.16	51	0.24	14	0.16
Somalia	5	0.02	3	0.02	12	0.08	9	0.04	0	0.00
South Africa	12631	46.28	4373	31.08	3840	24.80	4303	20.11	4287	48.98
Sudan	276	1.01	104	0.74	191	1.23	314	1.47	66	0.75
Swaziland	17	0.06	23	0.16	18	0.12	40	0.19	6	0.07
Tanzania	646	2.37	323	2.30	547	3.53	1213	5.67	256	2.92
Togo	17	0.06	27	0.19	51	0.33	99	0.46	0	0.00
Tunisia	742	2.72	833	5.92	572	3.69	478	2.23	630	7.20
Uganda	759	2.78	90	0.64	1044	6.74	1554	7.26	110	1.26
Zambia	376	1.38	92	0.65	375	2.42	559	2.61	45	0.51
Zimbabwe	721	2.64	323	2.30	459	2.96	582	2.72	83	0.95
Total	27291	100.00	14068	100.00	15486	100.00	21392	100.00	8753	100.00

countries together made up another 45 per cent. A few countries were actively involved in public, environmental and occupational health research in Africa: South Africa (14 per cent), Nigeria (13 per cent), Kenya (9 per cent), Egypt, Tanzania (6 per cent each) and Ethiopia (5 per cent).

In the area of infectious diseases, the most significant contribution was made by South Africa (one-fifth of all publications in Africa). Kenya, the second largest producer in the area, provided another

Table 4.1 (cont.)

Country	Materials Sci No.	%	Mathematics No.	%	Mechanics No.	%	Metallurgy No.	%	Microblgy No.	%
Algeria	3672	14.07	2404	11.05	1060	15.79	573	8.51	286	2.21
Angola	3	0.01	2	0.01	0	0.00	0	0.00	21	0.16
Benin	7	0.03	63	0.29	14	0.21	0	0.00	92	0.71
Botswana	37	0.14	146	0.67	5	0.07	7	0.10	62	0.48
Burkina Faso	44	0.17	68	0.31	7	0.10	2	0.03	203	1.57
Burundi	9	0.03	7	0.03	1	0.01	0	0.00	9	0.07
Cameroon	173	0.66	429	1.97	177	2.64	9	0.13	257	1.99
Cape Verde	1	0.00	6	0.03	1	0.01	2	0.03	2	0.02
Chad	2	0.01	8	0.04	0	0.00	0	0.00	12	0.09
Comoros	0	0.00	1	0.00	0	0.00	0	0.00	1	0.01
Congo	13	0.05	17	0.08	1	0.01	9	0.13	89	0.69
Djibouti	1	0.00	0	0.00	0	0.00	0	0.00	5	0.04
Egypt	10057	38.53	4862	22.35	2201	32.79	2036	30.25	1926	14.88
Eritrea	4	0.02	9	0.04	2	0.03	0	0.00	4	0.03
Ethiopia	117	0.45	105	0.48	12	0.18	6	0.09	320	2.47
Gabon	4	0.02	32	0.15	3	0.04	0	0.00	188	1.45
Gambia	2	0.01	0	0.00	0	0.00	0	0.00	175	1.35
Ghana	136	0.52	53	0.24	10	0.15	38	0.56	319	2.46
Guinea	12	0.05	40	0.18	18	0.27	4	0.06	196	1.51
Guinea-Bissau	0	0.00	1	0.00	0	0.00	0	0.00	40	0.31
Ivory Coast	7	0.03	4	0.02	2	0.03	4	0.06	13	0.10
Kenya	110	0.42	119	0.55	39	0.58	9	0.13	796	6.15
Lesotho	7	0.03	23	0.11	7	0.10	1	0.01	0	0.00
Liberia	2	0.01	4	0.02	0	0.00	0	0.00	12	0.09
Libya	147	0.56	109	0.50	55	0.82	34	0.51	42	0.32
Madagascar	9	0.03	24	0.11	7	0.10	1	0.01	86	0.66
Malawi	11	0.04	18	0.08	1	0.01	2	0.03	265	2.05
Mali	7	0.03	11	0.05	0	0.00	1	0.01	135	1.04
Mauritius	27	0.10	71	0.33	6	0.09	7	0.10	16	0.12
Mauritania	7	0.03	39	0.18	1	0.01	1	0.01	9	0.07
Morocco	2292	8.78	2629	12.09	551	8.21	370	5.50	329	2.54
Mozambique	8	0.03	11	0.05	0	0.00	1	0.01	81	0.63
Namibia	12	0.05	18	0.08	2	0.03	10	0.15	32	0.25
Niger	20	0.08	28	0.13	1	0.01	1	0.01	30	0.23
Nigeria	962	3.69	865	3.98	316	4.71	251	3.73	887	6.85
Rwanda	10	0.04	6	0.03	1	0.01	0	0.00	58	0.45
Reunion	8	0.03	66	0.30	7	0.10	1	0.01	63	0.49
Sao Tome and Principe	0	0.00	0	0.00	0	0.00	0	0.00	0	0.00
Senegal	98	0.38	166	0.76	10	0.15	10	0.15	347	2.68
Seychelles	1	0.00	1	0.00	0	0.00	0	0.00	4	0.03
Sierra Leone	9	0.03	5	0.02	1	0.01	1	0.01	24	0.19
Somalia	4	0.02	0	0.00	0	0.00	0	0.00	9	0.07
South Africa	4422	16.94	6080	27.95	1289	19.20	2726	40.50	3511	27.13
Sudan	109	0.42	91	0.42	23	0.34	7	0.10	145	1.12
Swaziland	14	0.05	43	0.20	11	0.16	0	0.00	1	0.01
Tanzania	90	0.34	54	0.25	19	0.28	20	0.30	364	2.81
Togo	34	0.13	18	0.08	10	0.15	2	0.03	21	0.16
Tunisia	3271	12.53	2835	13.03	819	12.20	538	7.99	875	6.76
Uganda	41	0.16	36	0.17	5	0.07	4	0.06	442	3.41
Zambia	33	0.13	22	0.10	1	0.01	15	0.22	139	1.07
Zimbabwe	38	0.15	102	0.47	16	0.24	28	0.42	0	0.00
Total	26104	100.00	21751	100.00	6712	100.00	6731	100.00	12943	100.00

10 per cent. Plant sciences was concentrated in three countries: South Africa (38 per cent), Egypt (12 per cent) and Nigeria (11 per cent). Contributions from other countries were insignificant. The gap between South Africa and other countries is notable.

In mathematics, a crucial area for Africa, most of the research publications originated from South Africa (28 per cent), Egypt (22 per cent), Tunisia (13 per cent), Morocco (12 per cent) and

Scientific Research Areas

Table 4.1 *(cont.)*

Country	Neuros & Neulgy No.	%	Nucl Sci & Tec No.	%	Nutr & Diet No.	%	Obst & Gyne No.	%	Oncology No.	%	Parasitology No.	%
Algeria	148	2.36	467	8.16	110	1.77	18	0.29	110	2.08	110	0.89
Angola	3	0.05	3	0.05	2	0.03	4	0.06	2	0.04	37	0.30
Benin	31	0.49	2	0.03	74	1.19	16	0.26	7	0.13	199	1.62
Botswana	6	0.10	14	0.24	30	0.48	11	0.18	12	0.23	21	0.17
Burkina Faso	16	0.26	4	0.07	81	1.31	32	0.52	11	0.21	427	3.47
Burundi	6	0.10	0	0.00	13	0.21	3	0.05	1	0.02	17	0.14
Cameroon	60	0.96	19	0.33	131	2.11	76	1.23	58	1.10	564	4.58
Cape Verde	0	0.00	2	0.03	2	0.03	3	0.05	0	0.00	1	0.01
Chad	0	0.00	0	0.00	6	0.10	0	0.00	0	0.00	12	0.10
Comoros	0	0.00	0	0.00	0	0.00	0	0.00	0	0.00	8	0.07
Congo	25	0.40	2	0.03	29	0.47	24	0.39	19	0.36	126	1.02
Djibouti	0	0.00	0	0.00	0	0.00	4	0.06	0	0.00	3	0.02
Egypt	1152	18.36	2598	45.40	889	14.34	1408	22.79	1639	31.04	1062	8.63
Eritrea	1	0.02	0	0.00	2	0.03	3	0.05	0	0.00	7	0.06
Ethiopia	75	1.20	7	0.12	169	2.73	158	2.56	20	0.38	462	3.76
Gabon	27	0.43	1	0.02	5	0.08	16	0.26	10	0.19	228	1.85
Gambia	6	0.10	0	0.00	70	1.13	22	0.36	21	0.40	171	1.39
Ghana	37	0.59	172	3.01	188	3.03	190	3.07	75	1.42	434	3.53
Guinea	18	0.29	1	0.02	57	0.92	49	0.79	21	0.40	247	2.01
Guinea-Bissau	2	0.03	0	0.00	15	0.24	8	0.13	2	0.04	15	0.12
Ivory Coast	7	0.11	6	0.10	5	0.08	3	0.05	1	0.02	33	0.27
Kenya	169	2.69	32	0.56	343	5.53	326	5.28	143	2.71	1910	15.52
Lesotho	2	0.03	0	0.00	3	0.05	3	0.05	0	0.00	4	0.03
Liberia	0	0.00	1	0.02	4	0.06	5	0.08	1	0.02	49	0.40
Libya	49	0.78	112	1.96	19	0.31	14	0.23	42	0.80	51	0.41
Madagascar	6	0.10	3	0.05	12	0.19	19	0.31	5	0.09	136	1.11
Malawi	30	0.48	0	0.00	148	2.39	68	1.10	43	0.81	170	1.38
Mali	21	0.33	1	0.02	33	0.53	23	0.37	17	0.32	214	1.74
Mauritius	1	0.02	1	0.02	24	0.39	2	0.03	4	0.08	12	0.10
Mauritania	5	0.08	0	0.00	5	0.08	5	0.08	1	0.02	24	0.20
Morocco	620	9.88	339	5.92	154	2.48	95	1.54	211	4.00	162	1.32
Mozambique	13	0.21	4	0.07	32	0.52	71	1.15	14	0.27	115	0.93
Namibia	8	0.13	11	0.19	1	0.02	3	0.05	2	0.04	13	0.11
Niger	5	0.08	0	0.00	17	0.27	25	0.40	5	0.09	81	0.66
Nigeria	465	7.41	321	5.61	1234	19.90	1006	16.28	283	5.36	977	7.94
Rwanda	7	0.11	3	0.05	6	0.10	35	0.57	10	0.19	43	0.35
Reunion	39	0.62	2	0.03	0	0.00	32	0.52	7	0.13	43	0.35
Sao Tome and Principe	0	0.00	0	0.00	0	0.00	0	0.00	0	0.00	13	0.11
Senegal	49	0.78	19	0.33	104	1.68	38	0.61	49	0.93	439	3.57
Seychelles	23	0.37	0	0.00	16	0.26	0	0.00	0	0.00	4	0.03
Sierra Leone	2	0.03	1	0.02	0	0.00	17	0.28	3	0.06	55	0.45
Somalia	2	0.03	0	0.00	3	0.05	5	0.08	0	0.00	33	0.27
South Africa	2145	34.19	1317	23.02	1131	18.24	1514	24.50	1581	29.94	1224	9.95
Sudan	33	0.53	89	1.56	137	2.21	63	1.02	73	1.38	406	3.30
Swaziland	5	0.08	4	0.07	6	0.10	3	0.05	1	0.02	9	0.07
Tanzania	88	1.40	22	0.38	234	3.77	208	3.37	71	1.34	890	7.23
Togo	11	0.18	0	0.00	7	0.11	2	0.03	6	0.11	66	0.54
Tunisia	620	9.88	130	2.27	372	6.00	175	2.83	457	8.66	0	0.00
Uganda	122	1.94	1	0.02	98	1.58	167	2.70	172	3.26	527	4.28
Zambia	46	0.73	8	0.14	84	1.35	93	1.51	24	0.45	251	2.04
Zimbabwe	68	1.08	3	0.05	96	1.55	114	1.84	46	0.87	223	1.81
Total	6274	100.00	5722	100.00	6201	100.00	6179	100.00	5280	100.00	12303	100.00

Algeria (11 per cent). They together produced 86 per cent of all mathematics publications for Africa. South Africa produced 40 per cent of the publications in metallurgy research and another 30 per cent came from Egypt. Algeria (9 per cent) and Tunisia (8 per cent) also played important parts in this subject.

Apart from the above ten prominent research areas, there were some other research areas that are crucial for African science. They are important not only in terms of the percentage of publications to total publications but also from the developmental perspective.

In tropical medicine, Nigeria was the top producer in Africa (16 per cent). About one-third of all publications in biochemistry and

Table 4.1 *(cont.)*

Country	Pediatrics No.	%	Pharamacology No.	%	Physics No.	%	Plant Sci No.	%	Polymer Sci No.	%
Algeria	78	1.03	488	2.37	5837	14.59	414	1.91	702	9.64
Angola	15	0.20	10	0.05	3	0.01	25	0.12	0	0.00
Benin	19	0.25	70	0.34	205	0.51	236	1.09	0	0.00
Botswana	31	0.41	90	0.44	95	0.24	108	0.50	8	0.11
Burkina Faso	57	0.75	104	0.50	49	0.12	148	0.68	1	0.01
Burundi	5	0.07	6	0.03	40	0.10	21	0.10	2	0.03
Cameroon	83	1.09	675	3.27	791	1.98	761	3.51	44	0.60
Cape Verde	1	0.01	2	0.01	0	0.00	1	0.00	0	0.00
Chad	1	0.01	6	0.03	4	0.01	5	0.02	0	0.00
Comoros	1	0.01	1	0.00	0	0.00	4	0.02	0	0.00
Congo	47	0.62	50	0.24	39	0.10	72	0.33	1	0.01
Djibouti	0	0.00	4	0.02	1	0.00	3	0.01	0	0.00
Egypt	1034	13.64	7330	35.55	12318	30.78	2540	11.71	3945	54.15
Eritrea	1	0.01	9	0.04	8	0.02	17	0.08	0	0.00
Ethiopia	138	1.82	337	1.63	196	0.49	569	2.62	23	0.32
Gabon	40	0.53	100	0.48	20	0.05	48	0.22	0	0.00
Gambia	113	1.49	30	0.15	1	0.00	1	0.00	0	0.00
Ghana	120	1.58	278	1.35	144	0.36	307	1.42	8	0.11
Guinea	152	2.00	157	0.76	35	0.09	212	0.98	1	0.01
Guinea-Bissau	42	0.55	2	0.01	1	0.00	2	0.01	0	0.00
Ivory Coast	15	0.20	25	0.12	17	0.04	104	0.48	0	0.00
Kenya	297	3.92	463	2.25	169	0.42	990	4.56	42	0.58
Lesotho	7	0.09	1	0.00	29	0.07	11	0.05	0	0.00
Liberia	2	0.03	2	0.01	2	0.00	4	0.02	1	0.01
Libya	55	0.73	172	0.83	252	0.63	83	0.38	40	0.55
Madagascar	28	0.37	116	0.56	26	0.06	183	0.84	6	0.08
Malawi	241	3.18	54	0.26	15	0.04	90	0.41	1	0.01
Mali	27	0.36	102	0.49	2	0.00	141	0.65	8	0.11
Mauritius	3	0.04	53	0.26	30	0.07	84	0.39	26	0.36
Mauritania	5	0.07	4	0.02	1	0.00	10	0.05	5	0.07
Morocco	270	3.56	677	3.28	4231	10.57	806	3.72	361	4.96
Mozambique	50	0.66	50	0.24	8	0.02	52	0.24	0	0.00
Namibia	3	0.04	19	0.09	43	0.11	67	0.31	1	0.01
Niger	20	0.26	32	0.16	21	0.05	117	0.54	1	0.01
Nigeria	1020	13.45	2767	13.42	1097	2.74	2411	11.11	344	4.72
Rwanda	39	0.51	55	0.27	18	0.04	57	0.26	0	0.00
Reunion	62	0.82	26	0.13	10	0.02	173	0.80	2	0.03
Sao Tome and Principe	0	0.00	0	0.00	0	0.00	0	0.00	0	0.00
Senegal	82	1.08	74	0.36	181	0.45	136	0.63	27	0.37
Seychelles	7	0.09	16	0.08	0	0.00	10	0.05	0	0.00
Sierra Leone	8	0.11	9	0.04	7	0.02	52	0.24	0	0.00
Somalia	2	0.03	16	0.08	1	0.00	17	0.08	0	0.00
South Africa	2081	27.45	4346	21.08	9888	24.71	8179	37.70	1121	15.39
Sudan	83	1.09	279	1.35	179	0.45	254	1.17	10	0.14
Swaziland	6	0.08	10	0.05	27	0.07	14	0.06	8	0.11
Tanzania	161	2.12	258	1.25	81	0.20	334	1.54	7	0.10
Togo	42	0.55	33	0.16	28	0.07	99	0.46	1	0.01
Tunisia	574	7.57	810	3.93	3756	9.39	1142	5.26	524	7.19
Uganda	205	2.70	179	0.87	21	0.05	258	1.19	1	0.01
Zambia	112	1.48	42	0.20	47	0.12	69	0.32	2	0.03
Zimbabwe	97	1.28	180	0.87	46	0.11	253	1.17	11	0.15
Total	7582	100.00	20619	100.00	40020	100.00	21694	100.00	7285	100.00

molecular biology in Africa were produced in South Africa. Egypt produced a quarter of publications in the same area. One-fourth of all African publications in immunology were drawn from South Africa. Kenya (12 per cent), Egypt (8 per cent) and Uganda (7 per cent) were also involved in immunological research. In microbiology, the major contributors were South Africa (27 per cent), Egypt (15 per cent), Nigeria, Tunisia (7 per cent each) and Kenya (6 per cent). Nigeria claimed first position, publishing 16 per cent of papers in tropical medicine. In biotechnology and applied microbiology, South Africa played the leading part producing 26 per cent of the total publications for Africa, followed by Egypt, Nigeria (19 per cent each) and Tunisia

Table 4.1 *(cont.)*

Country	Pub Env Occu Hea No.	%	Sci, Tech & Oth No.	%	Surgery No.	%	Thermody No.	%	Trop Med No.	%
Algeria	178	0.77	1099	5.11	98	1.06	769	13.12	87	0.45
Angola	42	0.18	33	0.15	7	0.08	1	0.02	60	0.31
Benin	188	0.82	0	0.00	11	0.12	8	0.14	239	1.23
Botswana	97	0.42	155	0.72	6	0.06	7	0.12	46	0.24
Burkina Faso	481	2.09	190	0.88	15	0.16	19	0.32	548	2.81
Burundi	69	0.30	22	0.10	4	0.04	0	0.00	97	0.50
Cameroon	682	2.97	414	1.93	68	0.74	54	0.92	775	3.97
Cape Verde	8	0.03	6	0.03	0	0.00	0	0.00	1	0.01
Chad	44	0.19	27	0.13	0	0.00	1	0.02	43	0.22
Comoros	8	0.03	1	0.00	0	0.00	0	0.00	14	0.07
Congo	259	1.13	103	0.48	15	0.16	0	0.00	261	1.34
Djibouti	9	0.04	8	0.04	2	0.02	0	0.00	9	0.05
Egypt	1480	6.44	2848	13.25	2096	22.69	2127	36.29	829	4.25
Eritrea	20	0.09	13	0.06	3	0.03	0	0.00	14	0.07
Ethiopia	1160	5.05	451	2.10	35	0.38	16	0.27	844	4.33
Gabon	225	0.98	116	0.54	23	0.25	0	0.00	333	1.71
Gambia	333	1.45	146	0.68	4	0.04	0	0.00	325	1.67
Ghana	1002	4.36	398	1.85	100	1.08	23	0.39	801	4.11
Guinea	512	2.23	156	0.73	48	0.52	0	0.00	550	2.82
Guinea-Bissau	79	0.34	23	0.11	0	0.00	0	0.00	52	0.27
Ivory Coast	59	0.26	80	0.37	28	0.30	2	0.03	115	0.59
Kenya	2014	8.76	1311	6.10	126	1.36	26	0.44	2022	10.37
Lesotho	41	0.18	37	0.17	1	0.01	8	0.14	16	0.08
Liberia	85	0.37	13	0.06	7	0.08	1	0.02	79	0.41
Libya	87	0.38	88	0.41	63	0.68	63	1.07	83	0.43
Madagascar	89	0.39	136	0.63	12	0.13	6	0.10	145	0.74
Malawi	803	3.49	269	1.25	65	0.70	2	0.03	593	3.04
Mali	277	1.21	128	0.60	16	0.17	1	0.02	408	2.09
Mauritius	39	0.17	42	0.20	2	0.02	7	0.12	22	0.11
Mauritania	28	0.12	17	0.08	3	0.03	1	0.02	31	0.16
Morocco	251	1.09	796	3.70	655	7.09	461	7.87	107	0.55
Mozambique	245	1.07	122	0.57	11	0.12	1	0.02	231	1.18
Namibia	50	0.22	116	0.54	15	0.16	2	0.03	22	0.11
Niger	163	0.71	74	0.34	10	0.11	1	0.02	192	0.98
Nigeria	2938	12.78	1388	6.46	747	8.09	292	4.98	3203	16.43
Rwanda	135	0.59	90	0.42	22	0.24	2	0.03	120	0.62
Reunion	49	0.21	97	0.45	47	0.51	4	0.07	55	0.28
Sao Tome and Principe	6	0.03	0	0.00	0	0.00	0	0.00	14	0.07
Senegal	552	2.40	298	1.39	66	0.71	17	0.29	611	3.13
Seychelles	57	0.25	24	0.11	1	0.01	0	0.00	10	0.05
Sierra Leone	94	0.41	27	0.13	32	0.35	3	0.05	87	0.45
Somalia	54	0.23	3	0.01	3	0.03	0	0.00	47	0.24
South Africa	3248	14.13	7365	34.25	3846	41.64	1185	20.22	1199	6.15
Sudan	674	2.93	152	0.71	62	0.67	19	0.32	703	3.61
Swaziland	35	0.15	26	0.12	4	0.04	5	0.09	16	0.08
Tanzania	1425	6.20	548	2.55	73	0.79	28	0.48	1412	7.24
Togo	109	0.47	36	0.17	9	0.10	4	0.07	122	0.63
Tunisia	419	1.82	1033	4.80	563	6.10	679	11.59	197	1.01
Uganda	1108	4.82	561	2.61	113	1.22	5	0.09	912	4.68
Zambia	518	2.25	177	0.82	41	0.44	1	0.02	451	2.31
Zimbabwe	459	2.00	238	1.11	59	0.64	10	0.17	344	1.76
Total	22987	100.00	21501	100.00	9237	100.00	5861	100.00	19497	100.00

(10 per cent). In entomology, one-third of the publications in Africa were by South Africans, which was much higher than the share of the second top producer Egypt (17 per cent). Kenya had 15 per cent publications in the area, while Nigeria owned 7 per cent.

South Africa produced one-third of all publications in Africa in the field of geology, which was double the figure of the second highest producer (Egypt). Morocco and Nigeria were among other countries that also published a good number of papers in geology. For research in marine and fresh water biology, South Africa produced the largest share, half of all publications in the continent.

Table 4.1 *(cont.)*

Country	Vet Sci No.	Vet Sci %	Virology No.	Virology %	Water Reso No.	Water Reso %	Zoology No.	Zoology %	All Res Areas No.	All Res Areas %
Algeria	212	1.45	26	0.44	788	7.58	239	1.88	42998	5.14
Angola	4	0.03	5	0.08	0	0.00	11	0.09	744	0.09
Benin	77	0.53	31	0.52	39	0.38	79	0.62	4501	0.54
Botswana	127	0.87	71	1.19	159	1.53	79	0.62	4385	0.52
Burkina Faso	147	1.00	97	1.62	74	0.71	32	0.25	5981	0.71
Burundi	12	0.08	8	0.13	13	0.13	18	0.14	748	0.09
Cameroon	162	1.11	264	4.42	88	0.85	156	1.23	14699	1.76
Cape Verde	3	0.02	1	0.02	5	0.05	9	0.07	216	0.03
Chad	40	0.27	2	0.03	28	0.27	7	0.06	516	0.06
Comoros	0	0.00	1	0.02	0	0.00	3	0.02	94	0.01
Congo	30	0.20	50	0.84	18	0.17	106	0.83	3115	0.37
Djibouti	2	0.01	1	0.02	3	0.03	3	0.02	149	0.02
Egypt	1802	12.30	388	6.49	1396	13.44	748	5.88	172822	20.65
Eritrea	15	0.10	4	0.07	6	0.06	7	0.06	515	0.06
Ethiopia	764	5.21	89	1.49	255	2.45	108	0.85	15868	1.90
Gabon	44	0.30	116	1.94	2	0.02	131	1.03	3458	0.41
Gambia	64	0.44	84	1.41	0	0.00	3	0.02	3139	0.38
Ghana	147	1.00	109	1.82	262	2.52	77	0.61	12167	1.45
Guinea	59	0.40	90	1.51	23	0.22	133	1.05	6542	0.78
Guinea-Bissau	2	0.01	49	0.82	4	0.04	3	0.02	764	0.09
Ivory Coast	38	0.26	14	0.23	22	0.21	22	0.17	1802	0.22
Kenya	1543	10.53	499	8.35	334	3.21	799	6.28	36501	4.36
Lesotho	8	0.05	2	0.03	12	0.12	6	0.05	521	0.06
Liberia	8	0.05	15	0.25	1	0.01	5	0.04	504	0.06
Libya	71	0.48	8	0.13	80	0.77	31	0.24	4528	0.54
Madagascar	55	0.38	34	0.57	11	0.11	311	2.44	3291	0.39
Malawi	55	0.38	161	2.69	90	0.87	51	0.40	6990	0.84
Mali	84	0.57	34	0.57	34	0.33	27	0.21	3709	0.44
Mauritius	13	0.09	5	0.08	21	0.20	47	0.37	1940	0.23
Mauritania	24	0.16	7	0.12	13	0.13	13	0.10	599	0.07
Morocco	406	2.77	73	1.22	482	4.64	268	2.11	40267	4.81
Mozambique	79	0.54	34	0.57	58	0.56	21	0.17	2892	0.35
Namibia	101	0.69	6	0.10	49	0.47	191	1.50	2449	0.29
Niger	45	0.31	19	0.32	73	0.70	22	0.17	2711	0.32
Nigeria	1551	10.59	221	3.70	504	4.85	447	3.51	66600	7.96
Rwanda	24	0.16	80	1.34	21	0.20	39	0.31	1982	0.24
Reunion	51	0.35	41	0.69	17	0.16	62	0.49	3246	0.39
Sao Tome and Principe	0	0.00	0	0.00	0	0.00	5	0.04	60	0.01
Senegal	202	1.38	193	3.23	71	0.68	163	1.28	8989	1.07
Seychelles	1	0.01	3	0.05	9	0.09	50	0.39	730	0.09
Sierra Leone	12	0.08	14	0.23	6	0.06	21	0.17	1067	0.13
Somalia	16	0.11	5	0.08	0	0.00	8	0.06	364	0.04
South Africa	4400	30.03	1936	32.40	3722	35.82	6884	54.09	238381	28.48
Sudan	541	3.69	54	0.90	79	0.76	89	0.70	9105	1.09
Swaziland	6	0.04	9	0.15	28	0.27	61	0.48	818	0.10
Tanzania	484	3.30	223	3.73	252	2.43	325	2.55	16528	1.97
Togo	26	0.18	13	0.22	8	0.08	0	0.00	1594	0.19
Tunisia	355	2.42	150	2.51	835	8.04	328	2.58	56210	6.72
Uganda	246	1.68	492	8.23	135	1.30	247	1.94	13732	1.64
Zambia	0	0.00	0	0.00	0	0.00	0	0.00	5318	0.64
Zimbabwe	494	3.37	145	2.43	260	2.50	232	1.82	10185	1.22
Total	14652	100.00	5976	100.00	10390	100.00	12727	100.00	837034	100.00

Note: Research areas of at least 5,000 included.

Tunisia, which had the second largest contribution, provided 7 per cent. In genetics and heredity research, the highest share of publications came from South Africa. South Africa was able to publish 32 per cent of the total. Other major contributors were Egypt (15 per cent) and Tunisia (10 per cent). In computer science, South Africa was at the forefront with the highest percentage (27 per cent) of research publications for Africa. Egypt (32 per cent), South Africa (24 per cent) and Nigeria (14 per cent) were the main sources of research in energy and fuels for Africa.

During 1945–2015, Tunisia produced a total of 6.7 per cent of publications towards African science. Tunisia's key contribution to Africa was in the fields of computer science (15 per cent of all publications in the area in Africa and the third country after South Africa and Egypt), mathematics (13 per cent), materials science, mechanics and thermodynamics (12 per cent each), biotechnology and applied microbiology, energy and fuels, engineering, genetics and heredity, and neurosciences and neurology (10 per cent each), biochemistry and molecular biology, food science and technology, physics, oncology (9 per cent each).

Kenya is the sixth largest producer of publications in Africa. Its major contributions to Africa were in the scientific fields of parasitology (16 per cent and first in Africa), entomology (15 per cent), immunology (12 per cent and second in Africa), infectious diseases (10 per cent and second in Africa), tropical medicine (10 per cent and second in Africa), veterinary science (11 per cent and third in Africa), and public, environmental and occupational health (9 per cent). The strengths of Kenya were evident in medicine (parasitology, entomology, immunology, infectious diseases, tropical medicine and health).

Algeria contributed 5 per cent to the total production of Africa. In comparison to other African countries, Algeria emerged as the third largest producer in physics, after Egypt and South Africa. It had 15 per cent of all physics publications in Africa. The contribution of Algeria to Africa in the field of engineering was 13 per cent. In mathematics Algeria (11 per cent) was the fifth country in Africa after South Africa, Egypt, Tunisia and Morocco. Algeria published 16 per cent of the publications in mechanics, and filled third position after Egypt and South Africa. Thermodynamics is another research area to which Algeria contributed significantly (13 per cent of the total).

Altogether, Tanzania's share in African science for the period was less than 2 per cent. In terms of percentile contribution of publications, there were not many research areas in which Tanzania could participate actively. However, they included publications in both parasitology and tropical medicine which were 7 per cent each of the total publications in Africa, and 6 per cent for public, environmental and occupational health. In the research area of infectious diseases, Tanzania made up 5.7 per cent of the total. These areas were generally in the broad area of medicine related to diseases.

Ethiopia ranked eighth in scientific publications in Africa. It had 1.9 per cent of the total publications for Africa during the period. In agriculture, Ethiopia produced 6 per cent of the total African publications. It contributed 5 per cent each to the fields of public, environmental and occupational health, and veterinary sciences in Africa. About 4 per cent each of the publications for Africa in general and internal medicine, infectious diseases, and tropical medicine also came from Ethiopia.

Strengths of Individual Countries, 1945–2015

In the count of publications, according to classified research areas, a few countries made a sizable contribution to science in Africa during 1945–2015. South Africa (28 per cent), Egypt (21 per cent), Nigeria (8 per cent), Tunisia (7 per cent), Algeria (5 per cent), Morocco (5 per cent) and Kenya (4 per cent), Tanzania (2 per cent) and Ethiopia (2 per cent) were among them. The analysis therefore starts with these main contributors.

North African countries made the largest contribution of publications categorised under the research areas. Four countries (Algeria, Egypt, Tunisia and Morocco) jointly produced 37 per cent of the total publications in Africa for the period of analysis. Their individual share for the whole of Africa, on average, was 9 per cent each. Sub-Saharan Africa had the remaining 63 per cent of all production. South Africa was the single most productive country in publications with a 28 per cent share of all the publications in Africa. South Africa alone produced more than half (55 per cent) of the total publications in sub-Saharan Africa.

South Africa's major scientific production was in the areas of chemistry (5.8 per cent of all its publications), general and internal medicine (5.3 per cent), environmental sciences and ecology (4.4 per cent), engineering (4.2 per cent), physics (4.1 per cent), plant sciences (3.4 per cent), science, technology and others (3.1 per cent), zoology (2.9 per cent), mathematics (2.6 per cent), agriculture, biochemistry and molecular biology (2.4 per cent each). Some of the least published areas were materials science, veterinary sciences (1.9 per cent each), geology, infectious diseases, marine and freshwater biology, pharmacology (1.8 per cent each), immunology, surgery, water resources (1.6 per cent each), microbiology (1.5 per cent), public, environmental and occupational health (1.4 per cent), biotechnology and applied microbiology (1.3 per cent), entomology (1.2 per cent), metallurgy (1.1 per cent), computer science (1 per cent), food science and technology, genetics and heredity, paediatrics, and neurosciences and neurology (0.9 per cent eaxh), virology (0.8 per cent), oncology (0.7 per cent), energy and fuels, nuclear science and technology, obstetrics and gynaecology (0.6 per cent each), mechanics, nutrition and dietetics, parasitology, polymer science, thermodynamics, and tropical medicine (0.5 per cent each).

Egypt followed South Africa in the total number of publications classified under research areas. Its key research areas were chemistry (15 per cent of all its publications), engineering (7.5 per cent), physics (7.1 per cent), materials science (5.8 per cent) and pharmacology (4.2 per cent). Agriculture, biochemistry and molecular biology, environmental sciences and ecology, mathematics and polymer science made 2–3 per cent of its publications. The least productive areas for Egypt were general and internal medicine (0.05 per cent), virology (0.2 per cent), marine and freshwater biology (0.3 per cent), zoology, infectious diseases (0.4 per cent each), tropical medicine, nutrition and dietetics (0.5 per cent each), genetics and heredity, parasitology, paediatrics (0.6 per cent each), and immunology (0.7 per cent).

Of all the publications produced by Nigeria, 9 per cent were in agriculture, 5.5 per cent in general and internal medicine, 5.1 per cent in chemistry, 4.8 per cent in tropical medicine, 4.2 per cent each in entomology, food science and technology, and pharmacology, 3.6 per cent in plant sciences, 3.5 per cent each in biotechnology and applied microbiology, engineering, and environmental sciences and ecology, 2.3 per cent in veterinary sciences, 2.1 per cent in science, technology and other topics, 1.9 per cent in nutrition and dietetics, 1.6 per cent in physics, 1.5 per cent each in biochemistry and molecular biology, geology, obstetrics and gynaecology, parasitology, and paediatrics, 1.4 per cent each in energy and fuels, and materials science, 1.3 per cent each in mathematics and microbiology, 1.2 per cent in infectious diseases, 1.1 per cent in surgery, 1 per cent in entomology,

0.8 per cent in water resources, 0.7 per cent each in immunology, and zoology, 0.5 per cent each in genetics and heredity, mechanics, nuclear sciences and technology, and 0.4 per cent in computer science.

With one-fifth of the total publications, Nigeria was the top producer in nutrition and dietetics in Africa. In obstetrics and gynaecology, paediatrics, plant sciences, and pharmacology, Nigeria was third in Africa. Publications in public, environmental and occupational health made Nigeria an equal partner with South Africa, 13 per cent against 14 per cent respectively. In veterinary science, also Nigeria played a significant role by having one-tenth of all publications in Africa.

The focus of Tunisia was limited to chemistry (8.8 per cent of all publications in the country), engineering (7 per cent), physics (6.7 per cent), materials science (5.8 per cent) and mathematics (5 per cent). Other less-published areas were biochemistry and molecular biology (2.9 per cent), agriculture (2.6 per cent), biotechnology and applied microbiology, computer science (2.2 per cent each), food science and technology (2.1 per cent), plant sciences (2 per cent), science, technology and other topics (1.8 per cent), microbiology (1.6 per cent), geology, mechanics, and water resources (1.5 per cent each), genetics and heredity (1.2 per cent), energy and fuels, marine and fresh water biology, neurosciences and neurology, and paediatrics (1.1 per cent each), immunology, and metallurgy (1 per cent each).

Algeria had more publications in physics than in any other research area, about 14 per cent of the total publications in the country. Engineering (12.6 per cent), chemistry (10.5 per cent), materials science (8.6 per cent), mathematics (5.6 per cent), computer science (2.8 per cent), and science, technology and others (2.6 per cent) were other important research areas for Algeria. The poorly researched areas included virology (0.06 per cent), obstetrics and gynaecology (0.04 per cent), general and internal medicine, paediatrics, surgery, tropical diseases (0.2 per cent each), immunology, marine and freshwater biology, neurosciences and neurology, nutrition and dietetics, oncology, parasitology (0.3 per cent each), genetics and heredity, infectious diseases, public, environmental and occupational health (0.4 per cent each), veterinary sciences (0.5 per cent), zoology (0.6 per cent), plant sciences (1 per cent), pharmacology (1.1 per cent), polymer science (1.6 per cent), thermodynamics, and water resources (1.8 per cent each). Obviously, medical subjects were not the preferred research areas for Algeria. None of the medical research areas were found to be among the highest published ones in the country. For instance, there were fewer publications in virology, obstetrics and gynaecology, general and internal medicine, paediatrics, surgery and tropical diseases than others. Algeria worked more intensely in physical sciences and in technology. These preferential areas of research were a feature of North Africa.

Morocco produced 11 per cent of all its publications in the field of chemistry, 10.5 per cent in physics, 6.5 per cent in mathematics, 5.7 per cent in materials science, 4.5 per cent in engineering, 2.6 per cent in geology, 2.5 per cent in the area of agriculture, 2.3 per cent in environmental sciences and ecology, 2 per cent each in plant sciences, science, technology and other topics, and in zoology, 1.9 per cent in biochemistry and molecular biology, 1.7 per cent in pharmacology, 1.6 per cent in surgery, 1.5 per cent in neurosciences and neurology, 1.4 per cent in mechanics, 1.3 per cent each in food science and technology, and general and internal medicine, 1.2 per cent in water resources, 1.1 per cent in thermodynamics, 1 per cent each in biotechnology and applied microbiology, computer science, and veterinary sciences, 0.9 per cent each in metallurgy, polymer science, 0.8 per cent each in microbiology, nuclear sciences and technology, 0.7 per cent each in energy and fuels, genetics and heredity, infectious diseases, marine and fresh water biology, paediatrics, 0.6 per cent in public, environmental and occupational health, 0.5 per cent each in immunology, oncology, 0.4 per cent each in nutrition and dietetics, parasitology, 0.3 per cent each in entomology, and tropical medicine, and 0.2 per cent each in obstetrics and gynaecology, and virology. Unlike sub-Saharan Africa, medical and related research was not very significant in this North African country.

Kenya's highest number of publications were in agriculture (7.8 per cent), infectious diseases (6.1 per cent), environmental sciences and ecology (6 per cent), general and internal medicine (5.9 per cent), public, environmental and occupational health, and tropical medicine (5.5 per cent each), parasitology (5.2 per cent), immunology (4.9 per cent), veterinary sciences (4.2 per cent), entomology (4.1 per cent), plant sciences (2.7 per cent), microbiology and zoology (2.2 per cent each) and biochemistry and molecular biology (2 per cent).

Tanzania's strengths in terms of the number of publications were seen in areas of public, environmental and occupational health (8.6 per cent of all its publications), tropical medicine (8.5 per cent), infectious diseases (7.3 per cent), environmental science and ecology

(5.9 per cent), agriculture (5.2 per cent), general and internal medicine (3.9 per cent), science, technology and others, and immunology (3.3 per cent each), veterinary sciences (2.9 per cent), microbiology (2.2 per cent), and engineering, plant sciences, and zoology (2 per cent each). Some areas such as computer science, mathematics, nuclear science and technology, polymer science, thermodynamics, mechanics, physics and metallurgy did not produce a significant number of publications.

Agriculture was a priority research area for Ethiopia. Twelve per cent of the total publications in the country were in agriculture. Following this were public, environmental and occupational health (7.3 per cent), general and internal medicine (6.4 per cent), tropical medicine (5.3 per cent), infectious diseases (4.9 per cent), veterinary sciences (4.8 per cent), environmental sciences and ecology (4.5 per cent), plant sciences (3.6 per cent), chemistry (3 per cent), parasitology (2.9 per cent), science, technology and others (2.8), and geology (2.5 per cent). The poorly produced areas were computer science, energy and fuels, mechanics, metallurgy, oncology, surgery, neurosciences and neurology, polymer science and virology.

As this analysis shows, there were a few countries in Africa that played a significant part in the production of science in Africa during 1945–2015. These countries also had their areas of strength. In many scientific research areas, South Africa was the undisputed leader. South Africa enjoyed dominance in marine and freshwater biology, geology, immunology, microbiology, infectious diseases, paediatrics, plant sciences, science, technology and other topics, surgery, virology, water resources and zoology.

South Africa also had capacity in research areas such as chemistry, general and internal medicine, environmental sciences and ecology, engineering and physics. In general and internal medicine, the share of publications for South Africa was higher than the share of publications for Africa. In environmental sciences and ecology, South Africa also performed better than Africa with a higher percentage of publications. In other areas of medicine (infectious diseases, immunology, surgery or microbiology), the production of scientific knowledge in South Africa was not very prominent within the country. Similarly, in neurosciences and neurology, obstetrics and gynaecology, paediatrics, plant sciences, public, environmental and occupational health, science, technology and other topics, surgery, virology, water resources and in zoology, South Africa led Africa. In zoological research, South Africa produced more than half of all publications on the continent. The contrast is that within the country the percentages of publications in infectious diseases and immunology were only less than 2 per cent of its total publications.

Some degree of correspondence was visible between the number of all publications and number of publications in key areas. Most of the countries with a higher number of total publications also had a higher share in the total publications of Africa in certain research areas.

Nigeria focused its research mainly in the areas of agriculture, medicine, chemistry, entomology, food science and technology and in pharmacology. The emphasis on agriculture and medicine (both general and internal medicine and tropical medicine) was obvious. These were historical reasons for this.

Algeria's strengths were predominantly in the physical sciences (physics, chemistry and mathematics) and in technology areas (engineering, materials science, computer science, science and technology and other topics), and not in life sciences, biomedicine and other topics. Is this a feature of North Africa? Clearly, scientific strengths of other North African countries like Morocco were in chemistry, physics, mathematics, materials science and in engineering. The set of areas such as virology, tropical medicine, entomology, parasitology, and immunology were not well developed in the country. Some research areas, namely, agriculture, plant sciences, zoology, geology, and science, technology and other topics showed signs of growth.

Current Research Foci of Africa, 2011–15

In this section, the analysis of research areas in Africa is confined to five years from 2011 to 2015. The importance of the period, other than that it is recent, is that 2011 was an historical turning point for science in Africa. The ministers of science and technology in Africa declared 2011 as the decade of science in Africa, committing increased shares of budgets and efforts to the development of science in Africa. A firm commitment to the development of science and technology was thereafter forthcoming.

During 2011–15, there were a total of 278,567 publications in Africa, grouped under 141 research areas. The average number of publications per country was 5,462, which is 473 short of the average number of publications for the whole period of 1945–2015. Instead of examining all these 141 research areas, some of which do not have a real presence in many countries, a few leading areas were chosen. Research areas that had 2,000 or more publications have been included to accommodate the research areas that had a reasonable presence in terms of the number of publications. Forty-two research areas had more than 2,000 publications. They were agriculture; astronomy; biochemistry and molecular biology; biotechnology and applied microbiology; chemistry; computer science; electrochemistry; energy and fuels; engineering; entomology; environmental sciences and ecology; food science and technology; general and internal medicine; genetics and heredity; geology; immunology; infectious diseases; life sciences; biomedicine and other topics; marine and fresh water biology; materials science; mathematics; mechanics; microbiology; neurosciences and neurology; obstetrics and gynaecology; oncology; optics; parasitology; paediatrics; pharmacology; physics; plant sciences; polymer science; public, environmental and occupational health; science, technology and others; surgery; thermodynamics; tropical medicine; veterinary sciences; virology; water resources; and zoology.

With 17,000 plus publications, chemistry held the highest share (6.4 per cent) of all publications in Africa during 2011–15. The shares of other research areas in the order of percentages were: engineering (5.4 per cent), physics (4.9 per cent), environmental sciences and ecology (4 per cent), science, technology and others (3.8 per cent), infectious diseases (3.6 per cent), materials science (3.4 per cent), agriculture (3 per cent), pharmacology (2.8 per cent), public, environmental and occupational health (2.7 per cent), mathematics (2.6 per cent), immunology (2.1 per cent), biochemistry and molecular biology (2 per cent), plant sciences (1.9 per cent), geology and parasitology (1.8 per cent), microbiology and tropical medicine (1.7 per cent each), general and internal medicine (1.6 per cent), food science and technology (1.5 per cent), biotechnology and applied microbiology, genetics and heredity (1.4 per cent each), virology, and water resources (1.3 per cent each), astronomy, energy and fuels, veterinary sciences, and zoology (1.2 per cent each), computer science (1.1 per cent), life sciences, biomedicine and other topics, mechanics, neurosciences and neurology, and paediatrics (0.9 per cent each), electrochemistry, entomology, marine and fresh water biology, obstetrics and gynaecology, oncology, optics, polymer

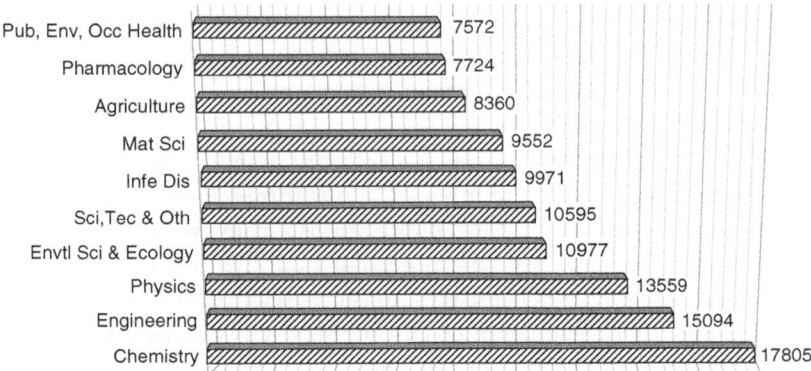

Figure 4.1 Top ten scientific research areas in contemporary Africa, 2011–15

science, surgery, and virology (0.8 per cent each), and thermodynamics (0.7 per cent).

Given the above data, the current research foci of Africa are chemistry; engineering; physics; environmental sciences and ecology; science, technology and others; infectious diseases; materials science; agriculture; pharmacology; and public, environmental and occupational health. This is now considered in relation to the contribution of countries (Figure 4.1).

Major Contributors in Africa, 2011–15

In line with the previous analysis for the period 1945–2015, the contribution of countries to the predominant research areas in Africa is examined below.

Scientific research in chemistry was centralised in two countries: Egypt (38 per cent) and South Africa (22 per cent). Tunisia (12 per cent) and Algeria (10 per cent) were the other two key players in chemistry. For research in engineering Africa relies on Egypt, South Africa and Algeria. They produced 32 per cent, 22 per cent and 17 per cent of publications respectively, or 71 per cent of all publications in Africa in engineering. Egypt (29 per cent), South Africa (24 per cent), Algeria (16 per cent), Tunisia (12 per cent) and Morocco (10 per cent) were the major countries in Africa for research in physics. Egypt and South Africa became the top publishers in physics (29 per cent and 24 per cent). In plant science research, South Africa was ahead of other countries owning 36 per cent of all publications in Africa. Other countries in this area, in descending order, were Egypt (14 per cent) and Tunisia (10 per cent). Nearly half of all the publications in polymer science were authored by scientists in Egypt, and another 20 per cent by South African scholars. Apart from these two countries, Tunisia (13 per cent) and Algeria (10 per cent) contributed considerably to polymer science research in Africa. Publications in environmental sciences and ecology came mostly from South Africa (31 per cent), Egypt (11 per cent), Tunisia (8 per cent), Kenya and Nigeria (7 per cent each). These countries altogether made up 64 per cent of the total publications in Africa in this particular research area.

Three countries in Africa are spearheading research in science and technology studies. Science, technology and others is an area which is taking root in Africa. South Africa produced 27 per cent of all publications in Africa, Egypt 17 per cent and Kenya 7 per cent. Infectious diseases research, another key research area for Africa of late, is carried out mainly in South Africa (21 per cent), Kenya (10 per cent) and in Uganda (8 per cent). Egypt (37 per cent), Algeria and South Africa (16 per cent each) produced most of the research publications in materials science. A few countries together produced most of Africa's publications in agriculture: South Africa (20 per cent), Egypt (13 per cent), Kenya, Nigeria, Tunisia (8 per cent each) and Ethiopia (6 per cent). For pharmacological research, Africa had good sources in Egypt (40 per cent) and South Africa (20 per cent). Nigeria also made a contribution of 10 per cent. For public, environmental and occupational health research, Africa turned to South Africa (19 per cent), Nigeria, Ethiopia, Kenya (8 per cent each), Uganda (7 per cent), Ghana and Tanzania (6 per cent each).

Tropical medicine research in Africa is not concentrated in one or two countries. Rather it is scattered among all countries. The top publisher in the area, namely Kenya, had only 10 per cent of the total production in Africa. Tanzania had 8 per cent while Nigeria and South Africa had 7 per cent each. Research in veterinary science was strong in four countries, namely, South Africa (27 per cent) and Egypt (20 per cent), Nigeria (8 per cent) and Ethiopia (7 per cent). For research in virology, Africa is indebted to South Africa for one-third of all its publications, while the next top countries, Egypt and Kenya, had only 9 per cent each and Uganda had 8 per cent. Water resources research was finding favour with South Africa (27 per cent), Egypt (16 per cent), Tunisia (11 per cent), and Algeria (8 per cent) than others. Zoological research in Africa was largely located in South Africa which produced almost half of all the publications for Africa. Egypt, the second country in this field, had only 7 per cent of the publications for the same period.

Immunological research in Africa continues to be centralised in South Africa (with a quarter of all publications in Africa), and some in Egypt and Uganda (9 per cent and 8 per cent). Entomological research in Africa originated largely from South Africa and Egypt. They jointly produced more than half of the total scientific publications (32 per cent and 21 per cent, respectively). In general and internal medicine, the major contributors were South Africa (31 per cent), Nigeria (20 per cent) and Egypt (10 per cent). Except for South Africa, which produced 19 per cent of research in public, environmental and occupational health, it cannot said to be located in any other single country. Rather it was widely dispersed with small contributions in all African countries. This could be a still-growing area in Africa.

Most of the geological research in Africa is conducted in two countries, namely, South Africa (27 per cent) and Egypt (22 per cent). In astronomy, South Africa has an unchallenged position, producing 64 per cent of all publications in Africa. Egypt, the second largest producer in astronomy, was way behind South Africa with 14 per cent. Except for Morocco and Algeria, no other country had made any worthwhile contribution to the field in Africa. In computer science, the major contributions came from Egypt (26 per cent), Tunisia (21 per cent), South Africa (19 per cent) and Algeria (16 per cent).

Seemingly, energy and fuels research was stronger in South Africa and Egypt than in any other countries; both had a joint share of half of all publications in Africa. Egypt declined in its production in the research area recently. Another quarter of publications were produced by Algeria (13 per cent) and Tunisia (11 per cent). Egypt (30 per cent), South Africa (23 per cent), Algeria (17 per cent) and Tunisia (15 per cent) were the strong publishers in the field of thermodynamics.

Research in genetics and heredity is not well developed in Africa. Three countries led genetics and heredity research in Africa: South Africa (26 per cent publications), Egypt (17 per cent) and Tunisia (13 per cent). The leading countries for life sciences, biomedicine and other topics were Egypt (41 per cent) and South Africa (25 per cent). South Africa advanced ahead of other African countries in marine and fresh water biology, if the count of the publications is any indication. It produced 43 per cent of all publications in Africa. Tunisia was the second largest contributor with 14 per cent. Egypt produced the lion's share of the publications in biochemistry and molecular biology (34 per cent of all publications in Africa). South Africa made another 24 per cent while Tunisia published 14 per cent in this area. South Africa and Egypt each produced a quarter of all papers in biotechnology and applied microbiology. Tunisia produced 14 per cent. In food science and technology most of the research in Africa occurred in Egypt, South Africa, Tunisia and in Nigeria.

More than half the publications in neurosciences and neurology were produced by two countries, namely, South Africa (29 per cent) and Egypt (26 per cent). Egypt (24 per cent), South Africa (15 per cent), Nigeria (11 per cent) and Kenya (6 per cent) were prominent countries where publications in obstetrics and gynaecology were concerned. In oncology, 40 per cent of the publications in Africa came from Egypt and 16 per cent from South Africa. Three countries, Egypt, South Africa and Algeria, contributed 36 per cent, 18 per cent and 16 per cent, respectively, to the field of optics. No country in Africa can claim to be a centre for research in parasitology, as the production was spread across all countries. The top publishers were South Africa (11 per cent) and Egypt (10 per cent). In paediatrics research, both South Africa and Egypt had a similar share, namely, 22 per cent and 21 per cent. Pharmacological research in Africa was mostly concentrated in Egypt, which produced 40 per cent of all the research publications. South Africa produced another 20 per cent of papers for Africa. In surgery, four major countries can be isolated: Egypt (34 per cent), South Africa (22 per cent), Morocco (8 per cent) and Nigeria (7 per cent).

Strengths of Countries in Africa, 2011–15

Country-wise statistics for 2011–15 are as follows: The most important countries that played a significant role in the production of science in recent years in Africa were South Africa (25 per cent of the total production in Africa), Egypt (22 per cent), Tunisia (9 per cent), Algeria (6 per cent), Nigeria (5.5 per cent), Morocco (4.4 per cent), Kenya

(3.8), Ethiopia (2.3 per cent), Uganda (2.1 per cent), Tanzania (2.0), and Ghana (1.7 per cent).

The production of science within these specific countries that led to the development of African science will now be explored. The facts are contained in Table 4.2.

South Africa's science seems to be focused on a few research areas that showed larger shares of publications during 2011–15. Chemistry topped the research areas with a 5.6 per cent of all publications in the country. Environmental sciences and ecology followed chemistry with 4.9 per cent while engineering had another 4.7 per cent of all publications produced in South Africa. Physics, and science, technology and other topics were the other two research areas that had a considerable number of publications (4.5 per cent and 4 per cent). Astronomy and infectious diseases had about 3 per cent each of the publications, mathematics 2.8 per cent, plant sciences 2.7 per cent, biochemistry and molecular biology, general and internal medicine, immunology, pharmacology, materials science, and zoology about 2 per cent each. The least published research areas (1 per cent or less) were computer science, electrochemistry, entomology, genetics and heredity, life sciences, biomedicine and other topics, mechanics, neurosciences and neurology, obstetrics and gynaecology, oncology, optics, parasitology, paediatrics, polymer science, surgery, thermodynamics, and tropical medicine.

Egypt, the second largest country in the production of science in Africa, produced most of its scientific publications in the field of chemistry (11 per cent of all its publications). Engineering (8 per cent), physics (6 per cent), materials science (6 per cent), pharmacology (5 per cent), and biochemistry and molecular biology (3 per cent) were other prominent areas for Egypt. The less significant areas in the country included mathematics (2.2 per cent), environmental sciences and ecology (2 per cent), agriculture (1.8 per cent), geology (1.7 per cent), life sciences, biomedicine and other topics (1.6 per cent), biotechnology and applied microbiology, and polymer science (1.5 per cent each). In the number of publications to its total output, Egypt did not do well in water resources (0.9 per cent), obstetrics and gynaecology, parasitology, public, environment and occupational health (0.8 per cent each), astronomy, entomology, genetics and heredity (0.7 per cent each), infectious diseases (0.6 per cent), and marine and fresh water biology, tropical medicine, virology and zoology

(0.3 per cent each). Egypt demonstrated its strengths in chemistry and engineering rather than in any other fields of research.

Chemistry, engineering, physics, materials science and mathematics were the three flourishing research areas for Tunisia. In these areas the country produced respectively 9 per cent, 8 per cent, 7 per cent, 6 per cent, and 5 per cent of all its publications. Following these three areas, Tunisia also had 3.5 per cent of its papers in environmental sciences and ecology, 3.1 per cent in biochemistry and molecular biology, 2.8 per cent in agriculture, 2.6 per cent each in computer science, and food science and technology, 2.1 per cent in plant sciences and 2 per cent in science, technology and other topics. It did not produce much in astronomy (0.1 per cent), tropical medicine (0.2 per cent), obstetrics and gynaecology, virology (0.3 per cent each), entomology (0.4 per cent), general and internal medicine, parasitology (0.5 per cent each), paediatrics, surgery, zoology (0.6 per cent each), electrochemistry, and life sciences, biomedicine and other topics (0.8 per cent each).

Algeria produced its highest shares of publications in engineering (14.2 per cent), physics (12.4 per cent), chemistry (9.8 per cent), materials science (8.6 per cent) and mathematics (5.7 per cent). It also produced modestly in computer science, mechanics (3 per cent each), science, technology and others (2.5 per cent), energy and fuels (2.4 per cent), optics, thermodynamics (2 per cent each), environmental sciences and ecology (1.9 per cent) and in water resources (1.6 per cent). Some areas were not well researched in Algeria including microbiology (0.9 per cent), biotechnology and applied microbiology (0.8 per cent), astronomy (0.7 per cent), zoology (0.5 per cent), entomology, immunology, infectious diseases (0.4 per cent), life sciences, biomedicine and other topics, marine and fresh water biology, neurosciences and neurology, parasitology (0.3 per cent each), oncology (0.2 per cent), genetic and heredity (0.2 per cent), general and internal medicine, paediatrics, surgery, tropical medicine (0.1 per cent each), obstetrics and gynaecology, public, environmental and occupational health, and virology (0 per cent each).

The focus areas for Nigeria were distributed among general and internal medicine (6.1 per cent), pharmacology (4.9 per cent), environmental sciences and ecology (4.8 per cent), agriculture (4.6 per cent), engineering (4.2 per cent), chemistry (4.1 per cent) and public, environmental and occupational health (3.8 per cent). There were other

Table 4.2 *Major scientific research areas of Africa in contemporary Africa, 2011–15*

Country	Agriculture No.	%	Astronomy No.	%	Bioch & Mol Bio No.	%	Biote Appl Micro No.	%	Chemistry No.	%
Algeria	223	2.67	126	3.90	189	3.38	145	3.66	1725	9.69
Angola	2	0.02	0	0.00	2	0.04	2	0.05	5	0.03
Benin	226	2.70	29	0.90	24	0.43	39	0.99	27	0.15
Botswana	39	0.47	0	0.00	31	0.55	9	0.23	79	0.44
Burkina Faso	137	1.64	10	0.31	22	0.39	34	0.86	47	0.26
Burundi	12	0.14	0	0.00	4	0.07	0	0.00	9	0.05
Cameroon	141	1.69	10	0.31	140	2.50	62	1.57	273	1.53
Cape Verde	4	0.05	0	0.00	0	0.00	0	0.00	2	0.01
Chad	0	0.00	3	0.09	2	0.04	0	0.00	0	0.00
Comoros	0	0.00	0	0.00	0	0.00	0	0.00	0	0.00
Congo	28	0.33	0	0.00	13	0.23	11	0.28	14	0.08
Djibouti	0	0.00	0	0.00	2	0.04	0	0.00	5	0.03
Egypt	1117	13.36	443	13.72	1921	34.32	973	24.58	6759	37.96
Eritrea	6	0.07	0	0.00	4	0.07	0	0.00	3	0.02
Ethiopia	577	6.90	38	1.18	48	0.86	69	1.74	169	0.95
Gabon	8	0.10	0	0.00	14	0.25	11	0.28	10	0.06
Gambia	4	0.05	0	0.00	18	0.32	11	0.28	0	0.00
Ghana	218	2.61	0	0.00	46	0.82	45	1.14	120	0.67
Guinea	23	0.28	0	0.00	30	0.54	7	0.18	24	0.13
Guinea-Bissau	2	0.02	0	0.00	0	0.00	0	0.00	0	0.00
Ivory Coast	0	0.00	0	0.00	0	0.00	0	0.00	0	0.00
Kenya	685	8.19	12	0.37	142	2.54	153	3.86	110	0.62
Lesotho	12	0.14	0	0.00	2	0.04	0	0.00	11	0.06
Liberia	0	0.00	0	0.00	0	0.00	0	0.00	0	0.00
Libya	25	0.30	0	0.00	18	0.32	9	0.23	114	0.64
Madagascar	68	0.81	5	0.15	43	0.77	15	0.38	44	0.25
Malawi	82	0.98	0	0.00	13	0.23	21	0.53	15	0.08
Mali	120	1.44	0	0.00	13	0.23	16	0.40	12	0.07
Mauritius	30	0.36	0	0.00	25	0.45	18	0.45	71	0.40
Mauritania	9	0.11	0	0.00	2	0.04	0	0.00	3	0.02
Morocco	274	3.28	229	7.09	171	3.05	115	2.90	965	5.42
Mozambique	41	0.49	0	0.00	14	0.25	11	0.28	13	0.07
Namibia	17	0.20	56	1.73	11	0.20	12	0.30	16	0.09
Niger	85	1.02	3	0.09	14	0.25	9	0.23	11	0.06
Nigeria	694	8.30	135	4.18	227	4.06	327	8.26	628	3.53
Rwanda	48	0.57	3	0.09	7	0.13	14	0.35	11	0.06
Reunion	65	0.78	6	0.19	55	0.98	25	0.63	19	0.11
Sao Tome and Principe	0	0.00	0	0.00	0	0.00	0	0.00	0	0.00
Senegal	127	1.52	2	0.06	29	0.52	39	0.99	88	0.49
Seychelles	0	0.00	0	0.00	10	0.18	0	0.00	0	0.00
Sierra Leone	9	0.11	0	0.00	3	0.05	2	0.05	0	0.00
Somalia	3	0.04	0	0.00	0	0.00	2	0.05	0	0.00
South Africa	1688	20.19	2069	64.10	1317	23.53	952	24.05	3934	22.09
Sudan	89	1.06	5	0.15	71	1.27	44	1.11	175	0.98
Swaziland	8	0.10	0	0.00	3	0.05	0	0.00	7	0.04
Tanzania	205	2.45	4	0.12	49	0.88	54	1.36	89	0.50
Togo	34	0.41	0	0.00	0	0.00	0	0.00	0	0.00
Tunisia	696	8.33	36	1.12	786	14.04	578	14.60	2140	12.02
Uganda	224	2.68	0	0.00	34	0.61	77	1.94	27	0.15
Zambia	53	0.63	2	0.06	12	0.21	24	0.61	4	0.02
Zimbabwe	202	2.42	2	0.06	17	0.30	24	0.61	27	0.15
Total	8360	100.00	3228	100.00	5598	100.00	3959	100.00	17805	100.00

areas in which Nigeria produced in modest terms: food science and technology (3.2 per cent), infectious diseases (2.9 per cent), science, technology and others (2.8 per cent), plant sciences, tropical medicine (2.3 per cent each), biotechnology and applied microbiology (2.2 per cent), materials science, physics (2.1 per cent each), geology (2 per cent), energy and fuels, and veterinary sciences (1.9 per cent each). Nigeria did not make much inroad into optics (0.2 per cent),

Table 4.2 *(cont.)*

Country	Comp Sci No.	%	Electrochem No.	%	Ener & Fuels No.	%	Engg No.	%	Entmlgy No.	%
Algeria	524	16.49	227	10.70	421	13.14	2501	16.57	72	3.26
Angola	2	0.06	0	0.00	0	0.00	6	0.04	0	0.00
Benin	7	0.22	0	0.00	4	0.12	19	0.13	67	3.04
Botswana	6	0.19	0	0.00	7	0.22	43	0.28	8	0.36
Burkina Faso	6	0.19	4	0.19	35	1.09	64	0.42	34	1.54
Burundi	0	0.00	0	0.00	0	0.00	0	0.00	3	0.14
Cameroon	30	0.94	10	0.47	41	1.28	139	0.92	51	2.31
Cape Verde	2	0.06	0	0.00	0	0.00	5	0.03	0	0.00
Chad	0	0.00	0	0.00	0	0.00	0	0.00	0	0.00
Comoros	0	0.00	0	0.00	0	0.00	0	0.00	0	0.00
Congo	0	0.00	0	0.00	0	0.00	10	0.07	2	0.09
Djibouti	0	0.00	3	0.14	0	0.00	3	0.02	0	0.00
Egypt	827	26.03	771	36.35	787	24.56	4841	32.07	457	20.72
Eritrea	0	0.00	0	0.00	0	0.00	0	0.00	2	0.09
Ethiopia	17	0.54	11	0.52	20	0.62	93	0.62	37	1.68
Gabon	0	0.00	0	0.00	0	0.00	5	0.03	4	0.18
Gambia	0	0.00	0	0.00	0	0.00	7	0.05	0	0.00
Ghana	19	0.60	0	0.00	41	1.28	115	0.76	28	1.27
Guinea	4	0.13	0	0.00	4	0.12	14	0.09	22	1.00
Guinea-Bissau	0	0.00	0	0.00	0	0.00	0	0.00	0	0.00
Ivory Coast	0	0.00	0	0.00	0	0.00	0	0.00	0	0.00
Kenya	27	0.85	0	0.00	54	1.68	81	0.54	197	8.93
Lesotho	0	0.00	0	0.00	8	0.25	5	0.03	4	0.18
Liberia	0	0.00	0	0.00	2	0.06	0	0.00	3	0.14
Libya	16	0.50	10	0.47	23	0.72	113	0.75	7	0.32
Madagascar	0	0.00	0	0.00	8	0.25	14	0.09	14	0.63
Malawi	11	0.35	0	0.00	15	0.47	16	0.11	3	0.14
Mali	2	0.06	0	0.00	3	0.09	10	0.07	10	0.45
Mauritius	13	0.41	0	0.00	14	0.44	24	0.16	7	0.32
Mauritania	0	0.00	0	0.00	0	0.00	3	0.02	11	0.50
Morocco	218	6.86	176	8.30	104	3.24	671	4.45	29	1.31
Mozambique	0	0.00	0	0.00	9	0.28	11	0.07	19	0.86
Namibia	7	0.22	3	0.14	2	0.06	19	0.13	6	0.27
Niger	4	0.13	0	0.00	5	0.16	19	0.13	21	0.95
Nigeria	72	2.27	101	4.76	295	9.20	643	4.26	92	4.17
Rwanda	4	0.13	0	0.00	3	0.09	11	0.07	2	0.09
Reunion	10	0.31	3	0.14	19	0.59	17	0.11	30	1.36
Sao Tome and Principe	0	0.00	0	0.00	0	0.00	0	0.00	0	0.00
Senegal	18	0.57	6	0.28	8	0.25	40	0.27	28	1.27
Seychelles	0	0.00	0	0.00	0	0.00	0	0.00	2	0.09
Sierra Leone	0	0.00	0	0.00	2	0.06	2	0.01	0	0.00
Somalia	0	0.00	0	0.00	0	0.00	0	0.00	0	0.00
South Africa	612	19.26	586	27.63	811	25.30	3313	21.95	709	32.14
Sudan	25	0.79	5	0.24	18	0.56	72	0.48	18	0.82
Swaziland	3	0.09	2	0.09	0	0.00	10	0.07	5	0.23
Tanzania	11	0.35	3	0.14	39	1.22	85	0.56	56	2.54
Togo	0	0.00	0	0.00	0	0.00	0	0.00	0	0.00
Tunisia	665	20.93	197	9.29	351	10.95	1964	13.01	99	4.49
Uganda	9	0.28	0	0.00	33	1.03	52	0.34	32	1.45
Zambia	2	0.06	0	0.00	5	0.16	17	0.11	6	0.27
Zimbabwe	4	0.13	3	0.14	14	0.44	17	0.11	9	0.41
Total	3177	100.00	2121	100.00	3205	100.00	15094	100.00	2206	100.00

polymer science (0.3 per cent), computer science, mechanics, thermodynamics (0.5 per cent each), zoology (0.6 per cent), oncology, and virology (0.7 per cent each).

Morocco paid more attention to physics (11.4 per cent), chemistry (7.9 per cent), mathematics (6.4 per cent), engineering (5.5 per cent), materials science (4.9 per cent), environmental sciences and ecology (2.7 per cent), and agriculture and geology (2.3 per cent each) than other subjects. Science, technology and

Table 4.2 *(cont.)*

Country	Envmtl Sci & Ecolgy No.	%	Food Sci Tech No.	%	Gen Int Med No.	%	Gene & Here No.	%	Geology No.	%
Algeria	341	3.11	211	5.10	16	0.35	36	1.49	254	5.17
Angola	8	0.07	0	0.00	4	0.09	2	0.08	16	0.33
Benin	89	0.81	84	2.03	15	0.33	24	0.99	21	0.43
Botswana	122	1.11	27	0.65	43	0.94	5	0.21	57	1.16
Burkina Faso	90	0.82	22	0.53	22	0.48	28	1.16	29	0.59
Burundi	16	0.15	2	0.05	0	0.00	0	0.00	5	0.10
Cameroon	238	2.17	129	3.12	68	1.49	50	2.07	118	2.40
Cape Verde	27	0.25	3	0.07	0	0.00	2	0.08	8	0.16
Chad	6	0.05	3	0.07	0	0.00	0	0.00	9	0.18
Comoros	0	0.00	0	0.00	0	0.00	2	0.08	0	0.00
Congo	54	0.49	14	0.34	25	0.55	16	0.66	19	0.39
Djibouti	5	0.05	0	0.00	2	0.04	0	0.00	2	0.04
Egypt	1277	11.63	816	19.71	460	10.06	415	17.19	1101	22.40
Eritrea	8	0.07	0	0.00	0	0.00	3	0.12	10	0.20
Ethiopia	381	3.47	113	2.73	78	1.71	68	2.82	151	3.07
Gabon	47	0.43	6	0.14	23	0.50	18	0.75	23	0.47
Gambia	12	0.11	2	0.05	32	0.70	25	1.04	0	0.00
Ghana	312	2.84	130	3.14	90	1.97	47	1.95	86	1.75
Guinea	46	0.42	5	0.12	78	1.71	16	0.66	18	0.37
Guinea-Bissau	5	0.05	0	0.00	13	0.28	5	0.21	0	0.00
Ivory Coast	0	0.00	0	0.00	0	0.00	0	0.00	0	0.00
Kenya	774	7.05	0	0.00	261	5.71	171	7.08	114	2.32
Lesotho	3	0.03	2	0.05	3	0.07	0	0.00	0	0.00
Liberia	6	0.05	0	0.00	7	0.15	0	0.00	0	0.00
Libya	31	0.28	10	0.24	38	0.83	8	0.33	58	1.18
Madagascar	123	1.12	27	0.65	8	0.17	31	1.28	32	0.65
Malawi	59	0.54	42	1.01	0	0.00	0	0.00	32	0.65
Mali	42	0.38	15	0.36	19	0.42	22	0.91	7	0.14
Mauritius	60	0.55	20	0.48	6	0.13	11	0.46	12	0.24
Mauritania	9	0.08	0	0.00	2	0.04	0	0.00	4	0.08
Morocco	332	3.02	164	3.96	64	1.40	97	4.02	284	5.78
Mozambique	54	0.49	18	0.43	43	0.94	10	0.41	28	0.57
Namibia	105	0.96	5	0.12	14	0.31	11	0.46	49	1.00
Niger	51	0.46	8	0.19	22	0.48	21	0.87	22	0.45
Nigeria	730	6.65	486	11.74	936	20.47	92	3.81	297	6.04
Rwanda	48	0.44	14	0.34	42	0.92	6	0.25	7	0.14
Reunion	98	0.89	31	0.75	7	0.15	43	1.78	37	0.75
Sao Tome and Principe	3	0.03	0	0.00	0	0.00	0	0.00	0	0.00
Senegal	136	1.24	41	0.99	25	0.55	28	1.16	36	0.73
Seychelles	44	0.40	0	0.00	0	0.00	9	0.37	0	0.00
Sierra Leone	12	0.11	4	0.10	15	0.33	0	0.00	0	0.00
Somalia	0	0.00	0	0.00	2	0.04	0	0.00	0	0.00
South Africa	3429	31.24	802	19.37	1422	31.10	636	26.35	1318	26.82
Sudan	72	0.66	66	1.59	43	0.94	38	1.57	38	0.77
Swaziland	22	0.20	6	0.14	4	0.09	0	0.00	3	0.06
Tanzania	305	2.78	52	1.26	122	2.67	54	2.24	98	1.99
Togo	18	0.16	0	0.00	0	0.00	0	0.00	0	0.00
Tunisia	885	8.06	663	16.01	128	2.80	301	12.47	404	8.22
Uganda	203	1.85	52	1.26	277	6.06	31	1.28	39	0.79
Zambia	75	0.68	23	0.56	48	1.05	11	0.46	11	0.22
Zimbabwe	164	1.49	23	0.56	45	0.98	21	0.87	58	1.18
Total	10977	100.00	4141	100.00	4572	100.00	2414	100.00	4915	100.00

others (2.1 per cent), astronomy, pharmacology (1.9 per cent each), computer science (1.8 per cent), optics, surgery (1.6 per cent each), electrochemistry, water resources (1.5 per cent), and biochemistry and molecular biology, food science and technology, neurosciences and neurology, physics and plant sciences (1.4 per cent each) were other areas in which Morocco published in considerable numbers. However, areas such as entomology, virology (0.2 per cent), obstetrics and gynaecology, parasitology, tropical medicine (0.3 per cent

Table 4.2 *(cont.)*

Country	Immulgy No.	%	Infec Dis No.	%	Life Sci No.	%	Mari & Fres No.	%	Materials Sci No.	%
Algeria	72	1.22	72	0.72	47	1.89	45	1.91	1519	15.90
Angola	13	0.22	40	0.40	0	0.00	10	0.43	3	0.03
Benin	52	0.88	131	1.31	7	0.28	13	0.55	5	0.05
Botswana	65	1.10	109	1.09	12	0.48	14	0.60	15	0.16
Burkina Faso	112	1.90	256	2.57	13	0.52	8	0.34	18	0.19
Burundi	2	0.03	5	0.05	0	0.00	8	0.34	7	0.07
Cameroon	93	1.58	283	2.84	37	1.49	20	0.85	68	0.71
Cape Verde	2	0.03	3	0.03	3	0.12	8	0.34	0	0.00
Chad	7	0.12	11	0.11	0	0.00	0	0.00	0	0.00
Comoros	0	0.00	0	0.00	0	0.00	0	0.00	0	0.00
Congo	85	1.44	125	1.25	6	0.24	13	0.55	4	0.04
Djibouti	0	0.00	5	0.05	0	0.00	0	0.00	0	0.00
Egypt	559	9.49	363	3.64	1020	41.10	179	7.61	3517	36.82
Eritrea	2	0.03	5	0.05	2	0.08	5	0.21	0	0.00
Ethiopia	98	1.66	379	3.80	29	1.17	47	2.00	61	0.64
Gabon	47	0.80	148	1.48	9	0.36	3	0.13	2	0.02
Gambia	104	1.77	148	1.48	6	0.24	0	0.00	2	0.02
Ghana	121	2.05	319	3.20	21	0.85	13	0.55	55	0.58
Guinea	105	1.78	201	2.02	7	0.28	13	0.55	3	0.03
Guinea-Bissau	45	0.76	48	0.48	0	0.00	3	0.13	0	0.00
Ivory Coast	0	0.00	0	0.00	0	0.00	0	0.00	0	0.00
Kenya	577	9.80	990	9.93	52	2.10	102	4.34	42	0.44
Lesotho	7	0.12	13	0.13	0	0.00	0	0.00	0	0.00
Liberia	6	0.10	19	0.19	0	0.00	0	0.00	0	0.00
Libya	12	0.20	19	0.19	20	0.81	11	0.47	49	0.51
Madagascar	30	0.51	113	1.13	8	0.32	25	1.06	4	0.04
Malawi	241	4.09	350	3.51	8	0.32	13	0.55	5	0.05
Mali	105	1.78	176	1.77	5	0.20	2	0.09	6	0.06
Mauritius	3	0.05	7	0.07	6	0.24	9	0.38	15	0.16
Mauritania	6	0.10	16	0.16	0	0.00	9	0.38	0	0.00
Morocco	82	1.39	104	1.04	52	2.10	48	2.04	601	6.29
Mozambique	89	1.51	158	1.58	9	0.36	20	0.85	2	0.02
Namibia	20	0.34	36	0.36	7	0.28	25	1.06	8	0.08
Niger	16	0.27	45	0.45	4	0.16	2	0.09	11	0.12
Nigeria	166	2.82	437	4.38	144	5.80	67	2.85	320	3.35
Rwanda	68	1.15	100	1.00	3	0.12	4	0.17	4	0.04
Reunion	19	0.32	60	0.60	3	0.12	57	2.42	5	0.05
Sao Tome and Principe	0	0.00	2	0.02	0	0.00	0	0.00	0	0.00
Senegal	111	1.88	286	2.87	24	0.97	37	1.57	32	0.34
Seychelles	0	0.00	3	0.03	2	0.08	31	1.32	0	0.00
Sierra Leone	11	0.19	34	0.34	3	0.12	0	0.00	0	0.00
Somalia	5	0.08	4	0.04	0	0.00	0	0.00	0	0.00
South Africa	1469	24.94	2100	21.06	625	25.18	1002	42.62	1535	16.07
Sudan	45	0.76	155	1.55	10	0.40	10	0.43	60	0.63
Swaziland	13	0.22	25	0.25	4	0.16	2	0.09	3	0.03
Tanzania	193	3.28	588	5.90	22	0.89	61	2.59	40	0.42
Togo	26	0.44	46	0.46	0	0.00	0	0.00	22	0.23
Tunisia	215	3.65	170	1.70	203	8.18	338	14.38	1478	15.47
Uganda	450	7.64	748	7.50	17	0.68	29	1.23	21	0.22
Zambia	152	2.58	266	2.67	4	0.16	17	0.72	8	0.08
Zimbabwe	169	2.87	250	2.51	28	1.13	28	1.19	2	0.02
Total	5890	100.00	9971	100.00	2482	100.00	2351	100.00	9552	100.00

each), life sciences, biomedicine and other topics, and marine and freshwater biology (0.4 per cent each) did not receive much attention in the country.

Ethiopia had its highest number of publications in public, environmental and occupational health (9.5 per cent), agriculture (9.2 per cent), environmental sciences and ecology (6.1 per cent), infectious diseases (6 per cent), science, technology and others (5.5 per cent), tropical medicine (4.3 per cent), paediatrics (4.2 per cent), veterinary

Table 4.2 *(cont.)*

Country	Mathematics No.	%	Mechanics No.	%	Microblgy No.	%	Neuros & Neulgy No.	%	Obst & Gyna No.	%	Oncology No.	%
Algeria	996	13.76	527	20.49	153	3.28	50	2.09	5	0.24	38	1.86
Angola	0	0.00	0	0.00	7	0.15	0	0.00	0	0.00	2	0.10
Benin	20	0.28	0	0.00	41	0.88	15	0.63	8	0.38	4	0.20
Botswana	70	0.97	2	0.08	28	0.60	0	0.00	8	0.38	9	0.44
Burkina Faso	31	0.43	5	0.19	100	2.14	9	0.38	15	0.71	6	0.29
Burundi	2	0.03	0	0.00	0	0.00	0	0.00	2	0.10	0	0.00
Cameroon	175	2.42	77	2.99	97	2.08	24	1.00	29	1.38	27	1.32
Cape Verde	3	0.04	0	0.00	0	0.00	0	0.00	0	0.00	0	0.00
Chad	2	0.03	0	0.00	7	0.15	0	0.00	0	0.00	0	0.00
Comoros	0	0.00	0	0.00	0	0.00	0	0.00	0	0.00	0	0.00
Congo	5	0.07	0	0.00	43	0.92	17	0.71	15	0.71	13	0.64
Djibouti	0	0.00	0	0.00	0	0.00	0	0.00	3	0.14	0	0.00
Egypt	1408	19.46	855	33.24	737	15.78	620	25.95	509	24.22	841	41.27
Eritrea	0	0.00	0	0.00	2	0.04	0	0.00	0	0.00	0	0.00
Ethiopia	26	0.36	5	0.19	103	2.21	40	1.67	103	4.90	14	0.69
Gabon	13	0.18	2	0.08	69	1.48	6	0.25	5	0.24	2	0.10
Gambia	0	0.00	0	0.00	70	1.50	0	0.00	5	0.24	4	0.20
Ghana	24	0.33	5	0.19	90	1.93	27	1.13	85	4.04	44	2.16
Guinea	7	0.10	0	0.00	76	1.63	5	0.21	21	1.00	4	0.20
Guinea-Bissau	0	0.00	0	0.00	16	0.34	2	0.08	5	0.24	0	0.00
Ivory Coast	0	0.00	0	0.00	0	0.00	0	0.00	0	0.00	0	0.00
Kenya	0	0.00	7	0.27	266	5.70	80	3.35	140	6.66	53	2.60
Lesotho	6	0.08	4	0.16	0	0.00	0	0.00	0	0.00	0	0.00
Liberia	0	0.00	0	0.00	3	0.06	0	0.00	4	0.19	0	0.00
Libya	23	0.32	10	0.39	25	0.54	0	0.00	5	0.24	27	1.32
Madagascar	7	0.10	2	0.08	48	1.03	3	0.13	10	0.48	0	0.00
Malawi	4	0.06	0	0.00	119	2.55	17	0.71	36	1.71	20	0.98
Mali	4	0.06	0	0.00	71	1.52	5	0.21	17	0.81	9	0.44
Mauritius	20	0.28	4	0.16	3	0.06	0	0.00	0	0.00	3	0.15
Mauritania	3	0.04	0	0.00	2	0.04	2	0.08	0	0.00	0	0.00
Morocco	783	10.82	153	5.95	129	2.76	167	6.99	39	1.86	118	5.79
Mozambique	4	0.06	0	0.00	45	0.96	5	0.21	15	0.71	9	0.44
Namibia	11	0.15	2	0.08	21	0.45	3	0.13	2	0.10	2	0.10
Niger	9	0.12	0	0.00	14	0.30	0	0.00	10	0.48	0	0.00
Nigeria	188	2.60	72	2.80	270	5.78	190	7.95	250	11.89	109	5.35
Rwanda	5	0.07	0	0.00	29	0.62	6	0.25	20	0.95	6	0.29
Reunion	16	0.22	2	0.08	32	0.69	11	0.46	20	0.95	4	0.20
Sao Tome and Principe	0	0.00	0	0.00	0	0.00	0	0.00	0	0.00	0	0.00
Senegal	82	1.13	2	0.08	91	1.95	15	0.63	11	0.52	16	0.79
Seychelles	0	0.00	0	0.00	2	0.04	7	0.29	0	0.00	0	0.00
Sierra Leone	0	0.00	0	0.00	12	0.26	0	0.00	9	0.43	3	0.15
Somalia	0	0.00	0	0.00	3	0.06	0	0.00	0	0.00	0	0.00
South Africa	1995	27.57	410	15.94	1062	22.74	696	29.13	322	15.32	319	15.65
Sudan	21	0.29	9	0.35	40	0.86	13	0.54	24	1.14	26	1.28
Swaziland	12	0.17	3	0.12	0	0.00	3	0.13	2	0.10	0	0.00
Tanzania	9	0.12	7	0.27	153	3.28	35	1.47	110	5.23	40	1.96
Togo	0	0.00	0	0.00	0	0.00	0	0.00	0	0.00	0	0.00
Tunisia	1222	16.89	397	15.44	436	9.34	223	9.33	68	3.24	189	9.27
Uganda	10	0.14	4	0.16	27	0.58	63	2.64	108	5.14	61	2.99
Zambia	0	0.00	0	0.00	59	1.26	23	0.96	46	2.19	10	0.49
Zimbabwe	21	0.29	6	0.23	69	1.48	7	0.29	16	0.76	6	0.29
Total	7237	100.00	2572	100.00	4670	100.00	2389	100.00	2102	100.00	2038	100.00

sciences (4.0 per cent) and virology (3.5 per cent). A few other areas including chemistry (2.7 per cent), pharmacology (2.6 per cent), water resources (2.5 per cent), geology (2.4 per cent), and plant sciences (2.3 per cent) had also done well. Ethiopia did not produce much in biotechnology and applied microbiology, genetics and heredity (1.1 per cent each), materials science (1 per cent), paediatrics (0.9 per cent), marine and fresh water biology (0.8 per cent), neurosciences and neurology (0.6 per cent), life sciences, biomedicine and other topics, zoology (0.5 per cent each), mathematics (0.4 per cent), computer science, energy and fuels (0.3 per cent each), electrochemistry, oncology, optics (0.2 per cent each), mechanics, polymer science and thermodynamics (0.1 per cent each).

Table 4.2 *(cont.)*

Country	Optics		Parasitology		Pediatrics		Pharmacology		Physics	
	No.	%	No.	%	No.	%	No.	%	No.	%
Algeria	347	16.35	52	1.02	14	0.55	234	3.03	2174	16.03
Angola	0	0.00	27	0.53	13	0.51	5	0.06	0	0.00
Benin	0	0.00	128	2.51	7	0.28	40	0.52	87	0.64
Botswana	4	0.19	12	0.24	21	0.83	39	0.50	22	0.16
Burkina Faso	4	0.19	187	3.66	23	0.91	46	0.60	19	0.14
Burundi	2	0.09	3	0.06	2	0.08	3	0.04	7	0.05
Cameroon	59	2.78	199	3.90	32	1.26	257	3.33	258	1.90
Cape Verde	0	0.00	0	0.00	0	0.00	2	0.03	0	0.00
Chad	0	0.00	3	0.06	0	0.00	5	0.06	2	0.01
Comoros	0	0.00	4	0.08	0	0.00	0	0.00	0	0.00
Congo	5	0.24	59	1.16	28	1.11	9	0.12	10	0.07
Djibouti	0	0.00	3	0.06	0	0.00	0	0.00	0	0.00
Egypt	755	35.58	528	10.35	522	20.62	3087	39.97	3915	28.87
Eritrea	0	0.00	4	0.08	0	0.00	6	0.08	0	0.00
Ethiopia	12	0.57	263	5.15	59	2.33	162	2.10	86	0.63
Gabon	5	0.24	85	1.67	4	0.16	39	0.50	12	0.09
Gambia	0	0.00	64	1.25	20	0.79	10	0.13	0	0.00
Ghana	8	0.38	220	4.31	46	1.82	95	1.23	36	0.27
Guinea	0	0.00	96	1.88	29	1.15	57	0.74	0	0.00
Guinea-Bissau	0	0.00	7	0.14	12	0.47	0	0.00	0	0.00
Ivory Coast	0	0.00	0	0.00	0	0.00	0	0.00	0	0.00
Kenya	13	0.61	486	9.52	152	6.00	133	1.72	38	0.28
Lesotho	0	0.00	3	0.06	0	0.00	0	0.00	5	0.04
Liberia	0	0.00	11	0.22	0	0.00	0	0.00	0	0.00
Libya	9	0.42	15	0.29	5	0.20	51	0.66	52	0.38
Madagascar	2	0.09	75	1.47	12	0.47	47	0.61	11	0.08
Malawi	3	0.14	69	1.35	101	3.99	23	0.30	4	0.03
Mali	0	0.00	101	1.98	11	0.43	36	0.47	0	0.00
Mauritius	5	0.24	8	0.16	0	0.00	22	0.28	13	0.10
Mauritania	0	0.00	12	0.24	2	0.08	2	0.03	0	0.00
Morocco	195	9.19	39	0.76	110	4.34	234	3.03	1388	10.24
Mozambique	0	0.00	67	1.31	17	0.67	26	0.34	0	0.00
Namibia	2	0.09	12	0.24	0	0.00	16	0.21	19	0.14
Niger	0	0.00	32	0.63	6	0.24	13	0.17	9	0.07
Nigeria	27	1.27	243	4.76	221	8.73	745	9.65	322	2.37
Rwanda	0	0.00	26	0.51	19	0.75	11	0.14	7	0.05
Reunion	0	0.00	37	0.73	23	0.91	6	0.08	3	0.02
Sao Tome and Principe	0	0.00	2	0.04	0	0.00	0	0.00	0	0.00
Senegal	10	0.47	222	4.35	36	1.42	16	0.21	62	0.46
Seychelles	0	0.00	3	0.06	3	0.12	0	0.00	0	0.00
Sierra Leone	0	0.00	19	0.37	0	0.00	2	0.03	0	0.00
Somalia	0	0.00	0	0.00	0	0.00	0	0.00	0	0.00
South Africa	388	18.28	573	11.23	562	22.20	1482	19.19	3197	23.58
Sudan	7	0.33	126	2.47	19	0.75	93	1.20	73	0.54
Swaziland	0	0.00	4	0.08	3	0.12	4	0.05	8	0.06
Tanzania	12	0.57	404	7.92	66	2.61	82	1.06	37	0.27
Togo	0	0.00	19	0.37	0	0.00	0	0.00	0	0.00
Tunisia	243	11.45	124	2.43	154	6.08	436	5.64	1653	12.19
Uganda	5	0.24	254	4.98	113	4.46	95	1.23	13	0.10
Zambia	0	0.00	129	2.53	37	1.46	9	0.12	10	0.07
Zimbabwe	0	0.00	44	0.86	28	1.11	44	0.57	7	0.05
Total	2122	100.00	5103	100.00	2532	100.00	7724	100.00	13559	100.00

Uganda's strengths were obvious in infectious diseases (12.8 per cent), public, environmental and occupational health (8.5 per cent), immunology (7.7 per cent), science, technology and others (7.4 per cent), tropical medicine (5.5 per cent), general and internal medicine (4.8 per cent), parasitology (4.4 per cent), agriculture (3.8 per cent), environmental sciences and ecology (3.5 per cent), and virology (3.2 per cent). A few research areas such as obstetrics and gynaecology, paediatrics (1.9 per cent each), plant sciences (1.7 per cent) and veterinary sciences (1.6 per cent) were also becoming important for the country. The least produced areas were surgery (1 per cent), food science and technology (0.9 per cent), geology (0.7 per cent), biochemistry and applied

Table 4.2 *(cont.)*

Country	Plant Sci No.	%	Polymer Sci No.	%	Pub Env Occu Hea No.	%	Sci, Tech & Oth No.	%	Surgery No.	%	Thermody No.	%
Algeria	171	3.19	221	10.37	0	0.00	444	4.19	19	0.81	352	17.46
Angola	6	0.11	0	0.00	15	0.20	22	0.21	2	0.09	0	0.00
Benin	98	1.83	0	0.00	77	1.02	99	0.93	3	0.13	0	0.00
Botswana	25	0.47	3	0.14	39	0.52	73	0.69	6	0.26	2	0.10
Burkina Faso	53	0.99	0	0.00	172	2.27	130	1.23	12	0.51	12	0.60
Burundi	6	0.11	2	0.09	8	0.11	9	0.08	0	0.00	0	0.00
Cameroon	0	0.00	5	0.23	154	2.03	228	2.15	34	1.45	16	0.79
Cape Verde	0	0.00	0	0.00	7	0.09	5	0.05	0	0.00	0	0.00
Chad	2	0.04	0	0.00	21	0.28	5	0.05	0	0.00	0	0.00
Comoros	2	0.04	0	0.00	0	0.00	0	0.00	0	0.00	0	0.00
Congo	24	0.45	0	0.00	82	1.08	52	0.49	10	0.43	0	0.00
Djibouti	2	0.04	0	0.00	0	0.00	3	0.03	2	0.09	0	0.00
Egypt	737	13.73	956	44.84	529	6.99	1699	16.04	802	34.14	601	29.81
Eritrea	0	0.00	0	0.00	0	0.00	3	0.03	2	0.09	0	0.00
Ethiopia	146	2.72	9	0.42	598	7.90	344	3.25	19	0.81	7	0.35
Gabon	16	0.30	0	0.00	31	0.41	64	0.60	4	0.17	0	0.00
Gambia	0	0.00	0	0.00	68	0.90	86	0.81	0	0.00	0	0.00
Ghana	71	1.32	5	0.23	422	5.57	283	2.67	53	2.26	10	0.50
Guinea	41	0.76	0	0.00	97	1.28	89	0.84	6	0.26	2	0.10
Guinea-Bissau	2	0.04	0	0.00	20	0.26	17	0.16	0	0.00	0	0.00
Ivory Coast	0	0.00	0	0.00	0	0.00	0	0.00	0	0.00	0	0.00
Kenya	299	5.57	18	0.84	618	8.16	752	7.10	63	2.68	5	0.25
Lesotho	3	0.06	0	0.00	13	0.17	20	0.19	0	0.00	0	0.00
Liberia	0	0.00	0	0.00	27	0.36	8	0.08	7	0.30	0	0.00
Libya	13	0.24	15	0.70	0	0.00	45	0.42	11	0.47	9	0.45
Madagascar	81	1.51	2	0.09	0	0.00	100	0.94	2	0.09	0	0.00
Malawi	14	0.26	0	0.00	306	4.04	195	1.84	37	1.58	0	0.00
Mali	43	0.80	3	0.14	71	0.94	82	0.77	5	0.21	0	0.00
Mauritius	31	0.58	7	0.33	18	0.24	25	0.24	2	0.09	4	0.20
Mauritania	0	0.00	0	0.00	10	0.13	9	0.08	0	0.00	0	0.00
Morocco	167	3.11	113	5.30	119	1.57	256	2.42	195	8.30	120	5.95
Mozambique	22	0.41	0	0.00	104	1.37	106	1.00	7	0.30	0	0.00
Namibia	26	0.48	0	0.00	31	0.41	56	0.53	12	0.51	2	0.10
Niger	35	0.65	0	0.00	34	0.45	37	0.35	5	0.21	0	0.00
Nigeria	345	6.43	44	2.06	578	7.63	431	4.07	163	6.94	81	4.02
Rwanda	18	0.34	0	0.00	66	0.87	73	0.69	16	0.68	0	0.00
Reunion	69	1.29	0	0.00	19	0.25	76	0.72	13	0.55	2	0.10
Sao Tome and Principe	0	0.00	0	0.00	0	0.00	3	0.03	0	0.00	0	0.00
Senegal	40	0.75	4	0.19	154	2.03	149	1.41	13	0.55	2	0.10
Seychelles	7	0.13	0	0.00	16	0.21	19	0.18	0	0.00	0	0.00
Sierra Leone	6	0.11	0	0.00	36	0.48	22	0.21	24	1.02	0	0.00
Somalia	0	0.00	0	0.00	4	0.05	0	0.00	0	0.00	0	0.00
South Africa	1919	35.75	429	20.12	1463	19.32	2815	26.57	511	21.75	466	23.12
Sudan	48	0.89	7	0.33	134	1.77	64	0.60	18	0.77	13	0.64
Swaziland	0	0.00	5	0.23	14	0.18	16	0.15	2	0.09	0	0.00
Tanzania	64	1.19	4	0.19	413	5.45	363	3.43	46	1.96	8	0.40
Togo	26	0.48	0	0.00	39	0.52	17	0.16	0	0.00	0	0.00
Tunisia	524	9.76	280	13.13	164	2.17	508	4.79	159	6.77	297	14.73
Uganda	98	1.83	0	0.00	495	6.54	433	4.09	57	2.43	3	0.15
Zambia	18	0.34	0	0.00	169	2.23	126	1.19	4	0.17	0	0.00
Zimbabwe	50	0.93	0	0.00	117	1.55	134	1.26	3	0.13	2	0.10
Total	5368	100.00	2132	100.00	7572	100.00	10595	100.00	2349	100.00	2016	100.00

microbiology, energy and fuels, entomology (0.6 per cent each), genetics and heredity, marine and fresh water biology (0.5 per cent each), materials science (0.4 per cent), life sciences, biomedicine and other topics (0.3 per cent), computer science, mathematics, physics (0.2 per cent each), mechanics, optics and thermodynamics (0.1 per cent). No publications were produced in astronomy, electrochemistry and polymer science.

In Tanzania, infectious diseases (10.4 per cent), tropical medicine (7.4 per cent), public, environmental and occupational health (7.3 per cent), parasitology (7.1 per cent), science, technology and

Table 4.2 *(cont.)*

Country	Trop Med No.	%	Vet Sci No.	%	Virology No.	%	Water Reso No.	%	Zoology No.	%	All Res Areas No.	%
Algeria	23	0.48	92	2.67	5	0.21	285	7.78	87	2.71	17594	6.32
Angola	29	0.60	0	0.00	0	0.00	0	0.00	7	0.22	306	0.11
Benin	113	2.35	27	0.78	11	0.47	25	0.68	29	0.90	2022	0.73
Botswana	9	0.19	21	0.61	35	1.49	50	1.37	25	0.78	1445	0.52
Burkina Faso	185	3.85	22	0.64	35	1.49	43	1.17	12	0.37	2458	0.88
Burundi	7	0.15	0	0.00	2	0.09	8	0.22	5	0.16	166	0.06
Cameroon	159	3.31	38	1.10	73	3.11	39	1.07	60	1.87	5048	1.81
Cape Verde	0	0.00	3	0.09	0	0.00	4	0.11	6	0.19	130	0.05
Chad	10	0.21	4	0.12	0	0.00	2	0.05	0	0.00	135	0.05
Comoros	0	0.00	0	0.00	0	0.00	0	0.00	3	0.09	13	0.00
Congo	75	1.56	6	0.17	13	0.55	10	0.27	44	1.37	1160	0.42
Djibouti	4	0.08	0	0.00	0	0.00	0	0.00	0	0.00	48	0.02
Egypt	163	3.39	696	20.23	211	8.98	586	16.01	213	6.65	63255	22.71
Eritrea	3	0.06	3	0.09	0	0.00	0	0.00	3	0.09	99	0.04
Ethiopia	268	5.57	250	7.27	22	0.94	155	4.23	34	1.06	6283	2.26
Gabon	82	1.71	10	0.29	36	1.53	2	0.05	25	0.78	1011	0.36
Gambia	82	1.71	2	0.06	27	1.15	0	0.00	0	0.00	948	0.34
Ghana	262	5.45	33	0.96	38	1.62	123	3.36	24	0.75	4817	1.73
Guinea	127	2.64	7	0.20	26	1.11	6	0.16	37	1.15	1550	0.56
Guinea-Bissau	16	0.33	0	0.00	13	0.55	3	0.08	0	0.00	269	0.10
Ivory Coast	0	0.00	0	0.00	0	0.00	0	0.00	0	0.00	0	0.00
Kenya	458	9.52	207	6.02	203	8.64	125	3.41	175	5.46	10668	3.83
Lesotho	3	0.06	5	0.15	0	0.00	3	0.08	0	0.00	149	0.05
Liberia	11	0.23	2	0.06	3	0.13	0	0.00	2	0.06	134	0.05
Libya	9	0.19	29	0.84	4	0.17	14	0.38	8	0.25	1236	0.44
Madagascar	75	1.56	24	0.70	15	0.64	7	0.19	132	4.12	1550	0.56
Malawi	162	3.37	11	0.32	67	2.85	48	1.31	3	0.09	2604	0.93
Mali	124	2.58	18	0.52	10	0.43	10	0.27	5	0.16	1384	0.50
Mauritius	9	0.19	3	0.09	0	0.00	5	0.14	10	0.31	703	0.25
Mauritania	11	0.23	0	0.00	3	0.13	6	0.16	5	0.16	169	0.06
Morocco	38	0.79	50	1.45	24	1.02	180	4.92	73	2.28	12157	4.36
Mozambique	84	1.75	26	0.76	15	0.64	23	0.63	8	0.25	1337	0.48
Namibia	11	0.23	31	0.90	3	0.13	18	0.49	38	1.19	944	0.34
Niger	38	0.79	8	0.23	5	0.21	25	0.68	7	0.22	795	0.29
Nigeria	350	7.28	287	8.34	101	4.30	147	4.02	95	2.96	15235	5.47
Rwanda	46	0.96	10	0.29	26	1.11	10	0.27	19	0.59	985	0.35
Reunion	33	0.69	28	0.81	24	1.02	12	0.33	25	0.78	1388	0.50
Sao Tome and Principe	0	0.00	0	0.00	0	0.00	0	0.00	2	0.06	12	0.00
Senegal	193	4.01	33	0.96	46	1.96	25	0.68	47	1.47	2940	1.06
Seychelles	0	0.00	0	0.00	0	0.00	3	0.08	19	0.59	268	0.10
Sierra Leone	27	0.56	0	0.00	7	0.30	0	0.00	5	0.16	311	0.11
Somalia	3	0.06	0	0.00	0	0.00	0	0.00	0	0.00	26	0.01
South Africa	352	7.32	919	26.71	792	33.72	976	26.66	1534	47.86	70628	25.35
Sudan	126	2.62	74	2.15	21	0.89	31	0.85	7	0.22	2548	0.91
Swaziland	4	0.08	2	0.06	4	0.17	6	0.16	25	0.78	265	0.10
Tanzania	420	8.73	130	3.78	67	2.85	89	2.43	91	2.84	5679	2.04
Togo	31	0.64	0	0.00	0	0.00	0	0.00	0	0.00	278	0.10
Tunisia	51	1.06	106	3.08	62	2.64	415	11.34	140	4.37	25299	9.08
Uganda	323	6.72	92	2.67	184	7.83	62	1.69	67	2.09	5829	2.09
Zambia	144	2.99	64	1.86	74	3.15	11	0.30	16	0.50	1974	0.71
Zimbabwe	56	1.16	68	1.98	42	1.79	69	1.88	33	1.03	2315	0.83
Total	4809	100.00	3441	100.00	2349	100.00	3661	100.00	3205	100.00	278567	100.00

others (6.4 per cent), and environmental sciences and ecology (5.4 per cent) made more progress than other areas. The key areas for Tanzania were medicine and diseases. Notable other research areas were agriculture (3.6 per cent), immunology (3.4 per cent), microbiology (2.7 per cent), veterinary sciences (2.3 per cent), general and internal medicine (2.2 per cent) and obstetrics and gynaecology (1.9 per cent). The least developed research areas among others included biotechnology and applied microbiology, entomology, genetics and heredity (1 per cent each), biochemistry and molecular biology, food science and technology (0.9 per cent each), surgery (0.8 per cent), materials science, oncology, physics

(0.7 per cent each), neurosciences and neurology (0.6 per cent), life sciences, biomedicine and other topics (0.4 per cent), computer science, mathematics, optics (0.2 per cent each), electrochemistry, astronomy, mechanics, polymer science and thermodynamics (0.1 per cent each).

Kenya demonstrated its strengths in infectious diseases (9.3 per cent), environmental sciences and ecology (7.3 per cent), science, technology and others (7.1 per cent), agriculture (6.4 per cent), immunology (5.4 per cent), parasitology (4.6 per cent), tropical medicine (4.3 per cent), plant sciences (2.8 per cent), general and internal medicine, microbiology (2.5 per cent each), entomology, veterinary sciences and virology (1.9 per cent each). Not much attention was paid to areas such as chemistry, marine and fresh water biology (1 per cent each), engineering, neurosciences and neurology (0.8 per cent each), computer science (0.7 per cent), surgery (0.6 per cent), energy and fuels, life sciences, biomedicine and other topics, oncology (0.5 per cent each), materials science, physics (0.4 per cent each), polymer science (0.2 per cent), astronomy, mechanics, optics and thermodynamics (0.1 per cent each). Kenya, like Tanzania, worked mostly in the areas of medicine and disease control.

Research Areas in Africa over the Years

In this analysis two datasets were used, one referring to the entire duration (1945–2015) and the other referring to the recent five years (2011–15). They cannot however be treated as two completely different sets. Despite this limitation, trends can be gleaned from these two datasets.

Compared to the research areas chosen for the analysis for the period of 1945–2015, some have disappeared while new areas have emerged. Research areas such as nuclear sciences and technology, nutrition and dietetics, and metallurgy have not been found among the prominent research areas during 2011–15, whereas astronomy, electrochemistry, life sciences, biomedicine and other topics, and optics are new entrants in the 2011–15 dataset.

Although the production of publications in chemistry had declined by about 1 per cent, it remained the focus of scientific research in Africa in the recent years. Similarly, engineering and physics continue to be the other prominent research areas that have shown an increase during 2011–15. Science, technology and others grew recently by 1.2 per cent while agriculture declined by 1 per cent. Infectious diseases added another 1 per cent of publications. Environmental sciences and ecology was another area that grew in the recent years. When tropical medicine declined, immunology and parasitology improved their share of publications. General and internal medicine almost halved its production in Africa in the last five years.

In the production of publications for 2011–15, South Africa was the top publisher with a quarter of all publications in Africa. Its percentage share has however decreased by 3 per cent from the period of 1945–2015. The second highest contributor was Egypt, which was close to South Africa with 22 per cent (an increase of 1 per cent). Nigeria lost its third position and now holds the fifth position after Tunisia and Algeria. Other than South Africa and Nigeria, some sub-Saharan countries such as Ethiopia, Uganda, Tanzania and Ghana were among the leading countries that produce science in Africa.

In agriculture, Nigeria, South Africa and Egypt were the top three publishers during the first period of (1945–2015). In the second period (2011–15), a shift in the percentage of production in this area is seen. South Africa topped with 20 per cent, Egypt became the second biggest publisher (13 per cent) and Nigeria lost its first position (8 per cent). The difference between the two periods was larger for Nigeria than for the other two countries (dropped from 18 per cent to 8 per cent).

Research in biochemistry and molecular biology shows some patterns as well. South Africa produced most of the publications during the first period. Egypt and Tunisia were in the second and third positions. During 2011–15, Egypt became the top producer, South Africa got the second position losing 8 per cent of publications, and Tunisia gained 5 per cent over the first period. In the area of biotechnology and applied microbiology, South Africa was leading when the total count of publications in Africa for the period of 1945–2015 was taken into account. Egypt and Nigeria had a similar percentage. Recently this proportion has been amended for Egypt and South Africa which produced about a quarter of publications each. Nigeria lost about 10 per cent. While there has not been much difference in the contribution of the top producing countries in chemistry (Egypt and South Africa), Tunisia strengthened its capacity in the second period. Algeria has also improved its share, but only slightly.

As a developing research area, computer science is growing in Africa. The most important change is seen in the proportionate share of publications in specific countries. Compared to the first period (from 14 per cent) Tunisia gained strength in the second period increasing to one-fifth of the total publications in Africa. South Africa was leading in this area in the first period with 27 per cent of the publications in the continent. In the second period it declined, which pushed the country into second position after Egypt. With more than a quarter of the total publications in the continent, Egypt maintained its percentage in both periods. Algeria is another country that improved its share from 14 per cent to 16 per cent.

As in the first period, Egypt maintained its lead in engineering in 2011–15. It is now credited with about one-third of the total publications in Africa. When South Africa lost its share in the second period, both Algeria and Tunisia increased their respective shares of publication in the second time period.

In entomological research, some shifts are visible. South Africa and Egypt grew in production in recent years and Kenya declined its share by half. In immunology, Kenya and Egypt tended to lose ground while South Africa maintained its position. On the other hand, Uganda grew its number of publications in the second period. In infectious diseases research, both South Africa and Kenya lost some percentages in the second period. Uganda increased its share slightly. However, South Africa produced most of the publications in this area for Africa. In microbiology, both South Africa and Nigeria declined their shares from the first period while Egypt had a small improvement. South Africa continued to lead in microbiological research. Kenya declined its production in parasitology while South Africa and Egypt had some growth in the recent years. Egypt consolidated its first rank in pharmacological research with an increase of 4 per cent of the total publications in Africa. South Africa lost greatly by having reduced its share by about 10 per cent from the first period. Nigeria and Kenya were making progress in tropical medicine research in the first period. In recent years, Nigeria lost its position. The share for Nigeria declined from 16 per cent to 7 per cent. Kenya also lost about 1 per cent of its publications. One-third of the research in virology conducted in Africa came from South Africa, which had a similar contribution of publications in the first period. Not much change was reported from other countries during the two time periods of analysis.

In scientific research conducted in general and internal medicine in Africa, the participation of South Africa decreased in recent years (2011–15) from 46 per cent that was reported during 1945–2015. Nigeria improved its position from a share of 13 per cent to 20 per cent. In infectious diseases, South Africa also had a similar contribution to African science as during 1945–2015. Tunisia retained its domination in this area.

Mathematics has been given attention in countries such as South Africa (28 per cent), Egypt (19 per cent), Tunisia (17 per cent), Algeria (14 per cent) and Morocco (11 per cent). South Africa had a similar percentage during 1945–2015 as well. Egypt improved its contribution by 3 per cent while Tunisia lost by 4 per cent. There was not much difference in mathematics research for Morocco between the two periods. Except in Tunisia which improved its production in materials science, changes were not seen between the periods among the prominent countries. No significant change was observed in the production of publications in environmental sciences and ecology, particularly among the leading countries in this research area.

In the recent five years, publications in the sphere of mechanics originated mostly from Egypt (33 per cent), Algeria (20 per cent), South Africa (16 per cent) and Tunisia (15 per cent). Algeria strengthened this research area from 1945 to 2015. In the scientific area of microbiology South Africa made 23 per cent of all the publications in Africa for the recent period. Other countries with a considerable percentage of publications were Egypt (16 per cent), Tunisia (9 per cent), Nigeria and Kenya (6 per cent each). For the period of 1945–2015 South Africa had 27 per cent of the publications in Africa while Egypt had 15 per cent. South Africa was leading Africa in science, technology and other topics during 1945–2015. However, Egypt took over the first position as seen in the production figures for 2011–15.

At the individual country level also, shifts in the production of publications have occurred. General and internal medicine was a major research area for South Africa, producing 5.3 per cent of the publications during 1945–2015. This has been reduced to 2 per cent during 2011–15. While environmental sciences and ecology, and physics had improved their positions in recent years, biochemistry and molecular biology had weakened. Agriculture remained at the same level. Of late, astronomy has become one of the prominent areas of research in South Africa. Small increases in mathematics and immunology publications are evident. Science, technology and other topics has grown from 3.1 per cent to 4 per cent. The strong areas in South Africa were in chemistry, environmental sciences and ecology, engineering and physics. On the other hand oncology and obstetrics and gynaecology had yet to develop.

South Africa, in the production of publications in agriculture, had the highest percentage in Africa. It was second to Nigeria for during 1945–2015. The percentage also increased from 17 to 20 for South Africa. Egypt continued to be the second best producer of publications in agriculture while Nigeria, Tunisia and Kenya had a similar share of about 8 per cent each. Scientific research in astronomy in Africa was centralised more in South Africa than in any other country. More than two-thirds of the publications in astronomy originate in South Africa. Advancement in astronomy has been very notable in South Africa with its recent investments made in the field.

The data in Table 4.3 shows the condensed version of countries and research areas for the two time periods. The data is considered from two vantage points. One is to group the countries that had at least 20 per cent of publications, or the highest percentage of publications in Africa, in any given research area. The second point is to look from the perspective of individual countries. Here publications with a percentage of 5 or more for any research areas within the country was included. For the management and analysis of data, only the major science-producing countries are examined for the second point of analysis.

Africa now looks up to its own specific countries for its advancement in selected scientific research fields. These were the countries which have been in the forefront of publishing a considerable number of research publications (20 per cent or more, or the top publisher in Africa). For advancement in most scientific fields, Africa now (according to the publications during 2011–15) relies only on a few countries: South Africa, Egypt, Tunisia, Nigeria and Kenya. These are the key

Table 4.3 *Leading research areas of major science producing countries in Africa, 1945–2015 & 2011–15*

Research areas	Countries with 20% and above, or the highest share of publications in Africa		Countries with the highest share of publications (5% & above) within country	
	1945–2015	2011–15	1945–2015	2011–15
Agriculture	Nigeria	South Africa	Nigeria Kenya Tanzania Ethiopia	Ethiopia Kenya
Astronomy		South Africa		
Biochemistry and molecular biology	Egypt South Africa	Egypt South Africa		
Biotechnology and applied microbiology	Egypt South Africa	Egypt South Africa		
Chemistry	Egypt South Africa	Egypt South Africa	Egypt South Africa Nigeria Tunisia Algeria Morocco	Egypt South Africa Tunisia Algeria Morocco
Computer science	Egypt South Africa	Egypt Tunisia		
Electrochemistry		Egypt South Africa		
Energy and fuels	Egypt South Africa	Egypt South Africa		
Engineering	Egypt South Africa	Egypt South Africa	Egypt Tunisia	Egypt Tunisia Algeria Morocco
Entomology	South Africa	Egypt South Africa		
Environmental sciences and ecology	South Africa	South Africa	Kenya Tanzania Ethiopia	Ethiopia Tanzania Kenya
Food sciences and technology	Egypt Nigeria	Egypt		

Table 4.3 (*cont.*)

Research areas	Countries with 20% and above, or the highest share of publications in Africa		Countries with the highest share of publications (5% & above) within country	
	1945–2015	2011–15	1945–2015	2011–15
General and internal medicine	South Africa	Nigeria South Africa	Nigeria South Africa Kenya Ethiopia	Nigeria
Genetics and heredity	South Africa	South Africa		
Geology	South Africa	Egypt South Africa		
Immunology	South Africa	South Africa		Uganda Kenya
Infectious diseases	South Africa	South Africa	Kenya Tanzania	Ethiopia Uganda Tanzania Kenya
Life sciences, biomedicine and other topics		Egypt South Africa		
Marine and fresh water biology	South Africa	South Africa		
Materials science	Egypt	Egypt	Egypt Tunisia Algeria Morocco	Egypt Tunisia Algeria
Mathematics	Egypt South Africa	South Africa	Tunisia Algeria	Tunisia Algeria Morocco
Mechanics	Egypt	Egypt		
Metallurgy	Egypt South Africa			
Microbiology	South Africa	South Africa		
Neurosciences and neurology	South Africa	Egypt South Africa		
Nuclear sciences and technology	Egypt South Africa	Egypt		

Table 4.3 (*cont.*)

Research areas	Countries with 20% and above, or the highest share of publications in Africa		Countries with the highest share of publications (5% & above) within country	
	1945–2015	2011–15	1945–2015	2011–15
Nutrition and Dietetics	Nigeria		Nigeria	
Obstetrics and gynaecology	Egypt South Africa	Egypt		
Oncology	Egypt South Africa	Egypt		
Parasitology	Kenya	South Africa	Kenya	
Paediatrics	South Africa	Egypt South Africa		
Pharmacology	Egypt South Africa	Egypt		Egypt
Physics	Egypt South Africa	Egypt South Africa	Egypt Tunisia Algeria Morocco	Tunisia Algeria Morocco
Plant sciences	South Africa	South Africa		
Polymer science	South Africa	Egypt South Africa		
Public, environmental and occupational health	South Africa	South Africa	Kenya Tanzania	Ethiopia Uganda Tanzania
Science, technology and others	South Africa	South Africa		Ethiopia Uganda Tanzania Kenya
Surgery	Egypt South Africa	Egypt South Africa		
Thermodynamics	Egypt South Africa	Egypt South Africa		
Tropical medicine	Nigeria	Kenya	Kenya Tanzania Ethiopia	Uganda
Veterinary sciences	South Africa	Egypt South Africa		

Table 4.3 (*cont.*)

Research areas	Countries with 20% and above, or the highest share of publications in Africa		Countries with the highest share of publications (5% & above) within country	
	1945–2015	2011–15	1945–2015	2011–15
Virology	South Africa	South Africa		
Water resources	South Africa	South Africa		
Zoology	South Africa	South Africa		

Note: The data presented in the table refers only to countries that had a substantial number of publications in Africa.

producers of science among the 54 African countries. The research areas are shown in Table 4.3.

The major countries of science production in Africa (at least 5 per cent of the production in the research area within the country) are listed in Table 4.3. The areas in which countries produced significantly in the current period (2011–15) included agriculture (Ethiopia and Kenya), chemistry (Algeria, Egypt, South Africa, Morocco and Tunisia), engineering (Algeria, Egypt, Morocco, Tunisia), general and internal medicine (Nigeria), immunology (Kenya and Uganda), infectious diseases (Ethiopia, Kenya, Tanzania and Uganda), materials science (Algeria, Egypt and Tunisia), mathematics (Algeria, Morocco and Tunisia), pharmacology (Egypt), public, environmental and occupational health (Ethiopia, Kenya, Tanzania and Uganda) and tropical medicine (Uganda).

The attention and focus of research in any given country at any given time depends on several internal factors. The areas of research, as Moravcsik (1984) noted, are determined primarily by the interests and capabilities of the scientific manpower that is available in the country. Countries have specific areas of science to focus, develop and advance, which vary from time to time and are affected by several internal factors. One can also see some characteristic features in this matter. It should be remembered that production of science in not even across countries (Frame et al., 1977). This applies to Africa as well.

Some of these findings correspond to those of other studies. Waast and Rossi (2010) reported (based on the analysis of publications for

1987–2006) that Arab countries that included some North African countries have a focus on exact and material sciences (mathematics, physics, chemistry and engineering) and have less interest in natural sciences and life sciences. Algeria, Morocco and Tunisia have an exceptional interest in mathematics, while Algeria has an exceptional advantage in computer science and electronics. Tunisia has done very well in several branches of agricultural sciences, animal husbandry and internal medicine (Waast and Rossi, 2010).

In SADC (Southern African Development Community) countries, as the analysis by Pouris (2010) for the period of 2004–8 showed, the focus areas are immunology, environment/ecology, plant and animal sciences, and agriculture. The data for 2009 indicated that in the scientific fields of agronomy and food science, Nigeria is ahead of South Africa with its number of publications (Irikefe et al., 2011). The analysis of PASCAL data for the period of 1987–90 confirmed that Nigeria devoted a larger share of its research activity to medicine and agriculture (Chatelin et al., 1997), which agrees with the data used here. According to this analysis, Tanzania has the highest activity index in agricultural sciences followed by immunology and environment/ecology. In the 1990s Tanzania's contribution to Africa was 2 per cent (Gaillard, 2003b), which agrees with the analysis. Two sectors, namely agriculture and health, with the highest critical mass of human resources produced the largest volumes of research outputs in Tanzania (COSTECH, 2015). South Africa had high research activity in plant and animal sciences and environment/ecology. Cherry (2010) documented that South Africa, being the research base of the continent, had historical strengths in areas of astronomy, geology, botany, zoology, clinical medicine and mining technology. South Africa exhibits strengths in clinical medicine and plant and animal sciences (Boshoff, 2009b). Between 1988 and 1992, animal and plant sciences and veterinary sciences did not do well, both in the number of publication outputs and world share, and the trend continued during 1989–93 as well (Ingwersen and Jacobs, 2004; Jacobs and Ingwersen, 2000). The national priorities of South Africa include human capital development, knowledge generation and exploitation R&D, knowledge infrastructure, space science and technology, energy and renewable energy technologies, and climate change (ECA, 2016). Most of these findings were also found to be true in our data.

Africa has given greater attention to the fields of agriculture and medical science for quite some time now, as Seaborg (1970) reported, although this varies across Africa. In French-speaking African countries the focus was more on agricultural research than medical research (Gaillard, 1992a). In the past, as in the analysis by Pouris (1989) presented for 1973–84, South Africa produced a substantial quantity of research in agriculture, contributing an increased percentage from almost nil to 3.75 per cent to world agricultural literature in 1984. A study of publications during 2005–7 in the field of HIV/AIDS research showed that South Africa is one of the major countries in the world contributing to world research (Pouris and Pouris, 2011). South Africa's contribution was 3.15 per cent of world research in HIV/AIDS. Macías-Chapula and Mijangos-Nolasco (2002) analysed the publications in AIDS research in Central Africa for the period of 1980–2000 and found that AIDS research is mostly done in the Democratic Republic of the Congo and Cameroon.

The analysis of publications according to the research areas presented in this chapter provides insight into understanding science that originates in Africa. Data for two time periods (1945–2015 and 2011–15) revealed the predominant scientific research areas within the continent. The datasets also revealed the strengths of African countries in specific scientific research areas, and the countries that contributed to the development of specific branches of science in Africa.

The division between the North African and sub-Saharan countries is wide enough, showing the gaps in the production of scientific publications. For the period of 1945–2015 the four countries in North Africa were capable of producing more than one-third of publications in Africa. In sub-Saharan Africa, South Africa alone produced about one-third of all the publications in Africa. South Africa, to its credit, had more than half of the publications in the sub-Saharan region. The preference of countries to pursue research in selected areas in North Africa and sub-Saharan Africa is evident in the data for the two periods.

Scientists are sometimes in a dilemma regarding their choice of research areas. There are very few scientists whose incentive to do science is to contribute to the utilitarian needs of society (Moravcsik, 1973). Some are very passionate about their own specific research interests, which they want to pursue throughout their career. Over the years, they might have mastered the skills and knowledge to advance and make a significant contribution to that particular area of science. This is their personal interest in science and they become fascinated by the outcomes of the research. They may not be contemplating the broader implications of their research for their country and its current scientific needs. Moravcsik (1984) argues that the most important function of a scientist in a developing country is not personal work that, if successful, contributes only a tiny part of the total body of science a country utilises. Personal research, in his view, is a major concern as it serves mainly to enable the scientist to stay in a much larger segment of science, which is of relatively minor importance. Although achievements of scientists in their preferred areas of research are beneficial from the individual point of view of accomplishments, the wider use and applicability of that research cannot be overlooked. Science is done ultimately for the benefit of society and not for the narrow sense of individual benefits. Society invests its resources in scientists, to make one a good scientist who will one day be part of the scientific force in the country to find solutions to its national problems of development.

The priority of a scientist in a developing region like Africa is therefore to work in areas in which the country needs to develop its expertise and knowledge. While national funding agencies in some countries have chosen their focus areas of research, scientists need to accept those ideas couched in a few focus areas. Scientists should not stay indifferent to those focus areas that have been identified and codified in the national science policies. Personal and preferential areas are secondary to the main national scientific objectives. There is no point in working in areas and spending resources on research problems that will not make any worthwhile contribution to the current needs of the country or in the foreseeable future. Resources, including scientific manpower, are valuable and precious for any developing country and any wastage cannot be justified.

The debate on the use, relevance, application and spending in scientific research is quite rife, even in developed countries. Some take the view that developing countries cannot afford to invest in basic science, but only in applied science as they have innumerable problems waiting to be addressed. The persistent view is also that basic or fundamental research should be conducted in the North and not in the South. A few reasons are put forward to defend this stance. This includes the view that the educational system in the South is not good enough, and the countries in the South do not have the culture to carry out research (Peimbert, 2000).[3] Rather than deliberating the divide on this issue, it is worth examining why basic science is important for developing countries. No country can afford to do away totally with basic research. In order to conduct applied research, basic research is essential. Applied research has direct relevance and importance to developing countries and it requires a critical mass in several basic sciences (Virasoro, 2000). This view has also been well received at international conferences on international donor support for research in developing countries.[4] A long time ago, Worthington (1938) observed that Africa is a field of pure as opposed to applied science and their interrelations have practical applications. There have been concerns about the declining investment in basic research in countries like South Africa (Mouton and Hackmann, 1997).

Research Areas and Development

It is relevant to consider the relationships between research areas and levels of development in Africa. Similarly, it is worth examining why some countries have strengths in certain areas but not in others.

The strength of research areas in specific countries and the general development of such countries can be connected to related indicators. Two indicators are useful in this examination: GDP and GERD. For such an analysis, the data collected and processed by UNESCO, the World Bank and the IMF are useful. Using the same set of indicators, they provide standardised data across countries.

In the most recent decade, as the above analysis of the major contributing countries has shown, a few countries that had strengths in certain areas can be isolated. Let us start with Egypt,

---

[3] The issue of a relatively underdeveloped science and technology culture in Africa has also been noted by other scholars (Thisen, 1993).

[4] The International Conferences on Donor Support to Development-Oriented Research in the Basic Sciences held at Sweden in June 1995, which brought together a host of donor agencies, had made these conclusions for strengthening research in the basic sciences: A foundation in the basic sciences is essential for all research in the applied sciences and for long-term development; adequate funding for the basic sciences, both domestic and international, is necessary; capacity-building in basic sciences should be undertaken; and higher education in the basic sciences should be supported (Virasoro, 2000).

South Africa, Tunisia and Nigeria, which produced the highest number of publications in specific research areas during 2011–15 (as presented in Table 4.3).

In terms of the average annual growth of the GDP for the period 2011–15, Egypt recorded an average annual growth rate of 2.9 per cent, ranging from 1.77 per cent in 2011 to 4.37 per cent in 2015. Nigeria registered a growth rate of 4.7 per cent, within a range of 2.7–6.3 per cent during the period. South Africa achieved an annual growth rate of 2.22 per cent, varying from 1.28 per cent to 3.28 per cent. Tunisia had 1.56 per cent of its GDP growing annually since 2011. It had a negative growth in 2011, but made up for this in the later years.[5]

As evident from the data, the countries that had made substantial progress in the production of science in specific areas have been growing their GDP. GDP growth in recent years in Egypt, Nigeria and in South Africa was substantial.

The per capita GDP is another measure to understand the level of development in these countries. The per capita GDP (average for 2011–15) for Egypt was $3,423. It increased in the country between 2011 and 2015 (from $3,077 to $3,731). In Nigeria, the average per capita GDP for the same period was $2,890, with a range of $2,583 in 2011 to $2,763 in 2015. South Africa reported an average of $6,975, but between 2011 and 2015, it decreased to $5,802 in 2015 from $8,083 in 2011: the trend was one of decline since 2011. Tunisia, a north African country, had an average of $4,187 (2011–15). Since 2011 it registered a declining pattern in its per capita GDP ($4,292 in 2011 to $3,884 to 2015). The average per capita GDP (2011–15) was $2,887 for Africa and $2,571 for sub-Saharan Africa. All four of these countries had per capita averages of more than the average for Africa and sub-Saharan Africa.

GERD data for these countries is an indicator of the progress they had achieved on the research front. But the problem again is the lack of comparable data from reliable sources across all countries. Egypt, during 2011–15, had an average of 0.61 per cent of its GDP for GERD. Since 2011, it increased its percentage of GERD from 0.53 to 0.72. South Africa, whose data is available for 2011 to 2013, showed an average GERD of 0.73 per cent, which was in the region of 0.72 to 0.73 per cent. Tunisia set apart an average 0.67 per cent of its GDP for

---

[5] Sourced and calculated from the IMF data for sub-Sahara and North Africa.

GERD during 2011–15. Its share decreased from 0.7 in 2011 to 0.63 per cent in 2015. Nigeria spent 0.22 per cent of its GDP on GERD in 2007. These figures can be compared with those for sub-Saharan and North Africa. During 2011–15, sub-Sahara allocated an average of 0.41 per cent of its GDP for GERD. This was an increase from 0.40 per cent in 2011. North Africa (which includes Egypt, Tunisia and Algeria), had an annual average of 0.46 per cent. The increase was evident from 2011 (0.42 per cent) to 2015 (0.5 per cent).

The advancement in specific areas and development is also related to the investment the country makes in these specific fields. The percentage of GERD can give some information in this regard. For instance, data regarding the percentages of GERD given to agricultural and veterinary sciences, natural sciences, engineering and technology, and health and medical sciences for countries are available. This data can provide some insights into the relationships between the development of research areas and the country.

South Africa, during 2011–15, spent 30.17 per cent of GERD on the natural sciences, 26.57 per cent on engineering and technology, 17.2 per cent on the medical and health sciences, and 7.9 per cent on the agricultural and veterinary sciences. South Africa, among other African countries, had produced a large number of publications in the fields of agriculture; astronomy; biochemistry and molecular biology; biotechnology and applied microbiology; chemistry; electrochemistry; energy and fuels; engineering; entomology; environmental sciences and ecology; general and internal medicine; genetics and heredity; geology; immunology; infectious diseases; life sciences; biomedicine and others; mathematics; microbiology; neurosciences and neurology; parasitology; paediatrics; physics; plant sciences; polymer sciences; science, technology and others; surgery; thermodynamics; veterinary sciences; virology; water resources and zoology. The contribution of the agricultural sector to development, particularly in regional development, and for economic growth in South Africa, has been acknowledged (Rooyen and Machethe, 2010).

Kenya and Ethiopia are the two other countries that had a sizable count of publications in the agricultural sciences. These two countries had a larger share of GERD allocated to these areas (47 per cent by Ethiopia and 44.8 per cent by Kenya in 2010). Kenya, which did not make much progress in engineering and technology, had spent

13.25 per cent of GERD on engineering and technology. Ethiopia, on the other hand, had spent only 4.7 per cent of GERD on engineering and technology. The consequences of this are obvious. Ethiopia did not appear on the list of countries that produced a substantial number of publications in research areas such as engineering. Kenya is another case among the countries that produced a considerable number of publications in the research areas of science, technology and others. Kenya in 2010 had 13.3 per cent of the share of GERD allocated for engineering and technology. The same is the case with South Africa which spent 26.6 per cent of its GERD for engineering and technology, and became a leading publisher in the research areas of engineering, and in science, technology and others. Uganda, in recent years, made progress in research areas of tropical medicine, public, environmental and occupational health, and infectious diseases. In 2010, Uganda allocated 18.14 per cent of GERD to medical and health sciences alone. By having 12.18 per cent of GERD on engineering and technology, Uganda made contributions to the group of leading publishers in science, technology and others. At the same time, Uganda had only 9 per cent of its GERD for the natural sciences. Obviously, Uganda is not among the largest producers in natural science subjects such as chemistry and physics.

These data suggests that GDP, per capita GDP and GERD have a strong relationship with the production of science. The percentage of GERD allocated to areas influences the production of science in specific research areas as well.

Let us now examine why some countries have strengths in certain research areas while others do not. Although one cannot find answers to this question for all countries and research areas, an attempt can be made to investigate this.

The agriculture sector in South Africa in 2006 contributed about 2.8 per cent of the GDP, and 8.5 per cent of total employment (Department of Trade and Industry, 2007 cited in Chitiga et al., 2008). Historically, agricultural research in South Africa received support for its R&D. It established the Agricultural Research Council (ARC) which was in charge of basic research and technology development. Research funding for the agriculture sector grew at an average of 5.1 per cent during 1911–50, and rose to 7 per cent later until 1971 (Liebenberg et al., 2011). Although agricultural spending stabilised at about 2.5 per cent in 2007, the total of research publications was over

20 per cent (against 14 per cent for 1945–2015) of all the publications South Africa produced during 2011–15. Also, as recent studies showed, South Africa grew its capacity to conduct research and develop the sector (Liebenberg et al., 2011). A host of laws, ordinances and statutes guided the development of agriculture in the country (Viljoen, 2005). The thrust of agricultural and economy-wide policies also assisted South Africa (Liebenberg and Pardey, 2011). The policies were meant to build an efficient and internationally competitive agricultural sector (Viljoen, 2005). These were instrumental in advancing agricultural sciences in the country.

The total expenditure on agriculture had a positive effect on Nigeria's economic growth. Studies (Iganiga and Unemhilin, 2017; Tijani et al., 2015) have examined the impact of governmental expenditure in agriculture on economic growth. Tijani et al. (2015) found priority in budget allocation essential to promoting economic growth in Nigeria.

Nigeria identified agriculture as the mainstay of the economy and the key development priority (Ani et al., 2014). Nigeria registered increasing agricultural growth in recent years, from 4.1 per cent in 1996 to 7.4 per cent in 2004, which was the result of increased expenditure on agriculture (Tijani et al., 2015). Beginning from the 1970s, the government introduced programmes to develop its agricultural sector (Oluwatoyese et al., 2016). Programmes including agricultural development, national acceleration of food production, a river basin development authority, structural adjustment programme, Operation Feed the Nation, investment in research institutes, and new agricultural policies initiated since 2001 were responsible for the country achieving increasing rates of growth in the sector.

Ethiopia has long recognised the importance of agriculture for economic growth. Being one of the fastest growing economies in the world, Ethiopia's agricultural growth was a major contributor to its economic growth (Bachewe et al., 2018). Having experienced bouts of severe drought and famine, Ethiopia adopted agricultural development as part of its overall development strategy (Dorosh and Rashid, 2012). Learning from experience, Ethiopia developed policies that emphasised investment in agricultural research (Dorosh and Rashid, 2012) since the 1980s. Ethiopia followed pathways of expanding its labour and land, and expansion of modern input.

As a result, its productivity grew by 2.3 per cent and agricultural output by 7.6 per cent per year during 2004–14 (Bachewe et al., 2018). Agricultural growth in the past two decades made a significant contribution to the overall development of Ethiopia, increasing from 1.3 per cent in the 1980s to 6.2 per cent in the 2000s (Dorosh and Rashid, 2012). The level of economic growth (the highest GDP growth in Africa at 9.6 per cent) that Ethiopia achieved in the more recent years has been startling (Clapham, 2017). Ethiopia implemented several policy interventions in agriculture to make the necessary changes in growth and development. For an extended period, the government of Ethiopia put agriculture at the centre of its national policy priorities (Bachewe et al., 2018).

The Kenyan economy is characterised by its high dependence on the agricultural sector, which has a major share in the GDP and involves about 75 per cent of its population (Gow and Parton, 1992). The agricultural policy of post-independent Kenya supported various aspects of the agricultural sector (Walker and Chinigò, 2018). One other reason for Kenya's progress in agriculture was the extent of international collaboration that it attracted to agricultural research. Recent studies, Muriithi et al. (2018) for instance, reported a high level of collaborative research in the academic community in Kenya and that collaboration in agriculture was as much as six times higher than that in other research fields. The study also indicated how an increase in collaboration further increased productivity. In Kenya, as a study of its productivity and economic growth reported (Gerdin, 2002), the contribution of productivity growth to output growth grew from 10.2 per cent in 1964–73 to 26.8 per cent in 1988–96. Capital was the most important contributor for this growth, and Kenya focussed on its agriculture by investing in it.

Astronomy flourished in South Africa, a leader in astronomy in Africa. The latest development is the Square Kilometre Array (SKA), co-hosted by South Africa and Australia in 2012, which was a huge feat for South African astronomy. The SKA is aimed to test the limits of human engineering and scientific endeavour and is expected to answer fundamental questions of science and about the laws of nature (Walker and Chinigò, 2018). Through legislation, government supported astronomy in the country to provide measures to advance astronomy and related scientific endeavours (Walker and Chinigò, 2018).

The SKA and similar attempts have taken South Africa to new heights in astronomy in Africa.

In the most recent decade, Egypt has reorganised its research sector with the allocation of more funding (Koenig, 2007). Science is advancing in Egypt specifically after the popular uprising and the country is adopting a grand vision to make science and scientific research the engine of its economy (Lawler, 2011). Following the revolution, there was great interest in science, and in chemistry in particular (Lawler, 2011). Traditionally, Egypt possessed strengths in chemistry (Koenig, 2007). Egyptians were among the first practising chemists in the world and are credited with a history of being the masters of metallurgy; produced the first synthetic pigment, Egyptian blue. They were ahead of many other countries in the research and production of chemical components such as cosmetics, perfumes and pharmaceuticals (Loyson, 2011). Rich in natural resources, Egypt is gifted and this has played a big role in its development (Loyson, 2011).

Africa is known for its history in the medical sciences (Gathiram and Hänninen, 2014). This eventually led some countries to advance in the field. Kenya, for instance, was badly affected by diseases such as HIV/ AIDS, and communicable diseases that include tuberculosis, malaria, respiratory diseases and diarrheal diseases. It faced a great many challenges in providing proper health care to its population. This led the WHO to predict that Kenya will face the burden of communicable and chronic illnesses in the coming years (Wachira and Martin, 2011). This situation necessitated that Kenya deal with the issue by investing sufficient funds in the medical and health sciences. This is reflected in some areas of medicine in which Kenya produced substantial research outputs.

Uganda's progress in the medical sciences is the result of several factors. It has drastically increased its health sector spending by making policy changes. It received funds from global agencies to fight diseases such as HIV/AIDS, tuberculosis and malaria (Wendo, 2003). It began to establish centres for treatment and research and programmes (for instance, vaccination on cancer) with the assistance of international partners (Gulland, 2012; Wendo, 2001). Uganda's achievement in the control of diseases such as HIV/AIDS would not have been possible without related research in this area. Uganda has been recognised as the first and most dramatic African success story in the reduction of the prevalence of HIV (Schoepf, 2003).

In the production of science, international collaboration is indispensable. International partnership in Africa is explored in the next chapter. Not only international partnership but also the forms of scientific collaboration as evident in the publications of African researchers are central to the analysis.

5 | Collaboration: Importance for Africa

Science is no longer an individual activity but a collective endeavour in which scholars, equipment and laboratories, not necessarily from a single location, are involved. It is scientific collaboration. The value of collaborative research has been highly regarded, particularly when the outcomes of such research are taken into account. As Michael Polanyi believed, science is more than the defining activity of a group of individuals and the product of their coordinated actions. It produces results that are much more potent than would emerge if they were working alone (cited in Ziman, 2000b: 21). In this chapter, scientific collaboration is discussed both generally and in the context of Africa.

The Indispensability of Collaboration

Science is global. Science is not isolated but connected locally, nationally and internationally. It advances through national and international links. As it is practised today, scientists are receptive to ideas from any place in the world and are willing to pursue long-distance ties (Schott, 1993). Scientists work together from one point on the earth to the other, as much as they work at local, national and regional levels. They conduct experiments simultaneously and in real time in laboratories that are located in distant regions, countries and continents. Some of the research they conduct involves thousands of scientists and technicians, as is evident from the number of authors listed in a single publication stored in the Web of Science (WoS). Not only science but also the scientific community is global even though they are scattered far and wide, and work in small, medium or large groups. They now have the facilities to work together in either the same locale or distant locales. They can work on a single project while remaining in their own offices and laboratories and process data remotely.

Advancement in communication technologies has made this easy, fast, convenient and possible.

Collaboration in science has become increasingly necessary as the scale of budgets for R&D and challenges to research continue to grow (The Royal Society, 2011). A large volume of literature confirms the undeniable role of collaboration in science (Adams, 2012, 2013; Eslami et al., 2013; Gazni et al., 2016; He et al., 2009; Onyancha and Maluleka, 2011; Pouris and Ho, 2014; Schubert and Sooryamoorthy, 2010; Sooryamoorthy, 2009a, 2009b, 2011, 2013, 2014, 2015, 2016; Toivanen and Ponomariov, 2011, to cite a few).

Collaboration is universal and is a standard feature of science. Collaborations are expanding in every region of the world (Adams, 2012). As a consequence of the increasingly collaborative approach to science, co-authored publications in science outnumber single-authored publications. Examining some 14 million publications for the period 2000–9, Gazni et al. (2012) conclude that there has been a consistent increase in the percentage of collaborated publications that were produced at many different parts of the world. Not only was there an increase in the production of collaborated publications but there was also an increase in the number of authors per publication, from 3.3 in 2000 to 4.1 in 2009 (Gazni et al., 2012). The analysis of publications by countries all over the world reveals that the rise in the output of countries like the United States and the United Kingdom is due to international collaboration (Adams, 2013). A worldwide analysis showed that six countries, namely, the United States, the United Kingdom, Germany, France, Italy and Canada produced 82 per cent of the total multinational publications during 2000–9 (Gazni et al., 2012).

Over 35 per cent (from 25 per cent some 15 years ago) of the papers published in international journals had international collaboration (The Royal Society, 2011). In a study of a few African countries for the period of 1991–7, Narváez-Berthelemot et al. (2002) found that the increased production of publications in the selected countries was associated with increasing levels of international collaboration.

National and international agencies support collaborative efforts in scientific research simply for reasons of benefits to science and society. The UN agenda of 2030 for sustainable development includes the goal of enhancing regional and international cooperation on and access to science, technology and innovation to enhance knowledge sharing on mutually agreed terms (UN, 2015).

Three forms of collaboration are common in scientific publications: domestic, international and multi-country. Domestic collaboration is between scholars within the same country while international collaboration requires at least one author from another country. In one publication there can be more than one form of collaboration. For instance, in a publication authored by five scientists, two of them can be from the same institution (domestic) and three from three other countries (international). When a paper does not engage more than one author there is no collaboration. They are sole or single-authored papers.

Collaboration is measured in many different ways, including the collaboration index that takes into account the number of authors per paper or the degree of collaboration which considers the strength of collaboration in publication. The collaboration coefficient and the modified collaboration coefficient are other newer forms of measurement (Savanur and Srikanth, 2010). Internationalisation through collaboration can also be measured. Some relevant indicators for this measure are intensity (the number of cross-national publications), propensity (the ratio of cross-national publications to total publications), and amplitude (the number of nations involved in cross-national publications) (Abramo et al., 2011).

Collaboration can be facilitated and accelerated by scientific development. The rate of scientific development is one of the main pivots in international scientific collaboration, as countries would be attracted to cooperate with highly developed countries (Gazni et al., 2012). Institutions that do not engage in international collaboration risk progressive disenfranchisement in the world of knowledge (Adams, 2013). It is important for scientists in developing countries to be connected to countries elsewhere. This global connectedness upgrades local knowledge and science, accelerating the learning of local scientists by tapping into advanced knowledge (Barnard et al., 2012).

One of the many reasons for the increasing growth in collaboration in science is the plexus of research fields in science. It is the interwoven combination of research fields and sub-research fields, which provides an evolution of science and the development of emerging research fields (Coccia and Bozeman, 2016). As reported by Coccia and Bozeman (2016), international collaboration has been accelerated across all research fields, but it is particularly pronounced in applied disciplines that include medical sciences and allied medical fields. There has only been low relative growth in international collaboration in more basic research fields such as astronomy, chemistry, mathematics and physics (Coccia and Bozeman, 2016).

Science does not know boundaries, geographical or political. It moves in the direction it wants to move, for new inventions, discoveries and applications. Collaboration in research among scientists from different countries often reflects political relationships that exist between the countries. Stated unequivocally, collaboration is beneficial to science. Whether it is equally useful and advantageous for all its partners and partnering countries is debatable. One of the many by-products of collaboration is evident in the science itself. It is the emergence of new disciplines (or decline of others) which is critical to science (Sun et al., 2013). New disciplines emerge from splitting and merging of communities in a collaborative network (Sun et al., 2013).

Capacity building in scientific research that can take place at different levels is another outcome of collaboration. This can occur at three levels. The first level is the partial research capacity in a given field to perform research in cooperation with competent research partners. The second level is when complete research capacity in a field is achieved independently and at the international level. Third is the national research capacity that requires the ability to prioritise and support research, train and attract good researchers, create conducive research environments, and form a system of incentives for the benefit of national development (Thulstrup et al., 1996). If these levels are considered, African countries that are at various stages of scientific research will fall under any one or more of these levels. Either way, collaboration can be avoided only at the loss of these levels of research capacity.

The connection between collaboration and productivity has been documented. Research productivity and the quality of output have positive effects on the degree of a scientist's international collaboration (Abramo et al., 2011). In North African countries (Algeria, Egypt, Morocco and Tunisia) international collaboration has expanded, leading to both internationalisation and increased research output (Landini et al., 2015). The correlation between productivity and international collaboration however varies across subjects and disciplines. Some studies showed significant correlation in subjects such as physical sciences, chemical sciences, earth sciences and industrial and information engineering in the selected countries (Abramo et al., 2009). Domestic collaboration and productivity seem to be correlated more in biological sciences than in physical or earth sciences (Abramo et al., 2009). Several other studies have also investigated the relationship between collaboration and productivity (Abramo et al., 2009; Gupta and Karisiddippa, 1999; He et al., 2009; Hu et al., 2013; Lakitan et al., 2012; Lee and Bozeman, 2005; Sooryamoorthy, 2014; Sooryamoorthy and Shrum, 2007; Ynalvez and Shrum, 2011).

One of the benefits of collaboration is the impact and visibility of research outputs, which has significance for the visibility of African science in world science. The impact of publications occurs at varying levels, depending on the field of science, where the research was conducted, the publication country of journal and the partners in such research. In a comparative study of the impact factors generated by Brazilian publications, there has been a disadvantage for endogenous papers, that is, papers that have authors from the same country (Meneghini, 1992). The impact of international publications rests on factors such as rigorous editorial practices, internationality of the editorial board and publication standards (Cano, 1992). The analysis of Scopus data for the period 2005–10 pertaining to fifty-four African countries showed that growth in publication output was accompanied by an increased ability of African scholars to publish in highly cited journals (AOSTI, 2014). As measured in terms of research quality, impact and intensity, African countries have achieved higher than world-average levels in the fields of engineering, public health and health services (AOSTI, 2014).

Citations have become a yardstick of the quality and value of scientific research. As essential aids to scientists to verify statements or data and peer recognition, citations are essential for the effective functioning of science as a social activity (Merton, 1988). Studies on collaboration present evidence that supports the relationship between collaborated publications and higher levels of citations counts. Internationally collaborated papers generally get cited more than single-authored or domestically collaborated papers (Adams, 2013; Aman, 2016; Khor and Yu, 2016; Onyancha and Maluleka, 2011; Puuska et al., 2014; Sin, 2011; Sooryamoorthy, 2009a, 2009b, 2017a; Tijssen, 2015).

In sub-Saharan Africa, international collaboration has been instrumental in increasing the citation impact of publications to about three times as impactful as institutional collaborations within regions in sub-Saharan Africa (Blom et al., 2016). This beneficial effect is not just for science produced in Africa. International collaborators of sub-Saharan researchers have also benefitted. They gained a relative higher citation impact in their collaborated publications with sub-Saharan scholars than their overall average impact (Blom et al., 2016). On the citation count, collaboration varied across regional and international collaboration in South Africa, a leading country in science. South Africa's international collaborations received higher counts of citation than its continental collaborations (Onyancha, 2011).

Despite the extensive literature on scientific collaboration, all the complexities involved in collaboration have not yet been brought to light. Collaboration of scientists is the result of a complex set of factors and it occurs due to strategic, organisational and operational reasons (Traoré and Landry, 1997). All the underlying determinants of collaboration are not completely known either (Abramo et al., 2011).

Theoretical Basis

Scientists collaborate with their peers nationally, regionally and internationally. A theoretical position in this regard is that, as Schott (1998) delineates, such collaborations accumulate in a centre, not in proportion to the research performance of the scientists in the centre, but accrue in excess of predictions of the research performance of the centre. Scientists around the world are tied to the centre, and are more tied to their peers in the centre than should be expected. Keeping this framework, one can look at the collaborative ties of African scholars, which is undertaken in this chapter using WoS data for the longer period of 1945–2015.

The core–periphery model has been used to explain the dynamics of collaboration. The classical world-system theory stipulates that the economic and technological inequality between the centre and periphery works to the detriment of the periphery. It is a viable model to explain the networks in international collaboration (Choi, 2012). Scholars from the periphery are dependent on core countries for funding, equipment and infrastructure (Safonova and Sokolov, 2013). Scholars from core countries at the same time are able to implement their technological ideas, collect raw data, get the technical tasks done and attract the best students from peripheral countries. The model, based on empirical data, predicts the intensity of migrations and colla-boration between countries. According to this model, the volume of student flows between countries is positively correlated with the wealth of the destination country and negatively correlated with the wealth of the country of origin; and the core of the academic world system is fractured between older and newer colonial powers; co-authorships are essentially symmetric showing a tendency of academics from more prosperous parts of the world to look for other resourceful research partners; scholars from poorer countries tend to co-author with those in other countries formerly dependent from the same colonial centre and not from the centre itself.

Studies have employed the core–periphery model to explain the characteristics of collaboration in countries. Karlovčec et al. (2016) present the core–periphery structure of researchers between core and periphery in the collaboration networks of Slovenian scientists. The dynamics of the strength and length of intervals of permanent presence in the core is shown in this analysis. Borgatti and Everett (1999) theorise the notion of a dense cohesive core and a sparse unconnected periphery. This can be applied to the examination of collaboration. Choi (2012), referring to the period between 1995 and 2010, found that increasingly there is periphery–periphery collaboration rather than core–core and core–periphery types. This can be tested in the case of African countries as well. As we will see in the analysis that follows in this chapter, this has been the case in Africa.

There are other theoretical lenses to comprehend international col-laboration and networks. They include inter alia academic (neo)colo-nialism, the classical world-system and the world society (Safonova and Sokolov, 2013). According to classical modernisation theory, modern sciences originated in the West and were exported to other countries as part of colonialisation and that those formerly peripheral countries eventually pass the stage of colonial science (Basalla, 1967). Basalla (1967) seeks to find an answer to the question of the diffusion of modern science from Western Europe to the rest of the world. Basalla's theory premised that the original home of modern science was in the nations of Western Europe and these nations were the arena of the scientific revolution that altered the epistemological, institutional and social architecture of premodern science (Raina, 1999). Basalla identified three stages in the institutionalisation of modern science outside Europe. In phase 1, a non-scientific society or nation (referring to the absence of Western science) provides a source for European science. Phase 2 is marked by the period of colonial science in those countries. In phase 3, the process of transplantation of modern science starts with commencement of the struggle to achieve an independent tradition. These phases are not alien to many African countries.

Neocolonial theories premise that the former colonies can never attain fully fledged membership in the global academic system as they are still bound by ties to their metropolitan countries and the network ties will be clustered along the lines of former colonial allegiances (Safonova and Sokolov, 2013). When the old colonial relationships are recreated at the level of international agreements, the procedures and applications for education from the former colonies are simplified. However, processes like these lead to the dominance of the metropolitan countries even in the absence of direct political dependency (Safonova and Sokolov, 2013).

Scientific Collaboration in Africa: Initiatives, Reality and Trends

Africa attracted scholars from overseas to conduct scientific research in crucial areas of science. Whether it was during or after the colonial period, collaboration prevailed in Africa. Scholars from other countries, the USA and Europe in particular, found Africa a fruitful field for science, original data and scientific experiments that opened the doors to numerous discoveries and advancements in science. Africa received scholars and scientists from a large number of countries. Their presence in African science, however, ranged from meagre to very large. Some countries were involved in almost all research areas while some others were limited to a few areas of mutual interest. This is evident from the analysis that follows.

Collaboration in Africa does not show any single or homogenous pattern across the continent. Although similarities exist, some collaborations are unique in historical and contemporary trends. Collaboration patterns in Africa are far from universal and exhibit layers of internal clusters and external links (Adams et al., 2013). These patterns can be explained by regional geography, history, legacy, culture and language rather than global influences. In international collaboration, contacts countries had in the past and colonial legacies cannot be neglected (Narváez-Berthelemot et al., 2002). Partners of

Africa reflect the colonial legacy. The colonial past continues to influence research agendas and patterns of collaboration. Thus, for instance, the increased collaborative papers written by South Africa and the United Kingdom, and Central Africa and France and Belgium (Boshoff, 2009b).

Scientific collaboration that exists in Africa has not been detached from its colonial countries. Many African countries continue to maintain ties with their previous colonisers like the United Kingdom, France, Belgium and Portugal. Ex-colonial countries in sub-Saharan Africa have stronger ties with their former colonisers than with other countries (Nagtegaal and de Brun, 1994). In central African countries, 35 per cent of collaboration occurred with their past colonial rulers (Boshoff, 2009a). This historical legacy in collaboration is carried on in both Francophone and Anglophone groups of nations (Adams et al., 2010; Jonathan et al., 2010). In international collaboration, Arab countries (including Algeria, Egypt, Libya, Morocco and Tunisia) have prominent partners in the USA, France and the United Kingdom (Elalami et al., 1992). Some of countries in North Africa maintain cooperation with European countries such as France due to their past colonial relations (Nour, 2005). International partners of Algeria, Morocco and Tunisia are mostly France, followed by Europe and the USA, and for Egypt it is Saudi Arabia (Landini et al., 2015).

France, for instance, endures as a collaborative partner for its former colonies as shown in the Scopus publication data for 1996–2008 (The Royal Society, 2011). This has been confirmed in the analysis of publications presented in this chapter. A study by Pouris (2010) for 2004–8 showed that England is among the top collaborative countries for all countries of SADC, while the USA is the top collaborating country only for Namibia. For all SADC countries, the top five collaborating countries are from outside the borders of Africa (Portugal, England, France, USA and Italy). Through mutual agreements for scientific and technical cooperation, the ties with past colonisers are maintained.[1] Neocolonial

---

[1] Similar cases can be drawn from other countries as well. Australian science, while attached to the colonial mentality, developed an independent structure in the early years of the 20th century, a product of both local imperatives and colonial relations (Inkster, 1985).

patterns in scientific collaboration are often manifested between the West and the developing countries.[2]

Initiatives were taken early to establish and advance scientific collaboration in Africa. In colonial times, there were structures that promoted regional cooperation among colonies. During this time, governments (Belgium, the United Kingdom, France, Portugal, the Federation of Rhodesia and Nyasaland, and the Union of South Africa) jointly established the Commission for Technical Cooperation in Africa South of the Sahara (CCTA) in 1950 (Odhiambo, 1993). In the same year, the Scientific Council for Africa South of the Sahara (CSA) was formed by scientists in these countries (Ohiambo, 1993). With the independence of African countries, the Organisation of African Unity (OAU) recognised the role of SCA for the continent. CCTA was then transformed into the Scientific, Technical and Research Commission of the OAU in 1963. The Africa Science Board of the National Academy of Sciences was reconstituted in 1965 for African scientists to work with their colleagues in other countries (Dillon, 1966).[3] The African Regional Centre for Technology (ARCT) was soon to be established in 1977 to contribute to the development and use of technology within the member states. The African Regional Organisation for Standardisation (ARSO) was another body that was constituted in 1979 to support regional cooperation in Africa. Around this time, the African Network of Scientific and Technological Institutions (ANSTI) was formed to bring forward collaboration of higher learning institutions in S&T in Africa. Many of these were not successful in achieving the intended objectives. There were problems of lack of political support from member countries and failure to provide necessary physical and financial support for their operations (UNESCO, 1987).

---

[2] This can be measured on a scale called the Neo-Colonial Relations Index, to understand this neocolonial dependency (Nagtegaal and de Brun, 1994). This measure expresses the number of coauthored papers between a former colonial country and its former colony. The index can vary between 0 and 1 in which 1 denotes total dependence on the former colonial country and all papers in the former colony are produced with a former colonial country.

[3] The background for the establishment of the Board in the USA was to facilitate scientific cooperation that recognises the desire of African scientists to contribute to the production of knowledge and scientists in other countries to take advantage of opportunities for research in Africa (Dillon, 1966).

The UNESCO conference in 1974 suggested a few types of international cooperation in S&T:

- The creation and strengthening of international scientific organisations set up under international and intergovernmental agreements,
- Establishment of national scientific institutions with an international impact, and
- Regional cooperation through the development of international research projects.

Following these conferences, mostly under the aegis of UNESCO, the need for collaboration has been emphasised. The CASTAFRICA II conference held in Arusha in Tanzania in 1987 reiterated the promotion of regional and interregional co-operation in S&T (UNESCO, 1987). This is to create a strong foundation for increased production and application of S&T. As the conference conceded, the role of regional cooperation in S&T was warranted by some main features: namely, that there was a serious shortage of high-level scientific and technological manpower; that a high level of investment was necessary to provide an infrastructure for the application of S&T; that the existence of similar problems was found in African countries; and that due to the nature and extent of a number of problems these could be dealt with regionally. Science, Technology and Innovation Strategy for Africa 2024 (STISA-24) recommended a multidisciplinary and multisectoral approach to collaboration in Africa. It stressed the need for increasing networking and collaboration within sectors of education, research, public and private, and at both national and cross-border levels.

African countries exhibit substantially higher collaboration patterns (Pouris and Ho, 2014). In one recent study, Pouris and Ho (2014) noted that of all the African countries, 29 of them published more than 90 per cent of their publications in collaboration with other countries. Nigeria was the only country found to have a collaboration rate of less than 50 per cent (Pouris and Ho, 2014). Publications in the Scopus database (2005–10) for fifty-four African countries indicated that there has been an increase in the level of collaboration (AOSTI, 2014). In the data presented below, collaboration of African countries also emerges as a striking feature of science in Africa. Africa collaborates with both developed and developing countries in the world.

Sub-Saharan Africa, particularly East Africa and Southern Africa, relies heavily on international collaboration for its research output (Blom et al., 2016). The percentage of international co-publications in sub-Saharan Africa had increased from 27 to 34 during 1990–5 (Krishna et al., 1998). As shown in a bibliometric analysis of Scopus data by Blom et al. (2016), Southern Africa produced 79 per cent of its research output through international collaboration, East Africa 70 per cent and West and Central Africa 45 per cent. A large percentage of researchers in sub-Saharan Africa (39 per cent of all researchers in East Africa and 48 per cent of all researchers in Southern Africa) are non-local and transitory and normally spend less than two years at institutions in sub-Saharan Africa (Blom et al., 2016).

In some African countries where production of publications is weak, international collaboration was found to be more than domestic collaboration (Mêgnigbêto, 2013b). Publications that involved authors from outside Africa exceeded publications between authors in Africa (AOSTI, 2014). The situation in continental collaboration has improved in some African countries (AOSTI, 2014).

Studies on collaboration in Africa however present the features that are characteristic of regional collaboration (i.e., countries within the continent are showing increasing levels of collaboration among themselves). Regional collaboration within Africa is centred around certain key countries which are the leaders in scientific output. South Africa has become the lynchpin of collaborative efforts in Africa, while other countries such as Nigeria, Egypt, Morocco and Sudan are also connecting to other African countries (The Royal Society, 2011). As a central hub of collaboration, South Africa collaborates with many other countries on the continent (AOSTI, 2014), many of which are shown in our data as well. The AOSTI analysis revealed that the level of inter-African–country collaboration for each AU member country has increased only slightly from 26 per cent to 27 per cent between 2005 and 2010. Extra-African collaboration (i.e., publications co-authored with non-AU countries) was about 50 per cent. South Africa's collaboration for the period of 1986–2005 with other African countries was very prominent (Onyancha, 2011). It was followed by other African countries such as Namibia, Kenya, Nigeria, Botswana, Ethiopia and Zambia, in declining order of publications. In regard to countries in general, South Africa produced its highest percentage of publications with the USA, followed by England, Germany, Australia, Canada,

France and the Netherlands. There was no African country within the first forty highly collaborated countries for South Africa. In South Africa, scholars were mostly associated with the USA, the United Kingdom, Germany, Australia and France, in descending order during the same period (Kraemer–Mbula and Scerri, 2015). As studies vary in terms of countries and period of analysis, all the findings are not expected to be alike. Similarities and resemblances however are obvious.

Scientific collaboration within sub-Saharan Africa is reportedly lower than between sub-Saharan and foreign countries (Onyancha and Maluleka, 2011). Collaboration between countries in Africa was in the region of 4 per cent in the publications for 2005–10 as against around 40 per cent for collaboration with a non-African country (AOSTI, 2014). In another study of Scopus data, inter-sub-Saharan collaboration was found to be about 3 per cent (Blom et al., 2016). In Central African countries for the period of 2000–6 collaboration between former colonisers and former colonies continued to exist (Boshoff, 2009a). A high level of neocolonial ties in science in Central African countries was reported (Boshoff, 2009a). For Ethiopian scholars, their major partners, as a bibliometric analysis of publications during 2008–14 indicated, in the order of the number of publications were the USA, the United Kingdom, Germany and India (Urama et al., 2015). For Kenya, the partners were from the USA, the United Kingdom, South Africa, Germany and the Netherlands (Urama et al., 2015). Tanzanian collaboration was mainly with the USA, the United Kingdom, Kenya, Switzerland and South Africa during 2008–14 (Kraemer–Mbula and Scerri, 2015). The most collaborative partners of Africa, as shown in several previous studies, were the USA, France and the United Kingdom (Adams et al., 2010; Adams et al., 2014; Blom et al., 2016; Jonathan et al., 2010; Narváez-Berthelemot et al., 2002; Pouris and Ho, 2014; Sooryamoorthy, 2015; Wagner et al., 2001).

Collaborations within Africa and with other countries allowed African scientists to conduct cutting-edge research in science (Mutapi, 2012). International collaboration has also helped advance certain scientific fields of inquiry in Africa. For instance, clinical medicine, chemistry, chemical engineering and technology have benefitted from international collaboration to advance their critical mass (Tijssen, 2015). A relationship between collaboration and specialisation has also been observed. South Africa collaborates mostly in life sciences,

Nigeria in agriculture and Egypt in the chemical sciences (Wagner et al., 2001). In South Africa, the most preferred subject areas for international collaboration are astronomy and astrophysics (Onyancha, 2011). Continental collaboration is preferred in subjects such as veterinary sciences, ecology, public, environmental and occupational health, environmental sciences, plant sciences, zoology, infectious diseases, tropical medicine and biochemistry and molecular biology.

International Partners of Africa and Publications

Using the WoS data for the period of 1945–2015, international partnerships that were involved in the publications of African scholars were examined. The analysis was made around the countries with which Africa had scientific contacts and collaboration. This was done for the two separate time periods of 1945–2015 and 2011–15. A total of 561,217 scientific publications for 1945–2015 and 308,538 for 2011–15 that had international partnership were analysed.

Partnership During 1945–2015

As shown in Table 5.1, major international partners of Africa are listed according to the number of publications for 1945–2015. There were 561,217 publications for this period.[4] The table not only provides the proportionate representation of international partners in African science but also the count of countries in Africa with whom the international partners had research collaboration (Figure 5.2).

A consideration of the first few international partners with whom Africa produced the largest number of publications shows that they were mostly from North America and Europe. An exception to this is the presence of two countries from other continents: Saudi Arabia and Australia. The first ten countries together produced 50 per cent of all publications in Africa for the selected period. Among these, two (the United States and France) had more than 10 per cent each of all publications. They were both involved in a quarter of all publications.

France was the second most common international partner in Africa for the entire period of 1945–2015. This has something to do with the

---

[4] The total number of publications does not tally with this number as a publication might have engaged more than one international partner.

Table 5.1 *Major partners of Africa in science, 1945–2015*

Country	No. of publications	% of all publications	Mean	S.D.	No. of African countries
US	68,797	12.26	1,348.96	3,373.69	51
France	59,655	10.63	1,193.10	2,762.12	50
England	42,339	7.54	830.18	1,966.75	51
Germany	26,569	4.73	542.22	1,441.08	36
Saudi Arabia	16,553	2.95	376.20	1,874.38	44
Canada	14,648	2.61	292.96	772.14	50
Australia	13,503	2.41	277.55	980.29	49
Italy	13,492	2.40	269.84	610.75	50
Netherlands	13,264	2.36	276.33	679.85	48
Belgium	12,298	2.19	250.98	440.97	49
Switzerland	12,261	2.18	245.22	529.38	50
Spain	11,736	2.09	234.72	598.48	50
South Africa	11,475	2.04	244.15	330.46	47
Japan	10,272	1.83	223.30	611.31	46
Sweden	9,677	1.72	205.89	459.84	47
India	8,590	1.53	186.74	465.3	46
China	8,280	1.48	176.17	476.88	47
Scotland	7,360	1.31	156.60	391.78	47
Denmark	6,876	1.23	149.48	315.24	46
Kenya	6,507	1.16	138.45	234.97	47
Brazil	5,961	1.06	129.59	328.99	46
Norway	5,519	0.98	117.43	305.62	47
Austria	5,146	0.92	114.36	305.54	45
Nigeria	4,579	0.82	95.40	219.32	48
Poland	4,497	0.80	121.54	361.51	37
Portugal	4,313	0.77	95.84	239.78	45
Russia	3,967	0.71	88.16	295.89	45
Uganda	3,697	0.66	85.98	161.88	43
Tanzania	3,694	0.66	83.95	147.71	44
Czech Republic	3,584	0.64	87.41	240.99	41
Israel	3,557	0.63	75.68	264.08	47
Greece	3,203	0.57	82.13	223.04	39
Cameroon	3,089	0.55	65.72	96.26	47
South Korea	3,002	0.53	75.05	245.85	40
Mexico	2,968	0.53	64.52	169.15	46

Table 5.1 (*cont.*)

Country	No. of publications	% of all publications	Mean	S.D.	No. of African countries
Malaysia	2,947	0.53	75.56	145.97	39
Finland	2,870	0.51	70.00	178.46	41
New Zealand	2,868	0.51	65.18	212.45	44
Hungary	2,819	0.50	80.54	238.35	35
Turkey	2,802	0.50	65.16	194.04	43
Ghana	2,725	0.49	61.93	90.57	44
Taiwan	2,668	0.48	65.07	187.02	41
Thailand	2,615	0.47	60.81	111.09	43
Pakistan	2,608	0.46	65.20	150.89	40
Zimbabwe	2,553	0.45	63.83	163.44	40
Argentina	2,529	0.45	63.23	199.51	40
Senegal	2,485	0.44	51.77	61.63	48
Columbia	2,454	0.44	63.18	152.04	39
Chile	2,410	0.43	60.25	214.7	40
Romania	2,361	0.42	69.44	189.09	34
Morocco	2,343	0.42	55.79	138.96	42
Egypt	2,302	0.41	52.32	87.06	44
Burkino Faso	2,248	0.40	54.83	64.84	41
Ethiopia	2,087	0.37	47.43	85.36	44
Iran	2,051	0.37	50.02	142.57	41
Ireland	1,992	0.35	45.27	114.92	44
Serbia	1,982	0.35	60.06	160.5	33
Tunisia	1,969	0.35	48.02	116.06	41
Malawi	1,915	0.34	46.71	85.76	41
Zambia	1,887	0.34	43.88	77.2	43
Cote Ivore	1,869	0.33	39.77	47.25	47
Armenia	1,849	0.33	80.39	197.23	23
UAE	1,719	0.31	47.75	159.34	36
Wales	1,674	0.30	38.05	93.58	44
Slovakia	1,601	0.29	43.27	142.22	37
Benin	1,581	0.28	37.64	47.26	42
Rep of Georgia	1,561	0.28	53.83	144.29	29
Mali	1,515	0.27	36.07	47.34	42
Bylaurs	1,430	0.25	75.26	172.08	19
Algeria	1,404	0.25	36.95	96.99	38

Table 5.1 (*cont.*)

Country	No. of publications	% of all publications	Mean	S.D.	No. of African countries
Slovenia	1,241	0.22	47.73	145.9	26
Indonesia	1,230	0.22	29.29	38.48	42
Botswana	1,211	0.22	30.28	71.32	40
Bulgaria	1,186	0.21	34.88	101.22	34
Qatar	1,181	0.21	39.37	128.08	30
Gambia	1,153	0.21	27.45	34.75	42
Swaziland	1,134	0.20	45.36	159.82	25
Sudan	1,125	0.20	26.16	42.1	43
Mozambique	1,107	0.20	27.00	46.19	41
Kuwait	1,089	0.19	40.33	154.17	27
Ukraine	1,081	0.19	36.03	105.48	30
Niger	1,071	0.19	23.28	35.14	46
Namibia	1,060	0.19	28.65	110.64	37
Croatia	1,034	0.18	33.35	103.24	31
Lebanon	1,023	0.18	26.92	57.76	38
Azerbaijan	1,018	0.18	46.27	134.44	22
Peru	1,007	0.18	24.56	49.86	41
Singapore	1,007	0.18	25.18	49.24	40
Bangladesh	9,26	0.16	25.72	31.77	36
Vietnam	924	0.16	23.69	28.83	39
Guinea	916	0.16	24.76	82.35	37
Jordan	906	0.16	26.65	57.05	34
Gabon	876	0.16	23.05	29.74	38
Rwanda	869	0.15	22.87	32.09	38
Oman	850	0.15	23.61	54.89	36
Philippines	844	0.15	20.59	28.37	37
Estonia	818	0.15	23.37	78.86	35
Madagascar	817	0.15	19.00	25.6	43
Congo	789	0.14	18.35	21.32	43
Cyprus	737	0.13	24.57	86.55	30
North Ireland	715	0.13	23.06	65.86	31
Zaire	708	0.13	18.63	20.65	38
Sri Lanka	705	0.13	19.05	36.95	37
Guinea-Bissau	647	0.12	23.11	86.81	28
Lithuania	637	0.11	24.50	85.77	26

Table 5.1 (*cont.*)

Country	No. of publications	% of all publications	Mean	S.D.	No. of African countries
Syria	612	0.11	19.13	33.12	32
Togo	568	0.10	16.71	25.55	34
Baharain	513	0.09	16.03	32.03	32
Luxumberg	477	0.08	11.93	14.56	40
Yemen	473	0.08	16.31	55.18	29
Libya	452	0.08	15.07	47.15	30
Iraq	413	0.07	14.75	30.93	28
Cambodia	406	0.07	10.97	10.47	37
Congo People's Rep	375	0.07	23.44	86.56	16
Central African Republic	350	0.06	11.29	15.89	31
Cuba	347	0.06	10.21	26.45	34
Ecuador	313	0.06	7.83	12.73	40
USSR	156	0.03	6.50	11.09	24
Total for all authors	561,217	100.00	102.82	258.09	39.33

Note: Total includes all partners that are not shown in the table.

colonial connection of France with African countries. Although the data relates to the period 1945–2015, there were not many publications up until 1970. This means the data mostly covers the period of independent Africa. England and Germany were the other colonial partners of Africa that continue to keep their links alive with the new independent Africa.

All the first ten countries listed had associated with most of the countries in Africa. The USA and England associated with fifty-one African countries while Germany had collaborated with fewer (thirty-six) countries in Africa. On average, foreign partners associated with thirty-nine African countries for the production of scientific publications. There were only twenty-one countries that associated with Africa to produce 1 per cent or more of its publications in Africa. India, China (1.5 per cent each) and Brazil (1.1 per cent) were among them.

The data also provides information about regional collaboration. Of those countries that were involved in the production of at least 1 per cent of publications, there were only two African countries. South Africa had 2 per cent while Kenya was involved in 1.16 per cent of all publications. Both South Africa and Kenya worked with forty-seven other African countries.

The association between countries in Africa and international partners is presented in Table 5.2. This data shows which of the international partners are closely or loosely collaborated with which African country. Only the top ten international partners are analysed. The total publications of international partners included all the publications they had produced with Africa and not all of their publications.

The USA collaborated mostly with South Africa and Egypt, partnering with 31 per cent and 16 per cent of all the US collaborated publications. There were a few other countries with which the USA had scientific contacts that were evident in joint publications. The US partners produced 9 per cent of their publications with their counterparts in Kenya, 5 per cent with Nigeria, 4 per cent with Uganda, 3 per cent each with Morocco and Tanzania and about 2 per cent each with Ethiopia, Ghana and Malawi.

France was keen to associate with Tunisia (20 per cent of its publications in Africa), Algeria, Morocco (18 per cent each), South Africa (10 per cent), Cameroon, Egypt, Senegal (4 per cent each) and Burkina Faso (2 per cent). These countries were the French colonies in Africa. Except South Africa with whom France had 10 per cent of its publications with African countries, France did not have serious scientific collaboration with other African countries other than its former colonies. There were also other colonies of France including Gabon, Chad, Niger and Madagascar that did not have much joint scientific production with France. These are not scientifically strong countries in Africa. In other words, scientific association with France is not influenced by the colonial legacy of countries alone but also by their standing in science.

The principal collaborator of England in Africa was South Africa with which it produced one-third of all its publications with African countries. Kenya became England's second major partner with 10 per cent of publications. Following this were Egypt (9 per cent), Nigeria (6 per cent), Tanzania (5 per cent) and Uganda (4 per cent), all British colonies of colonial Africa. Notably, England did not have much association with most of the formerly French colonies such as Algeria, Morocco, Senegal and Tunisia. England had significant association going back to the period when Egypt was the condominium of both France and England.

Germany's leading scientific collaborators were South Africa and Egypt, producing 32 per cent and 21 per cent, respectively, of its publications in Africa. South Africa and Egypt were not colonies of Germany. This has to be explored in the second dataset referring to the recent five years. Germany had a long-standing relationship with South Africa, mainly during the apartheid period. Kenya, Morocco (5 per cent each), Nigeria (4 per cent), Algeria, Cameroon (3 per cent each), Ethiopia (2.5 per cent), Ghana, Tanzania and Tunisia (2 per cent each) were other countries that Germany favoured for joint science production. Cameroon and parts of Tanzania were part of Germany at some point in time history.

Saudi Arabia published 75 per cent of its publications in Africa with Egypt alone. It also published with Tunisia (6 per cent), Algeria (5 per cent), South Africa (4 per cent), Morocco and Sudan (3 per cent each). Saudi Arabia's collaboration with many of these countries is due to its contacts with Arab and Muslim countries.

Although Canada was not a coloniser of Africa, it established scientific contacts with several African countries. Canada produced its publications in Africa mostly with countries such as South Africa (33 per cent), Egypt (16 per cent), Morocco (8 per cent), Kenya (7 per cent), Tunisia (5 per cent) and Uganda (3 per cent). South Africa and Egypt were the two key countries in Africa that advanced in S&T. This might be a reason for Canada to work with scientists from these countries. Similarly, Australia worked with South Africa to produce more than half of its publications with African countries. Guinea (7 per cent), Egypt, Kenya, Morocco (5 per cent each), Nigeria (3 per cent) and Tanzania (2 per cent) were some other countries with which Australian scientists worked.

Italy had a few colonies in Africa (Eritrea, Ethiopia, Libya and Somalia). In joint scientific publications, Italy chose to work mostly with South Africa (producing 26 per cent of its publications in Africa), Egypt (15 per cent), Morocco (11 per cent), Tunisia (9 per cent), Algeria (6 per cent), Nigeria (5 per cent), Kenya (4 per cent) and Cameroon (3 per cent). Seemingly, Italy preferred to work with scientifically strong countries in Africa. The Netherlands produced one-

Collaboration: Importance for Africa

Table 5.2 *International partners and Africa, 1945–2015*

Country	US No.	US %	France No.	France %	England No.	England %	Germany No.	Germany %	Saudi Arabia No.	Saudi Arabia %
Algeria	831	1.21	10966	18.38	557	1.32	699	2.63	869	5.25
Angola	77	0.11	57	0.10	51	0.12	17	0.06	2	0.01
Benin	300	0.44	931	1.56	266	0.63	218	0.82	7	0.04
Botswana	672	0.98	61	0.10	265	0.63	114	0.43	47	0.28
Burkina Faso	487	0.71	1352	2.27	444	1.05	309	1.16	7	0.04
Burundi	48	0.07	72	0.12	19	0.04	15	0.06	0	0.00
Cameroon	1166	1.69	2607	4.37	705	1.67	826	3.11	29	0.18
Cape Verde	17	0.02	11	0.02	21	0.05	14	0.05	0	0.00
Chad	44	0.06	183	0.31	26	0.06	14	0.05	1	0.01
Comoros	9	0.01	21	0.04	6	0.01	1	0.00	0	0.00
Congo	376	0.55	593	0.99	158	0.37	150	0.56	6	0.04
Djibouti	12	0.02	70	0.12	6	0.01	1	0.00	1	0.01
Egypt	10866	15.79	2118	3.55	3655	8.63	5549	20.88	12438	75.14
Eritrea	69	0.10	27	0.05	34	0.08	19	0.07	6	0.04
Ethiopia	1460	2.12	351	0.59	971	2.29	661	2.49	66	0.40
Gabon	334	0.49	835	1.40	220	0.52	557	2.10	3	0.02
Gambia	426	0.62	126	0.21	1200	2.83	98	0.37	10	0.06
Ghana	1681	2.44	240	0.40	1307	3.09	618	2.33	31	0.19
Guinea	706	1.03	255	0.43	519	1.23	136	0.51	11	0.07
Guinea-Bissau	47	0.07	16	0.03	81	0.19	7	0.03	0	0.00
Ivory Coast	34	0.05	319	0.53	16	0.04	11	0.04	0	0.00
Kenya	6318	9.18	779	1.31	4097	9.68	1360	5.12	48	0.29
Lesotho	69	0.10	8	0.01	21	0.05	6	0.02	0	0.00
Liberia	179	0.26	23	0.04	23	0.05	34	0.13	3	0.02
Libya	190	0.28	155	0.26	333	0.79	94	0.35	100	0.60
Madagascar	695	1.01	822	1.38	260	0.61	252	0.95	8	0.05
Malawi	1442	2.10	151	0.25	1398	3.30	152	0.57	8	0.05
Mali	672	0.98	591	0.99	248	0.59	145	0.55	8	0.05
Mauritius	153	0.22	95	0.16	269	0.64	25	0.09	23	0.14
Mauritania	44	0.06	151	0.25	16	0.04	13	0.05	3	0.02
Morocco	2009	2.92	10632	17.82	1020	2.41	1406	5.29	431	2.60
Mozambique	420	0.61	101	0.17	229	0.54	72	0.27	6	0.04
Namibia	345	0.50	192	0.32	289	0.68	325	1.22	9	0.05
Niger	319	0.46	564	0.95	136	0.32	92	0.35	8	0.05
Nigeria	3440	5.00	462	0.77	2414	5.70	1109	4.17	152	0.92
Rwanda	454	0.66	99	0.17	158	0.37	95	0.36	23	0.14
Reunion	209	0.30	1271	2.13	91	0.21	68	0.26	5	0.03
Sao Tome and Principe	5	0.01	0	0.00	12	0.03	0	0.00	0	0.00
Senegal	821	1.19	2418	4.05	436	1.03	177	0.67	11	0.07
Seychelles	136	0.20	78	0.13	95	0.22	18	0.07	9	0.05
Sierra Leone	197	0.29	3	0.01	160	0.38	32	0.12	3	0.02
Somalia	28	0.04	5	0.01	16	0.04	7	0.03	1	0.01
South Africa	21348	31.03	6222	10.43	13149	31.06	8619	32.43	641	3.87
Sudan	558	0.81	172	0.29	594	1.40	405	1.52	425	2.57
Swaziland	117	0.17	6	0.01	77	0.18	4	0.02	3	0.02
Tanzania	2337	3.40	270	0.45	2247	5.31	628	2.36	20	0.12
Togo	96	0.14	344	0.58	45	0.11	96	0.36	3	0.02
Tunisia	1135	1.65	12140	20.35	518	1.22	593	2.23	1008	6.09
Uganda	3029	4.40	340	0.57	1827	4.32	445	1.67	27	0.16
Zambia	1233	1.79	82	0.14	709	1.67	104	0.39	17	0.10
Zimbabwe	1137	1.65	268	0.45	925	2.18	170	0.64	16	0.10
Total	68797	100.00	59655	100.00	42339	100.00	26580	100.00	16553	100.00

third of its publications in Africa with South Africa, followed by Kenya (9 per cent), Egypt (7 per cent), Morocco, Tanzania (5 per cent each), Ethiopia, Ghana (4 per cent each), Nigeria, Uganda and Zimbabwe (3 per cent each). South Africa had been in touch with the Dutch for a long time.

The Congo, Rwanda and Burundi were colonies of Belgium. The scientific partners of Belgium in Africa were, among others, South Africa (produced 22 per cent of its publications in Africa), Egypt (10 per cent), Kenya (7 per cent), Algeria, Morocco, Tunisia

Table 5.2 *(cont.)*

Country	Canada No.	Canada %	Australia No.	Australia %	Italy No.	Italy %	Netherlands No.	Netherlands %	Belgium No.	Belgium %
Algeria	406	2.77	107	0.79	779	5.77	118	0.89	579	4.71
Angola	7	0.05	10	0.07	28	0.21	12	0.09	12	0.10
Benin	126	0.86	44	0.33	91	0.67	245	1.85	379	3.08
Botswana	106	0.72	109	0.81	53	0.39	66	0.50	53	0.43
Burkina Faso	145	0.99	69	0.51	233	1.73	171	1.29	314	2.55
Burundi	2	0.01	6	0.04	13	0.10	33	0.25	166	1.35
Cameroon	252	1.72	187	1.38	415	3.08	217	1.64	524	4.26
Cape Verde	1	0.01	4	0.03	6	0.04	8	0.06	1	0.01
Chad	12	0.08	2	0.01	11	0.08	3	0.02	15	0.12
Comoros	3	0.02	0	0.00	4	0.03	0	0.00	4	0.03
Congo	47	0.32	44	0.33	58	0.43	53	0.40	338	2.75
Djibouti	9	0.06	1	0.01	5	0.04	0	0.00	4	0.03
Egypt	2332	15.92	671	4.97	2025	15.01	862	6.50	1185	9.64
Eritrea	12	0.08	9	0.07	34	0.25	51	0.38	0	0.00
Ethiopia	199	1.36	181	1.34	308	2.28	571	4.30	492	4.00
Gabon	38	0.26	41	0.30	41	0.30	166	1.25	142	1.15
Gambia	40	0.27	83	0.61	68	0.50	147	1.11	205	1.67
Ghana	300	2.05	242	1.79	145	1.07	484	3.65	134	1.09
Guinea	106	0.72	1041	7.71	78	0.58	105	0.79	79	0.64
Guinea-Bissau	13	0.09	13	0.10	24	0.18	33	0.25	11	0.09
Ivory Coast	2	0.01	1	0.01	0	0.00	4	0.03	37	0.30
Kenya	1092	7.45	706	5.23	482	3.57	1237	9.33	870	7.07
Lesotho	6	0.04	10	0.07	3	0.02	9	0.07	7	0.06
Liberia	8	0.05	9	0.07	5	0.04	3	0.02	6	0.05
Libya	97	0.66	65	0.48	92	0.68	31	0.23	31	0.25
Madagascar	68	0.46	77	0.57	81	0.60	57	0.43	101	0.82
Malawi	140	0.96	182	1.35	67	0.50	252	1.90	101	0.82
Mali	74	0.51	55	0.41	103	0.76	92	0.69	107	0.87
Mauritius	36	0.25	96	0.71	23	0.17	27	0.20	17	0.14
Mauritania	16	0.11	10	0.07	30	0.22	24	0.18	10	0.08
Morocco	1191	8.13	608	4.50	1484	11.00	645	4.86	613	4.98
Mozambique	39	0.27	112	0.83	93	0.69	6	0.05	95	0.77
Namibia	51	0.35	159	1.18	49	0.36	68	0.51	26	0.21
Niger	32	0.22	35	0.26	44	0.33	60	0.45	106	0.86
Nigeria	528	3.60	383	2.84	691	5.12	409	3.08	264	2.15
Rwanda	52	0.35	33	0.24	30	0.22	123	0.93	267	2.17
Reunion	38	0.26	73	0.54	79	0.59	50	0.38	65	0.53
Sao Tome and Principe	0	0.00	0	0.00	5	0.04	0	0.00	0	0.00
Senegal	128	0.87	74	0.55	171	1.27	131	0.99	321	2.61
Seychelles	17	0.12	45	0.33	14	0.10	28	0.21	14	0.11
Sierra Leone	29	0.20	19	0.14	15	0.11	40	0.30	15	0.12
Somalia	1	0.01	5	0.04	75	0.56	7	0.05	4	0.03
South Africa	4869	33.24	6843	50.68	3508	26.00	4552	34.32	2737	22.26
Sudan	62	0.42	88	0.65	110	0.82	235	1.77	52	0.42
Swaziland	20	0.14	17	0.13	5	0.04	8	0.06	12	0.10
Tanzania	343	2.34	270	2.00	257	1.90	716	5.40	431	3.50
Togo	33	0.23	7	0.05	19	0.14	43	0.32	48	0.39
Tunisia	770	5.26	170	1.26	1184	8.78	166	1.25	623	5.07
Uganda	469	3.20	216	1.60	206	1.53	420	3.17	359	2.92
Zambia	96	0.66	99	0.73	71	0.53	143	1.08	196	1.59
Zimbabwe	185	1.26	172	1.27	77	0.57	333	2.51	126	1.02
Total	14648	100.00	13503	100.00	13492	100.00	13264	100.00	12298	100.00

(5 per cent each), Cameroon, Ethiopia, Tanzania (4 per cent each), Benin, Burkina Faso, Congo, Senegal and Uganda (3 per cent each). Belgium had 2.2 per cent publications with Rwanda and 1.4 per cent with Burundi.

Individually, colonial legacy, religion or language affinity are not the determining factors that facilitate joint scientific endeavours between countries. The stage at which the countries are in scientific development is a further decisive condition for scientific association and joint publications. As seen in the data for 1945–2015, former colonial powers also had scientific contacts with other African countries, while maintaining links with their former colonies.

Partnership During 2011–15

Table 5.3 presents the figures of international partners of Africa in the recent five years, namely, 2011–15. For this period there were 308,538 publications. Compared to the publication trends revealed in the first dataset for the period of 1945–2015, new patterns are emerging in Africa in scientific collaboration. Still, Africa attracts most of its international partners from North America and Europe. Of the first ten international partners of Africa there were also countries such as Saudi Arabia, Australia and South Africa. These ten countries jointly produced 44 per cent of all publications that Africa produced with international partnerships. The USA alone had 10 per cent of the publications. In the order of the number of publications were France (8 per cent), England (6 per cent), Saudi Arabia, Germany (4 per cent each), and all of the other five countries had 2 per cent each. The share of the publications of these countries has been altered during recent years. South Africa is the only country in Africa among the top ten countries of international partnership.

International partners of Africa collaborated with an average of thirty-one African countries to produce joint publications during this period. The USA worked with fifty countries in Africa while Saudi Arabia associated with thirty-eight countries. More than 1 per cent of the joint publications with Africa came from twenty-four foreign countries, among them China, India, Brazil and Kenya. South Africa and Kenya were the only African countries that appeared on the list of countries with at least 1 per cent of publications. Since the dataset is for a shorter period of five years, countries with at least 0.5 per cent of their publications in Africa should be considered, giving a figure of 1,529 papers. There were fifty-one countries that had 0.5 per cent or more of their publications with African countries. Eight countries were from Africa that represented regional collaboration within the continent. Along with South Africa and Kenya, other African countries were Nigeria, Uganda, Tanzania, Ghana, Cameroon and Morocco. Some Asian countries such as Malaysia, Taiwan, South Korea, Pakistan and Thailand also partnered with Africa, producing 0.5–1 per cent of their publications with Africa. East and West European countries (Russia,

Table 5.3 *Major partners of Africa in science, 2011–15*

Country	No. of publications	% of all publications	Mean	S.D.	No. of African countries
US	30,883	10.01	617.66	1478.71	50
France	23,579	7.64	491.23	1083.08	48
England	18,442	5.98	384.21	913.32	48
Saudi Arabia	12,687	4.11	333.87	1550.81	38
Germany	12,472	4.04	277.16	699.1	45
Australia	7,342	2.38	159.61	512.73	46
Italy	7,340	2.38	163.11	376.39	45
Spain	7,311	2.37	166.16	390.98	44
Canada	7,298	2.37	162.18	414.17	45
South Africa	7,102	2.30	165.16	212.11	43
Netherlands	6,868	2.23	159.72	399.86	43
Switzerland	6,849	2.22	148.89	334.75	46
Belgium	5,969	1.93	132.64	234.24	45
China	5,920	1.92	144.39	366.22	41
India	5,721	1.85	136.21	348.2	42
Japan	5,363	1.74	138.50	344.62	38
Sweden	5,108	1.66	121.62	298.12	42
Brazil	4,043	1.31	98.61	244.68	41
Kenya	3,709	1.20	90.46	142.97	41
Scotland	3,597	1.17	89.93	228.24	40
Denmark	3,577	1.16	94.13	219.1	38
Norway	3,316	1.07	82.90	227.76	40
Portugal	3,226	1.05	78.68	190.86	41
Austria	3,195	1.04	91.29	217.26	35
Turkey	3,068	0.99	92.97	229.89	33
Poland	2,809	0.91	112.36	283.29	25
Russia	2,776	0.90	77.11	229.41	36
Nigeria	2,559	0.83	62.41	150.16	41
Czech Republic	2,555	0.83	79.84	207.63	32
Malaysia	2,318	0.75	66.23	123.96	35
Uganda	2,270	0.74	59.74	100.05	38
Taiwan	2,174	0.70	62.11	167.41	35
South Korea	2,110	0.68	60.29	195.22	35
Tanzania	2,101	0.68	53.87	83.92	39
Mexico	2,016	0.65	54.49	136.4	37

Table 5.3 (*cont.*)

Country	No. of publications	% of all publications	Mean	S.D.	No. of African countries
Columbia	2,009	0.65	59.09	146.88	34
Israel	1,959	0.63	55.97	160.17	35
Pakistan	1,902	0.62	55.94	129.24	34
Greece	1,871	0.61	60.35	179.71	31
Finland	1,856	0.60	50.16	130.75	37
Romania	1,853	0.60	71.27	178.38	26
Argentina	1,838	0.60	51.06	176.44	36
Chile	1,815	0.59	56.72	184.72	32
Serbia	1,767	0.57	67.96	170.3	26
New Zealand	1,764	0.57	47.68	134.88	37
Iran	1,726	0.56	52.30	138.79	33
Ghana	1,719	0.56	47.75	62.45	36
Cameroon	1,696	0.55	40.38	60.47	42
Thailand	1,681	0.54	44.24	82.17	38
Armenia	1,588	0.51	176.44	252.04	42
Morocco	1,529	0.50	46.33	109.3	33
Hungary	1,498	0.49	65.13	192.78	23
Rep of Georgia	1,490	0.48	64.78	154.17	23
Burkina Faso	1,389	0.45	38.58	36.58	36
Egypt	1,378	0.45	41.76	59.81	33
Belarus	1,360	0.44	123.64	205.64	11
Slovakia	1,334	0.43	53.36	148.73	25
Senegal	1,318	0.43	32.95	33.8	40
Ethiopia	1,223	0.40	31.36	46.99	39
Zimbabwe	1,221	0.40	35.91	85.59	34
Zambia	1,140	0.37	34.55	50.07	33
Tunisia	1,125	0.36	32.14	69.27	35
Slovenia	1,124	0.36	59.16	154.51	19
Malawi	1,118	0.36	34.94	56.11	32
Ireland	1,115	0.36	32.79	77.85	34
Cote Ivore	1,019	0.33	24.26	23.24	42
Azerbaijan	973	0.32	69.50	161.79	14
Qatar	890	0.29	40.45	100.93	22
UAE	888	0.29	32.89	79.58	27
Benin	854	0.28	23.72	23.67	36

Table 5.3 (*cont.*)

Country	No. of publications	% of all publications	Mean	S.D.	No. of African countries
Bulgaria	851	0.28	30.39	94.32	28
Mali	848	0.27	27.35	29.84	31
Ukraine	829	0.27	36.04	102.21	23
Croatia	808	0.26	42.53	113.97	19
Algeria	791	0.26	31.64	68.05	25
Wales	767	0.25	22.56	54.47	34
Papua New Guinea	756	0.25	26.07	103.54	29
Indonesia	736	0.24	21.65	23.41	34
Mozambique	733	0.24	20.36	28.16	36
Vietnam	696	0.23	20.47	21.96	34
Estonia	694	0.22	27.76	88.15	25
Peru	690	0.22	20.29	40.19	34
Lebanon	671	0.22	23.96	42.55	28
Bangladesh	665	0.22	20.15	22.76	33
Singapore	661	0.21	18.36	29.18	36
Gambia	634	0.21	21.13	18	30
Rwanda	633	0.21	20.42	24.71	31
Sudan	625	0.20	17.86	27.21	35
Cyprus	620	0.20	26.96	92.59	23
Botswana	611	0.20	21.07	40.22	29
Niger	594	0.19	16.05	19.3	37
Jordan	594	0.19	22.00	35.75	27
Namibia	550	0.18	18.97	55.28	29
Lithuania	537	0.17	35.80	108.69	15
Philippines	528	0.17	15.09	18.83	35
Madagascar	509	0.16	15.42	16.31	33
Sri Lanka	504	0.16	16.80	35.54	30
Oman	491	0.16	20.46	36.73	24
Gabon	478	0.15	15.93	16.66	30
Zaire	460	0.15	13.53	13.85	34
Congo	433	0.14	13.53	12.65	32
Guinea	429	0.14	15.89	40.33	27
Kuwait	407	0.13	45.22	103.03	9
North Ireland	358	0.12	17.05	40.42	21

Table 5.3 (*cont.*)

Country	No. of publications	% of all publications	Mean	S.D.	No. of African countries
Bahrain	320	0.10	14.55	19.63	22
Luxemburg	306	0.10	9.56	10.35	32
Syria	300	0.10	12.50	16.99	24
Yemen	299	0.10	17.59	41.88	17
Cambodia	284	0.09	10.52	6.58	27
Togo	271	0.09	11.78	13.3	23
Iraq	270	0.09	13.50	19.65	20
Libya	265	0.09	12.05	31.63	22
Guinea-Bissau	252	0.08	12.00	35.2	21
Uruguay	243	0.08	10.13	11.91	24
Reunion	242	0.08	11.52	18.44	21
Congo, Demo Rep	230	0.07	11.50	29.26	20
Panama	198	0.06	8.61	7.01	23
Ecuador	197	0.06	7.88	10.87	25
Cuba	193	0.06	13.79	32.26	14
Venezuela	188	0.06	8.55	10.37	22
Nepal	182	0.06	8.27	9.44	22
Swaziland	168	0.05	9.88	21.11	17
Sierra Leone	167	0.05	6.68	6.09	25
Costa Rica	164	0.05	7.45	8.58	22
Bolivia	138	0.04	5.75	3.92	24
Central African Republic	98	0.03	7.54	7.82	13
Rep of Congo	90	0.03	7.50	14.39	12
Total	308538	100.00			31

Note: Total includes all partners that are not shown in the table.

Czech Republic, Serbia and Romania) have also taken part in scientific collaboration with Africa. Three South American countries included among these twenty-four countries were Columbia, Argentina and Chile.

Scientific collaboration between specific countries in Africa and international partners is also examined to reveal further dimensions of collaboration (Table 5.4). All the top ten international partnering countries are included in the analysis (Figure 5.3).

In the recent five-year period (2011–15), the USA produced most of its publications in Africa with a few select countries. From this data, scientific association between African countries and the USA can be assumed and explained. South Africa was the major US partner, jointly producing 30 per cent of all publications that the USA had produced with African countries. Egypt and Kenya were the second and third research partners, accounting for 15 per cent and 9 per cent, respectively. With Uganda, the USA produced 5 per cent of the latter's publication. The interest of the USA in Nigeria was evident in 4.2 per cent of the publications. Tanzania had a similar percentage of publications with the USA (3.7 per cent). Other countries that the USA favoured for collaboration were Ethiopia, Ghana, Morocco (3 per cent each), Malawi and Zambia (2 per cent each).

France has chosen Tunisia as its major research partner in Africa, producing the highest number of publications in Africa (22 per cent). Collaboration with Algeria led to the production of 17 per cent of publications, 15 per cent with South Africa and 13 per cent with Morocco. Other African countries with whom the French associated were Egypt (5 per cent), Cameroon, Senegal (4 per cent each), Reunion (3 per cent), Benin, Burkina Faso, Kenya and Madagascar (2 per cent each).

Producing one-third of all its publications in Africa, England has found its best research partner in South Africa. No other country in Africa had published even half of South Africa's publications with England. England made 10 per cent publications with Egypt, 8 per cent with Kenya, 5 per cent each with Tanzania and Uganda, 4 per cent each with Nigeria and Morocco, and 3 per cent each with Malawi, Ghana and Ethiopia.

Saudi Arabia, in association with Egypt, published three-quarters of all its publications in Africa. Egypt has been the biggest African partner of Saudi Arabia. A few other countries with which Saudi Arabia worked in a significant way were Tunisia (7 per cent), Algeria (5 per cent), South Africa (4 per cent) and Morocco (3 per cent). The rest of the countries in Africa did not have much contact with Saudi Arabia for scientific research.

Germany's key partners in Africa, in order of percentage of publications, were South Africa (31 per cent), Egypt (22 per cent), Morocco

Table 5.4 *International partners and Africa, 2011–15*

Country	US No.	US %	France No.	France %	England No.	England %	Saudi Arabia No.	Saudi Arabia %	Germany No.	Germany %
Algeria	345	1.12	4004	16.98	221	1.20	583	4.60	261	2.09
Angola	39	0.13	19	0.08	24	0.13	0	0.00	5	0.04
Benin	161	0.52	466	1.98	115	0.62	7	0.06	95	0.76
Botswana	357	1.16	27	0.11	117	0.63	45	0.35	54	0.43
Burkina Faso	287	0.93	568	2.41	250	1.36	7	0.06	156	1.25
Burundi	20	0.06	12	0.05	16	0.09	0	0.00	7	0.06
Cameroon	502	1.63	907	3.85	281	1.52	26	0.20	396	3.18
Cape Verde	9	0.03	5	0.02	7	0.04	0	0.00	7	0.06
Chad	13	0.04	50	0.21	16	0.09	0	0.00	7	0.06
Comoros	4	0.01	8	0.03	3	0.02	0	0.00	0	0.00
Congo	231	0.75	174	0.74	99	0.54	5	0.04	86	0.69
Djibouti	3	0.01	32	0.14	5	0.03	0	0.00	0	0.00
Egypt	4570	14.80	1214	5.15	1842	9.99	9574	75.46	2748	22.03
Eritrea	16	0.05	6	0.03	12	0.07	2	0.02	8	0.06
Ethiopia	794	2.57	152	0.64	468	2.54	43	0.34	304	2.44
Gabon	126	0.41	271	1.15	86	0.47	3	0.02	222	1.78
Gambia	195	0.63	61	0.26	393	2.13	9	0.07	47	0.38
Ghana	852	2.76	0	0.00	582	3.16	19	0.15	290	2.33
Guinea	253	0.82	105	0.45	176	0.95	9	0.07	65	0.52
Guinea-Bissau	22	0.07	7	0.03	38	0.21	0	0.00	6	0.05
Ivory Coast	0	0.00	0	0.00	0	0.00	0	0.00	0	0.00
Kenya	2717	8.80	392	1.66	1566	8.49	34	0.27	621	4.98
Lesotho	28	0.09	4	0.02	7	0.04	0	0.00	0	0.00
Liberia	70	0.23	13	0.06	17	0.09	0	0.00	8	0.06
Libya	77	0.25	69	0.29	166	0.90	74	0.58	47	0.38
Madagascar	327	1.06	432	1.83	130	0.70	6	0.05	148	1.19
Malawi	736	2.38	73	0.31	594	3.22	6	0.05	90	0.72
Mali	333	1.08	227	0.96	0	0.00	6	0.05	62	0.50
Mauritius	80	0.26	48	0.20	86	0.47	21	0.17	17	0.14
Mauritania	12	0.04	54	0.23	4	0.02	2	0.02	2	0.02
Morocco	871	2.82	2952	12.52	665	3.61	380	3.00	828	6.64
Mozambique	262	0.85	58	0.25	125	0.68	6	0.05	47	0.38
Namibia	196	0.63	81	0.34	133	0.72	6	0.05	155	1.24
Niger	148	0.48	182	0.77	66	0.36	6	0.05	38	0.30
Nigeria	1283	4.15	192	0.81	731	3.96	107	0.84	384	3.08
Rwanda	297	0.96	36	0.15	104	0.56	12	0.09	69	0.55
Reunion	114	0.37	605	2.57	61	0.33	4	0.03	45	0.36
Sao Tome and Principe	3	0.01	0	0.00	5	0.03	0	0.00	0	0.00
Senegal	387	1.25	849	3.60	190	1.03	7	0.06	93	0.75
Seychelles	63	0.20	50	0.21	49	0.27	7	0.06	15	0.12
Sierra Leone	129	0.42	11	0.05	67	0.36	2	0.02	21	0.17
Somalia	10	0.03	3	0.01	10	0.05	0	0.00	0	0.00
South Africa	9348	30.27	3407	14.45	5998	32.52	471	3.71	3919	31.42
Sudan	179	0.58	70	0.30	167	0.91	267	2.10	155	1.24
Swaziland	60	0.19	4	0.02	36	0.20	3	0.02	2	0.02
Tanzania	1154	3.74	144	0.61	946	5.13	16	0.13	319	2.56
Togo	47	0.15	135	0.57	0	0.00	0	0.00	29	0.23
Tunisia	489	1.58	5071	21.51	228	1.24	870	6.86	283	2.27
Uganda	1634	5.29	168	0.71	918	4.98	20	0.16	195	1.56
Zambia	650	2.10	48	0.20	311	1.69	16	0.13	62	0.50
Zimbabwe	380	1.23	113	0.48	311	1.69	6	0.05	54	0.43
Total	30883	100.00	23579	100.00	18442	100.00	12687	100.00	12472	100.00

(7 per cent), Kenya (5 per cent), Nigeria, Cameroon and Tanzania (3 per cent each). Australia published mostly with South Africa (47 per cent), Egypt, Morocco (7 per cent each) and Kenya (5 per cent). The greatest African partner of Australia was South Africa, with which it produced nearly half of its publications in Africa. Italy chose to work mostly with South Africa (28 per cent), Egypt (17 per cent), Morocco (11 per cent), Tunisia (10 per cent), Algeria (4 per cent), Cameroon, Kenya and Nigeria (3 per cent each). Spain, which was not among the top ten countries of international

Table 5.4 *(cont.)*

Country	Australia No.	Australia %	Italy No.	Italy %	Spain No.	Spain %	Canada No.	Canada %	South Africa No.	South Africa %
Algeria	43	0.59	320	4.36	406	5.55	189	2.59	85	1.20
Angola	10	0.14	8	0.11	27	0.37	4	0.05	18	0.25
Benin	28	0.38	47	0.64	40	0.55	70	0.96	47	0.66
Botswana	38	0.52	33	0.45	17	0.23	55	0.75	218	3.07
Burkina Faso	51	0.69	82	1.12	39	0.53	81	1.11	84	1.18
Burundi	3	0.04	6	0.08	0	0.00	2	0.03	5	0.07
Cameroon	109	1.48	227	3.09	56	0.77	145	1.99	352	4.96
Cape Verde	3	0.04	5	0.07	25	0.34	0	0.00	0	0.00
Chad	2	0.03	6	0.08	3	0.04	8	0.11	3	0.04
Comoros	0	0.00	0	0.00	0	0.00	0	0.00	0	0.00
Congo	30	0.41	32	0.44	15	0.21	23	0.32	65	0.92
Djibouti	0	0.00	4	0.05	3	0.04	6	0.08	0	0.00
Egypt	488	6.65	1255	17.10	1020	13.95	1229	16.84	223	3.14
Eritrea	3	0.04	15	0.20	6	0.08	2	0.03	7	0.10
Ethiopia	117	1.59	158	2.15	112	1.53	82	1.12	233	3.28
Gabon	25	0.34	23	0.31	41	0.56	23	0.32	27	0.38
Gambia	51	0.69	35	0.48	38	0.52	17	0.23	45	0.63
Ghana	156	2.12	77	1.05	73	1.00	134	1.84	315	4.44
Guinea	373	5.08	43	0.59	152	2.08	37	0.51	35	0.49
Guinea-Bissau	12	0.16	15	0.20	7	0.10	5	0.07	4	0.06
Ivory Coast	0	0.00	0	0.00	0	0.00	0	0.00	0	0.00
Kenya	364	4.96	241	3.28	185	2.53	431	5.91	773	10.88
Lesotho	9	0.12	0	0.00	0	0.00	3	0.04	42	0.59
Liberia	4	0.05	3	0.04	2	0.03	5	0.07	4	0.06
Libya	43	0.59	49	0.67	25	0.34	43	0.59	11	0.15
Madagascar	55	0.75	50	0.68	22	0.30	38	0.52	74	1.04
Malawi	97	1.32	47	0.64	28	0.38	78	1.07	293	4.13
Mali	42	0.57	43	0.59	28	0.38	37	0.51	34	0.48
Mauritius	31	0.42	17	0.23	22	0.30	20	0.27	50	0.70
Mauritania	6	0.08	7	0.10	11	0.15	4	0.05	5	0.07
Morocco	540	7.35	837	11.40	1280	17.51	687	9.41	526	7.41
Mozambique	77	1.05	45	0.61	179	2.45	19	0.26	149	2.10
Namibia	88	1.20	36	0.49	27	0.37	33	0.45	304	4.28
Niger	29	0.39	17	0.23	15	0.21	18	0.25	35	0.49
Nigeria	206	2.81	238	3.24	134	1.83	170	2.33	958	13.49
Rwanda	25	0.34	17	0.23	14	0.19	44	0.60	91	1.28
Reunion	41	0.56	43	0.59	35	0.48	22	0.30	81	1.14
Sao Tome and Principe	0	0.00	0	0.00	0	0.00	0	0.00	0	0.00
Senegal	45	0.61	59	0.80	76	1.04	67	0.92	86	1.21
Seychelles	29	0.39	8	0.11	14	0.19	9	0.12	27	0.38
Sierra Leone	15	0.20	11	0.15	8	0.11	22	0.30	20	0.28
Somalia	4	0.05	2	0.03	2	0.03	0	0.00	0	0.00
South Africa	3463	47.17	2069	28.19	2008	27.47	2481	34.00	0	0.00
Sudan	54	0.74	57	0.78	26	0.36	33	0.45	111	1.56
Swaziland	11	0.15	0	0.00	0	0.00	7	0.10	91	1.28
Tanzania	162	2.21	146	1.99	100	1.37	122	1.67	356	5.01
Togo	0	0.00	0	0.00	0	0.00	0	0.00	0	0.00
Tunisia	97	1.32	706	9.62	864	11.82	394	5.40	50	0.70
Uganda	152	2.07	121	1.65	73	1.00	286	3.92	431	6.07
Zambia	56	0.76	44	0.60	24	0.33	47	0.64	245	3.45
Zimbabwe	55	0.75	36	0.49	29	0.40	66	0.90	489	6.89
Total	7342	100.00	7340	100.00	7311	100.00	7298	100.00	7102	100.00

publishers during 1945–2015, worked with countries such as South Africa (27 per cent), Morocco (18 per cent), Egypt (14 per cent), Tunisia (12 per cent) and Algeria (6 per cent). The percentages of publications with other countries were very small.

Canada chose two countries in Africa for most of its joint publications. It worked with South Africa (34 per cent) and Egypt (17 per cent) for its major share of (51 per cent) of publications in Africa. Some other partners of Canada were Morocco (9 per cent), Kenya (6 per cent), Tunisia (5.4 per cent), Uganda (4 per cent) and Algeria (3 per cent).

South Africa is the only African country to appear on the list of the first ten countries of international partners in Africa. The preferred countries for South Africa on the continent were Nigeria (13 per cent), Kenya (11 per cent), Morocco, Zimbabwe (7 per cent each), Uganda (6 per cent), Cameroon, Tanzania (5 per cent each), Ghana, Malawi and Namibia (4 per cent each).

Changes in International Partnership

There were a total of 561,217 publications during 1945–2015 which involved international partners. For the selected five-year period of 2011–15, there were a total of 308,538 publications. That is, 55 per cent of the publications, according to the classification of international partners, were published in the most recent five-year period. Were there any shifts in international partnership in Africa? Two data sets (1945–2015 and 2011–15) should be compared for this information.

The percentage of publications is an indicator in this analysis. The first ten countries can be considered for examination. The first ten countries with which Africa maintained contacts remained more or less the same in the second period. The only change was that two new countries, namely, Spain and South Africa, made it into the list of top ten. They were not among the first ten countries during 1945–2015 and moved the Netherlands and Belgium to eleventh and thirteenth positions respectively.

During 1945–2015, the USA produced 12 per cent of all publications in Africa that had international partnership. This had declined to 10 per cent in the past five years (2011–15). For France, the percentage had been reduced by 3 per cent. England also declined its publication count in the recent five years. But for Saudi Arabia there was an improvement from 3 per cent to 4 per cent. No significant change was observed in the percentage of publications of Australia or Italy but Canada lost slightly. The Netherlands and Belgium declined their percentages of publication during 2011–15. When the first twenty countries were examined, a few more changes were observed. Spain strengthened both its position (from twelfth to eighth) and share of publications. Switzerland, Japan, Sweden and Scotland decreased in percentage while China, India, Brazil and Kenya increased their respective shares of publications during 2011–15.

Have there been any increases or decreases in the average number of countries in Africa with which international collaborators published? The average has decreased from thirty-nine to thirty-one countries recently.

For the USA, its African partners with which most of its publications in Africa were published have not changed. South Africa, Egypt and Kenya remained the first three partners of the USA. The USA works more closely now with Uganda than Nigeria if the percentage of publications is any indication. The USA improved its association with Tanzania, Ethiopia and Ghana by producing slightly more publications in the second period. The key partners of France remained Tunisia, Algeria and Morocco. However, it increased its production with Tunisia and South Africa while it decreased publications with Morocco. With Cameroon, Egypt and Senegal, France maintained its share of publications in Africa as for the 1945–2015 period.

In the recent period, England was the third most common country published with African countries. With South Africa, its share of publications did not change between the two periods. In second place of partnership was Kenya and not Egypt as in the first period. With Nigeria, England's share declined by 2 per cent and with Uganda it increased by 1 per cent.

Saudi Arabia moved up one position, becoming the fourth country in terms of number of publications with other Africa countries. No change occurred in the percentage of publications of Saudi Arabia with Egypt, South Africa, Algeria and Morocco between the two periods. Germany occupied the fifth position. Germany also maintained its scientific association with South Africa, Egypt and Kenya but this slightly declined with Nigeria. It improved its links with Morocco and Tanzania.

South Africa continued to be the best partner of Australia, producing about half of the latter's publications in Africa. The new partners of Australia included Egypt, Morocco (7 per cent each) and Guinea (5 per cent) with a sizable percentage of publications. Italy continued to work mostly with their main partners such as South Africa, Egypt, Tunisia (increased in the recent years), Morocco, Cameroon (at the same level), Algeria, Nigeria, and Kenya (declined slightly).

Canada lagged behind, dropping from sixth top international partner of Africa during 1945–2015 to ninth top partner during 2011–15.

Canada increased its share of publications with South Africa, Egypt, Morocco and Uganda slightly and decreased with Kenya.

Collaboration in Major Sub-Saharan African Countries

The analysis of the scientometric data presented in Chapter 4 revealed that some African countries lead in the production of science in Africa: Algeria, Egypt, Ethiopia, Kenya, Nigeria, Morocco, South Africa, Tanzania, Tunisia and Uganda. In terms of their contributions to the total production in Africa, South Africa comes first with 25 per cent of the total production in Africa. Egypt had 22 per cent, Tunisia 9 per cent, Algeria 6 per cent, Nigeria 5.5 per cent, Morocco 4.4 per cent, Kenya 3.8 per cent, Ethiopia 2.3 per cent, Uganda 2.1 per cent and Tanzania 2 per cent. A discussion on these major countries can offer insights into scientific development within these countries on the continent.

Pursuing the analysis of collaboration further, a few major science-producing countries in sub-Saharan Africa were selected for a deeper level of analysis. The countries chosen were Ethiopia, Kenya, Nigeria, South Africa and Tanzania as having the highest number of publications in sub-Saharan Africa. A sampling procedure was adopted to make data analysis manageable. Publication records of every five years for each of these countries were selected. Going back from 2015, data was gathered and entered manually into a software programme. The sampled years were 2015, 2010, 2005, 2000, 1995, 1990, 1985, 1980 and 1975. Sampling was stopped at 1975, as not many publications were stored in the database prior to this date. The publication records captured for analysis for the sampled years were 2,482 for Ethiopia, 5,225 for Kenya, 9,955 for Nigeria, 35,112 for South Africa and 2,451 for Tanzania. Altogether 55,225 publications were analysed for the existing forms of collaboration in these countries (Map 5.1).

Before collaboration is examined, a brief introduction to relevant features of these countries is appropriate.

Ethiopia

Ethiopia is the oldest independent country in Africa which maintained its freedom from colonial rule during the scramble for Africa (Mouton

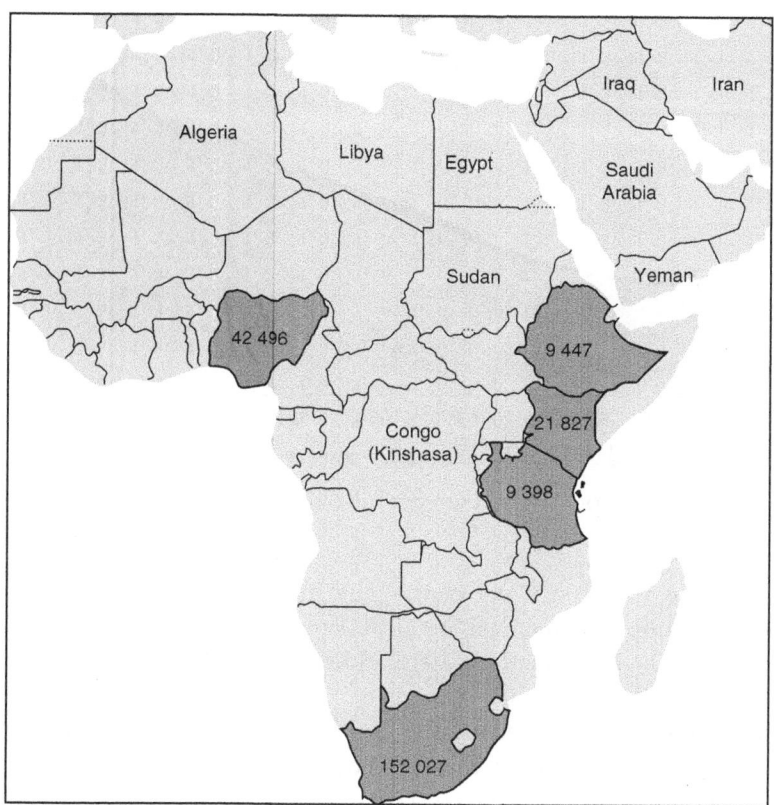

Map 5.1 Publications of top five countries in sub-Saharan Africa, 1945–2015

and Boshoff, n.d.). It became an independent country in 1941, the first country in Africa to obtain freedom.

Ethiopia has been ranked as one of the fastest-growing economies in the world (AfDB et al., 2015). The economy grew at 10.3 per cent during 2013–14, making Ethiopia one of the better-performing countries in Africa. The trend in growth was expected to continue in 2015 and 2016 (AfDB et al., 2015). The decadal GDP growth rate was 10.8 per cent per annum. In the last decade, Ethiopia witnessed some of the fastest growth in Africa (Urama et al., 2015). Its real per capita GDP growth rate increased from 1.4 per cent in the 1980s to 2.3 per cent in the 1990s, and 6.7 per cent in 2010–14 (UNCTAD, 2015). In 2017, the annual

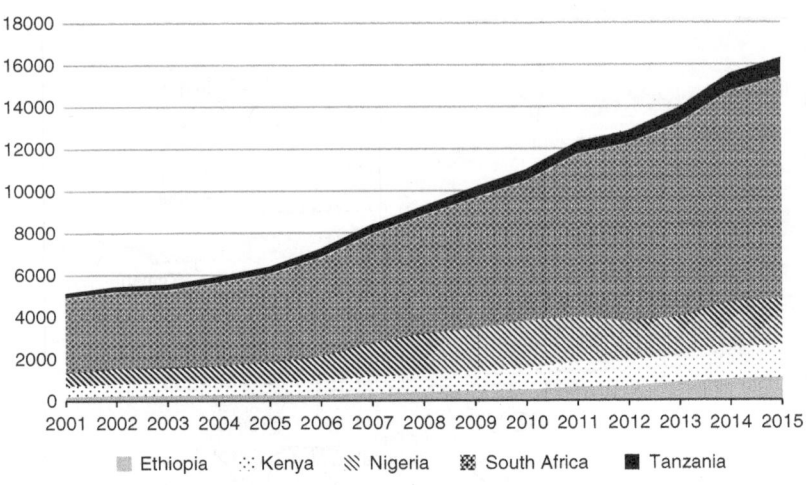

Figure 5.1 Publication trends in top five sub-Saharan countries, 2001–15

growth rate was 10.24 per cent.[5] The increasing rates of growth allowed Ethiopia to increase its GERD to GDP from 0.24 per cent in 2009 to 0.61 per cent in 2013 (Soete et al., 2015). Its growth rates are well above the average of sub-Saharan Africa (World Bank, 2016). In key economic indicators (KEI), Ethiopia has a poor ranking of 136 of the 140 countries (World Bank, 2007), declining between 1995 and 2007 from 134 (Figure 5.1). In 2010, Ethiopia had a total of 8,279 R&D personnel and 3,701 researchers, which translated into 100 R&D personnel per million inhabitants and 45 researchers per million inhabitants (NPCA, 2014). This has increased to 18,438 R&D personnel, 8,221 researchers and 194 R&D personnel per million inhabitants in 2013, as per the latest available data.[6]

With a modest beginning in 1950, Ethiopia built its universities to produce the technical and skilled manpower required for the country. Ethiopia also solicited private sector participation to contribute to its higher education system. The expansion in the higher education system was visible in the growing number of academic staff and students between 1996–7 and 2003–4 (Mouton and Boshoff, n.d.). The higher education system in Ethiopia has passed through an extensive reform

---

[5] http://data.uis.unesco.org. Accessed 2 December 2019.   [6] As in footnote 1.

since 2003 (Ayalew, 2012), which had its own effects on the production of knowledge in recent years at higher learning institutions in the country. Between 1994 and 2009, student enrolment increased sharply (2,782 per cent) at both undergraduate and graduate levels (Ayalew, 2012).

Ethiopia was among the few countries in Africa to invest hugely in its higher education system, which had a ripple effect on the production of science and knowledge in the country. The average higher education budget between 2005–6 and 2009–10 was 24 per cent of its education budget, and its GDP percentage for the education sector grew from 3.1 per cent to 7 per cent in the same period (Teferra, 2013). There has been a great surge in the production of university graduates as well. But the rapid expansion of the higher education system also has its effects manifested in the shortage of experienced academic staff, inadequate infrastructure, weak research output and declining quality of education (Ayalew, 2012).

In 2014, Ethiopia placed universities specialising in S&T under the new ministry of S&T to promote innovation and technology-driven enterprises (Urama et al., 2015).

Ethiopia also had its share of problems in regard to its capacity for scientific research. Adequate financial support for research, weak incentive structure, inefficient administration at universities and research centres, funding for equipment and supplies, and lack of effective international communication were among the constraints that affect research capacity (Thulstrup et al., 1996).

Kenya

In the production of scientific publications, Kenya is another major country in sub-Saharan Africa after South Africa and Nigeria. In a multi-wave study of agricultural scientists in Kenya, Schafer et al. (2016) showed that the publication productivity increased particularly between 2005 and 2010. However, in relation to population size, Kenya's productivity of publications was not great (Irikefe et al., 2011). Kenya became an independent country in 1963. It was the most politically stable country in East Africa in the 1970s, which was advantageous for the university system (Eisemon, 1986). Since the 1970s, the Kenyan university system has expanded its role in research (Schafer et al., 2016).

The Kenyan economy has grown in recent years. It had a GDP growth of 6.9 per cent in 2012 and 5.7 per cent in 2013 (AfDB et al., 2015). In 2017, it declined to 4.9 per cent (http://data.uis.unesco.org). In 2010, as the latest data that is available for Kenya, Kenya had a total of 42,566 R&D personnel and 9,305 researchers, which translated into 1,051 R&D personnel per million inhabitants (NPCA, 2014). This improved in the later years. The second *African Innovation Outlook* (NPCA, 2014) reported that Kenya is among the three top countries with the largest absolute numbers of R&D personnel per every million inhabitants in the population.

With the enacting of the Science and Technology Act of 1977, the role of S&T in the Kenyan economy became prominent and steps were taken to strengthen the country's S&T system. The ministry of S&T, which was renamed or amalgamated with other departments several times, takes care of the coordination of STI policy, research, and higher and technical education (Esau, n.d.). The national science policy in Kenya in 1980 prescribed 1 per cent of GDP for R&D but it remained at 0.5 per cent (Eisemon and Davies, 1997). In 2010, the figure was 0.8 per cent.[7]

The experience of Kenya brings into focus the challenge of not achieving expected economic growth despite investments in S&T (ROK, 2012). First of all, Kenya's National Council for Science and Technology (NCST) functioned only as a department in a ministry, which made the Council ineffective to coordinate various R&D activities in the country. Secondly, there was no national research agenda for researchers in the country to follow; as a consequence of which researchers and research institutions moved on with their research programmes without knowing what others were doing. This also caused non-alignment of the research programme with national development goals. Thirdly, the link between research institutions and industry was weak, resulting in the lack of commercialisation of research outcomes and knowledge. Fourthly, low levels of funding available at the national level made researchers rely on foreign funding and pursue a foreign agenda in scientific research. Finally, there were challenges pertaining to inadequate expertise, reliance on foreign experts, lower levels of science culture among the population, poor

---

[7] https://tradingeconomics.com/kenya/research-and-development-expenditure-pe rcent-of-gdp-wb-data.html. Accessed 2 December 2019.

performance in mathematics and science subjects at schools, weak and improper mechanisms for the implementation and evaluation of STI initiatives along with others.

The data referring to 1985–91, suggested that scientific research in universities and government institutions had declined from the previous years (Eisemon and Davies, 1997). Unlike many other African countries, scientific research in Kenya is undertaken at government-owned research institutes, such as Kenya Medical Research Institute (KEMRI) and Kenya Agricultural Research Institute (KARI), and a good share of their funding comes from international collaborators (Irikefe et al., 2011). Kenya, in relation to other African countries, received greater support from foreign sources for its scientific research and training (Eisemon and Davis, 1997). In the early 1990s, Kenya chose its research priorities in areas of agricultural, health, industrial, natural and physical and social sciences (Esau, n.d.). The priorities of Kenya, as seen from its national policies on STI, are agriculture, human resource development, industry and entrepreneurship, physical infrastructure, energy, environment and natural resources, education and training, ICT, and health and life sciences.

Nigeria

Nigeria is a major player in the production of scientific knowledge on the continent, particularly in the sub-Saharan region. As shown in our analysis of data presented in previous chapters, it comes after South Africa in the production of scientific publications. Nigeria has a strong scientific system. The commitment to the advancement of S&T for national development is enshrined in its 1999 constitution and its educational policy underlines this (Kashim and Adelabu, 2010). In the KEI ranking, Nigeria has not made any positive changes but declined between 1995 and 2007 (from 112 to 115) (World Bank, 2007). The Nigerian economy recorded sustained economic growth for the past decade with an annual increase of GDP around 7 per cent (AfDB et al., 2015). Despite the volatility of oil prices, the outlook for 2015 was a growth of 5 per cent in 2015 (AfDB et al., 2015). Nigeria is publishing far less in relation to its GDP than many other African countries, and it has fewer researchers (Irikefe et al., 2011).

As a consequence of colonial policies, Nigeria faced an acute shortage of manpower at all levels immediately after it gained political independence in 1960 (Ajeyalemi and Balyelo, 1990). In the earlier days, Nigeria did not have a dedicated ministry of S&T. A ministry was evolved later in 1980 from the Nigerian Council for Science and Technology and the Nigerian Council for Science and Technology. It had the infrastructure to support scientific journals and professional societies and the mechanism for national policymaking (Rabkin et al., 1979). It also had access to a large number of scientists and engineers in the 1990s (Gaillard, 1992a).

In British colonial Nigeria, scientific research for practical applications was centred around tropical medicine and agriculture. There were research institutions meant for research in medicine that were established during the colonial period. Close to independence, the Nigerian scientific system began to develop (Chatelin et al., 1997). This continued even in the 1980s. The scientific system and the working conditions of Nigerian scientists were dismal in the beginning years of its independence. These were featured by erratic government funding to universities and lack of basic facilities (including water, electricity, tools of communication and equipment) which had an impact on scientists' productivity (Adamson, 1992).

During the 1970s, the Nigerian scientific community, in relation to the oil boom, grew impressively (Chatelin et al., 1997). Publications of researchers based at major universities and research institutes doubled during this time (Lebeau et al., 2000). Nigeria thus became the giant of Africa. But later, in the mid-1980s, as a consequence of weakening currency, budgetary provisions for scientific equipment, books and journals were curtailed (Chatelin et al., 1997). The crisis in the country was aggravated by other institutional challenges. Nigerian research institutions were no longer in a position to devise research strategies and lost touch with the international scientific community (Lebeau et al., 2000). Between 1987 and 1995, Nigeria lost a substantial number of academics (Lebeau et al., 2000).

Nigerian academic scientists, as Ehikhamenor (1988) reported, were constrained by lack of essential equipment to conduct research, finding it hard to replace parts, maintain equipment and get reagents and chemicals for experiments due to dwindling budget allocation. The lack of updated resources in the libraries and the lack of necessary institutional support added to their woes. The working and living conditions of academics were deplorable (Gaillard, 2000; Hudu, 2015; Lebeau et al., 2000). This led the highly qualified academic staff at universities to leave the profession and seek employment in other sectors (Hudu, 2015).

During 1987–2001, there was a dramatic decline in the production of scientific publications that originated from Nigeria (Gaillard, 2003a). Gaillard's (2003a) analysis of the ISI publications showed that the decline was 50 per cent. According to him, this is due to the dilapidation of research infrastructure by budgetary cuts, absence of career prospects for scientists, and emigration or change of profession of scientists. Another empirical study of academics in science disciplines found that deficiencies in infrastructure, particularly ICTs, affected science and its internationalisation (Ani and Biao, 2005). The financial crisis that extended even to the 2000s caused great damage to scientific research and research outputs. Well-established and trained researchers could not look for public money to conduct their research and the higher education system in the country was in a state of abandonment (Lebeau, 2003).

To address all these issues, Nigeria had to concentrate on the education system outlined in its national development plans beginning in 1962 (Ajeyalemi and Balyelo, 1990). Nigeria had, from early on, focused on science education from the primary school level. Through programmes – among them the African Primary Science Programme, the Mid-west Primary Science Project, changes in the curricula for integrated science and mathematics subjects, the Nigeria Secondary School Science Project (Ajeyalemi and Balyelo, 1990) – the interest of the nation in science education and to produce a new generation of students keen on science subjects was generated. Nigeria's Integrated Science Curriculum was aimed at laying a basic foundation for literacy in science, scientific concepts, scientific skills and scientific attitudes (Ajeyalemi and Balyelo, 1990). The university system in Nigeria underwent expansion in the years following its independence in 1960 (Kolinsky, 1985). This expansion was to continue in the 1970s with the increase in oil prices. The system, in the view of Bamiro (2012), suffered from lack of quality caused by the shortfall in the number of academic staff and inadequate teaching and research facilities.

South Africa

Undeniably, South Africa is a giant in scientific research on the continent and the biggest science producer in Africa. In the 1990s,

South Africa produced about a quarter of Africa's scientific publications (Meyer, 1997). South Africa is the primary producer of science in Africa, contributing to 30 per cent of all publication output of Africa during 2000–4 (Pouris and Pouris, 2009).

South Africa maintained its position as a leading country in Africa, not only in the production of scientific publications but also in having a strong scientific system. It has a solid and consolidated scientific infrastructure, strong scientific communities, critical mass in several sectors of science, an effective incentive system to promote research and has pursued a sound S&T policy (Waast and Krishna, 2003a). In 2010, South Africa had a total of 29,486 R&D personnel and 18,719 researchers, which translated into 588 R&D personnel per million inhabitants and 373 researchers per million inhabitants (NPCA, 2014). In 2016, the figures were 80,029 R&D personnel, 56,761 researchers and 1,428 R&D personnel per million inhabitants.[8] In 1990, it spent only 0.61 per cent of its GDP on R&D but this percentage increased over the years although it has not yet reached the 1 per cent target. The economy has not been doing well for some time now. The growth rate declined to 1.5 per cent in 2014 and was expected to improve to 2 per cent in 2015 (AfDB et al., 2015). In 2014, South Africa had 1,108 R&D personnel per million inhabitants, placing it second after Egypt and Kenya with the highest number of R&D personnel proportionate to the population per country in Africa (NPCA, 2014). South Africa is the top African country in the KEI. However its rank dropped from 41 in 1995 to 50 in 2007 (World Bank, 2007). It spent 0.7 per cent of the GDP on R&D in 2012, which increased to 0.8 per cent in 2016.

South Africa has the most comprehensive and well-developed higher education system in sub-Saharan Africa (Atuahene, 2011). The country is credited with universities that have strong research capacity and a private sector with substantive and resourceful R&D capabilities (Scholes et al., 2008). However, South Africa has a serious shortage of skilled personnel in science and faces the challenge of attracting large number of students to science (Philander, 2009).

---

[8] All head count. http://data.uis.unesco.org. Accessed 2 December 2019.

Tanzania

Tanzania obtained freedom from the British in 1961. Of late, it has become one of the best-performing economies in Africa (THDR, 2015; UNCTAD, 2015).[9] The economy in the United Republic of Tanzania has performed well, with a growth rate of 7.3 per cent in 2013 (AfDB et al., 2015). As per UNESCO statistics, the annual rate in 2017 was 6.8 per cent (http://data.uis.unesco.org). The ideal target of spending 1 per cent of the GDP for R&D has not been accomplished yet. The percentage lingered around 0.35 in the early 1990s (Widstrand, 1992). The latest available figure indicates that it was 0.5 per cent in 2013 (http://data.uis.unesco.org). Tanzania is one of the few African countries that improved their standing in the KEI index between 1995 and 2007. This was due to the progress it made in one of sub-indices, namely, innovation (World Bank, 2007).

In 2010 Tanzania had a total of 2,928 R&D personnel and 1,599 researchers, which translated into 65 R&D personnel per million inhabitants and 37 researchers per million inhabitants (NPCA, 2014). According to the latest available data, there were 5,418 R&D personnel, 3,064 researchers and 107 R&D personnel per million inhabitants in 2013 (http://data.uis.unesco.org). The R&D institutions in the country have not been able to attract or retain skilled personnel due to lack of effective incentive systems in place (COSTECH, 2015). The research output of Tanzanian researchers remained low in relation to the size of its scientific community (Gaillard, 2000; 2003b). During 1987–2001, as the study of ISI publications by Gaillard (2003a) suggested, the production of scientific publications in Tanzania reported growing features. However, conditions are not conducive for scientific research. The current national priorities for the country are food and agriculture, industry, energy, natural resources, environment, health, transport and communication, science and innovation (ECA, 2016).

Tanzanian scientific production is heavily dependent on collaboration, both international and continental (within Africa) (Gaillard, 2000). Not only are they dependent on foreign countries for publications but also for foreign support to go for training, fund their research and contribute to their incomes (Gaillard, 2003b). Tanzanian researchers, to a large extent, depend on foreign aid and co-publications.

---

[9] The GDP grew from 0.9 per cent in the 1990s to 4 per cent in 2000s, and in 2010–14 it was 4.1 per cent (UNCTAD, 2015).

Gaillard's (2000) report on Tanzania makes this clear. Scientific cooperation with European countries and the USA has contributed to this growth in Tanzania wherein about three-quarters were co-authored with foreign authors (Gaillard, 2003a). This is in agreement with the data analysed here as well.

Having seen some of the features of the countries in terms of science production let us go back to collaboration and its forms that exist in these big five countries in sub-Saharan Africa.

No Collaboration

In Ethiopia, for all the selected years, single-authored papers formed 8.5 per cent of all its publications. This means that the rest of the publications were collaborations (panel All, Table 5.5). When all the five countries (Ethiopia, Kenya, Nigeria, Tanzania and South Africa) are considered, 14 per cent of the papers did not have any type of collaboration. Rather, they were all single-authored papers. The

Table 5.5 *Forms of collaboration in major sub-Saharan countries, 1975–2015*

Country	1975 No collaboration No.	%	Domestic No.	%	International No.	%	All No.	%	1980 No collaboration No.	%	Domestic No.	%
Ethiopia	7	26.9	7	26.9	12	46.2	26	100	12	28.6	20	47.6
Kenya	66	36.5	67	37.0	56	30.9	181	100	98	38.6	106	41.7
Nigeria	191	45.9	62	23.0	58	13.9	416	100	370	47.0	347	44.3
South Africa	391	32.3	734	65.2	108	8.9	1212	100	628	34.4	1044	62.4
Tanzania	24	49.0	14	28.6	15	30.6	49	100	25	43.1	21	36.2
Total	679		884		247		1884		1133		1538	

Country	1980 International No.	%	All No.	%	1985 No collaboration No.	%	Domestic No.	%	International No.	%	All No.	%
Ethiopia	12	28.6	42	100	20	31.3	32	50.0	14	21.9	64	100
Kenya	56	22.0	254	100	89	28.0	175	7.2	83	26.1	318	100
Nigeria	69	8.8	788	100	424	42.9	462	46.8	116	11.7	988	100
South Africa	169	9.2	1828	100	623	26.5	1470	70.2	291	12.4	2355	100
Tanzania	14	24.1	58	100	27	27.3	49	49.5	26	26.3	99	100
Total	319		2970		1183		2188		529		3824	

Country	1990 No collaboration No.	%	Domestic No.	%	International No.	%	All No.	%	1995 No collaboration No.	%	Domestic No.	%
Ethiopia	34	26.8	60	47.2	43	33.9	127	100	23	14.6	86	54.4
Kenya	91	23.5	205	53.0	119	30.7	387	100	64	14.1	210	46.2
Nigeria	298	32.0	532	57.1	121	13	932	100	150	23.8	368	58.4
South Africa	549	20.0	1875	77.4	392	14.3	2748	100	491	17.5	1728	77.9
Tanzania	23	18.7	57	46.3	55	44.7	123	100	17	9.6	82	46.3
Total	995		2729		729		4317		745		2474	

Table 5.5 (*cont.*)

Country	1995 International		All		2000 No collaboration		Domestic		International		All	
	No.	%	No.	%	No.	%	No.	%	No.	%	No.	%
Ethiopia	52	32.9	158	100	29	13.6	89	41.6	116	54.2	214	100
Kenya	218	47.9	455	100	53	10.8	199	40.4	317	64.4	492	100
Nigeria	123	19.5	630	100	147	19.0	467	60.5	214	27.7	773	100
South Africa	666	23.8	2801	100	502	14.9	1963	79.6	1172	34.8	3365	100
Tanzania	104	58.8	177	100	26	12.4	86	40.8	143	67.8	211	100
Total	1161		4221		757		2804		1102		5055	

Country	2005 No collaboration		Domestic		International		All		2010 No collaboration		Domestic	
	No.	%	No.	%	No.	%	No.	%	No.	%	No.	%
Ethiopia	19	6.8	98	34.9	200	71.4	281	100	29	5.6	265	51.5
Kenya	30	5.3	189	33.4	462	81.6	566	100	29	2.8	401	38.8
Nigeria	133	13.1	683	67.4	288	28.4	1014	100	204	9	1638	72.5
South Africa	479	11.5	2229	82.3	1803	43.1	4183	100	518	8.5	4197	87.7
Tanzania	19	6.0	120	37.5	274	85.6	320	100	14	2.6	230	42.7
Total	680		3319		1871		6364		794		6731	

Country	2010 International		All		2015 No collaboration		Domestic		International		All	
	No.	%	No.	%	No.	%	No.	%	No.	%	No.	%
Ethiopia	349	68.0	515	100	37	3.5	682	64.6	745	70.6	1055	100
Kenya	845	81.7	1034	100	22	1.4	895	58.2	1397	90.8	1538	100
Nigeria	646	28.6	2258	100	106	4.9	1420	65.9	1240	57.5	2156	100
South Africa	3077	50.8	6096	100	454	4.3	7169	94.0	6506	63.5	10524	100
Tanzania	467	86.6	539	100	19	2.2	550	62.9	778	89.0	875	100
Total	5414		10442		638		10716		10640		16148	

Country	All No collaboration		Domestic		International		All	
	No.	%	No.	%	No.	%	No.	%
Ethiopia	210	8.5	1339	53.9	1543	62.2	2482	100
Kenya	542	10.4	2447	46.8	3553	68.0	5225	100
Nigeria	2023	20.3	5979	61.0	2875	28.9	9955	100
South Africa	4635	13.2	22409	82.6	14184	40.8	35112	100
Tanzania	194	7.9	1209	49.3	1876	76.6	2451	100
Total	7604		33383		24031		55225	

Chi-square results

1. Ethiopia: Single authored publication (no collaboration), $p=0.000$, domestic collaboration $p=0.000$, international collaboration $p=0.000$, and multicountry collaboration $p=0.000$.

2. Kenya: Single authored publication (no collaboration), $p=0.000$, domestic collaboration $p=0.000$, international collaboration $p=0.000$, and multicountry collaboration $p=0.000$.

3. Nigeria: Single authored publication (no collaboration), $p=0.000$, domestic collaboration $p=0.000$, international collaboration $p=0.000$, and multicountry collaboration $p=0.000$.

4. South Africa: Single authored publication (no collaboration), $p=0.000$, domestic collaboration $p=0.000$, international collaboration $p=0.000$, and multicountry collaboration $p=0.000$.

5. Tanzania: Single authored publication (no collaboration), $p=0.000$, domestic collaboration $p=0.000$, international collaboration $p=0.000$, and multicountry collaboration $p=0.000$.

percentages for other countries were 10 (Kenya), 20 (Nigeria), 13 (South Africa) and 7 (Tanzania). The lowest percentage of single-authored publications was reported for Tanzania (8 per cent) and the highest percentage for Nigeria (20 per cent). In recent years (for 2015), the percentages of single-authored publications were even smaller, ranging between 1.4 per cent (Kenya) and 4.9 per cent (Nigeria). South Africa produced 4.3 per cent single-authored publications while Ethiopia had 3.5 per cent. Tanzania made only 2.2 per cent.

Within countries certain trends can be discerned. In Ethiopia, the number of single-author publications declined substantially after 1995. Until then, more than a quarter of all publications were single-authored ones. In 1995, it had almost halved to 15 per cent. It continued to decline, reaching 3.5 per cent in 2015. In Kenya, the percentage of single-authored publications began to reduce from 1985. It lost 10 per cent from the previous year of 1980. By 2015, Kenya had only 1.4 per cent single-authored publications. Nigeria, in 1975, had 46 per cent of publications that were single-authored. This proportion of single-authored papers to joint-authored papers showed a gradual decrease since 1990, reaching 5 per cent in 2015. This percentage in 2015 was however greater than that of Ethiopia and Kenya. For South Africa, the highest percentages of single-authored publications were produced in 1975 and 1980 (about one-third of all publications). In the following years, the percentage of single-authored publications had begun to shrink. By 2015, the percentage had become 4.3. Tanzania, when it began in 1975, had about half of its publications produced by single scholars. Among the five countries, Tanzania had the highest percentages of single-authored publications. By 1985, it had dropped to about one-third of all publications. In 2015, Tanzania made only 2.2 per cent single-authored publications. It became the country after Kenya in sub-Saharan Africa with the second-lowest percentage of single-authored publications. Association between year and joint-authored publications was statistically significant in a Chi-square test.

Domestic Collaboration

Collaboration within countries for these five countries for the selected period showed a major share of publications. Sixty per cent of the total publications for all the five countries for all the sampled years were the product of domestic collaboration. The highest percentage was found in South Africa (83 per cent), followed by Nigeria (61 per cent), Ethiopia (54 per cent), Tanzania (49 per cent) and Kenya (47 per cent).

The figures for the more recent years (in 2015) show a somewhat different trend. All the countries had about two-thirds or more of publications emerging out of domestic collaboration South Africa led with 94 per cent. Individual country-wise patterns are also worth examining. Ethiopia, in 1975, had 27 per cent domestically collaborated publications. This increased to 50 per cent in 1985 and 65 per cent in 2015. In between there were ups and downs for Ethiopia. Kenya began with 37 per cent in 1975, and moved ahead with 53 per cent in 1990 and 58 per cent in 2015. Between 1990 and 2010 the percentage declined. A quarter of publications in Nigeria in 1975 had domestic participation, which began to grow from 1980 and touched the mark of 66 per cent in 2015. In 2000, the percentage had declined to 28 per cent. In 2010, about two-thirds of Nigeria's publications had domestic collaboration. Compared to the other four countries on the list, South Africa had a head start in 1975. Of all its publications for this year, 65 per cent had domestic collaboration, which continued to increase in the later years. By 2015, most of the papers (94 per cent) published by South African authors had the element of domestic collaboration. About 40–50 per cent of publications of Tanzania in most of the years had domestic collaboration. In 2015, the percentage was 63.

International Collaboration

A total of 43 per cent of all publications for the selected years between 1975 and 2015 for all five countries had at least one foreign partner participating in the production of publications. How does it pan out across the five countries?

International collaboration in these countries has been rather prominent in the recent years. The percentage of international publications had been more than half of all the publications. It ranged between 58 per cent (Nigeria) and 89 per cent (Tanzania).

International collaboration varied across these countries (Figures 5.3 and 5.4). Nigeria had the lowest percentage (29 per cent) and Tanzania had the highest (77 per cent). In between were South Africa (41 per cent), Ethiopia (62 per cent) and Kenya (68 per cent).

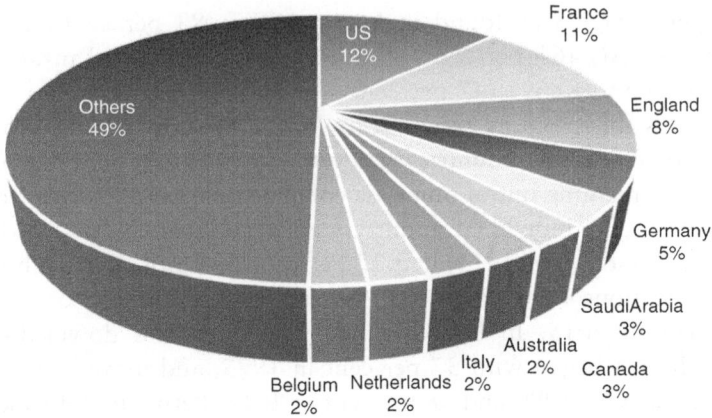

Figure 5.2 Top major international partners of Africa, 1945–2015

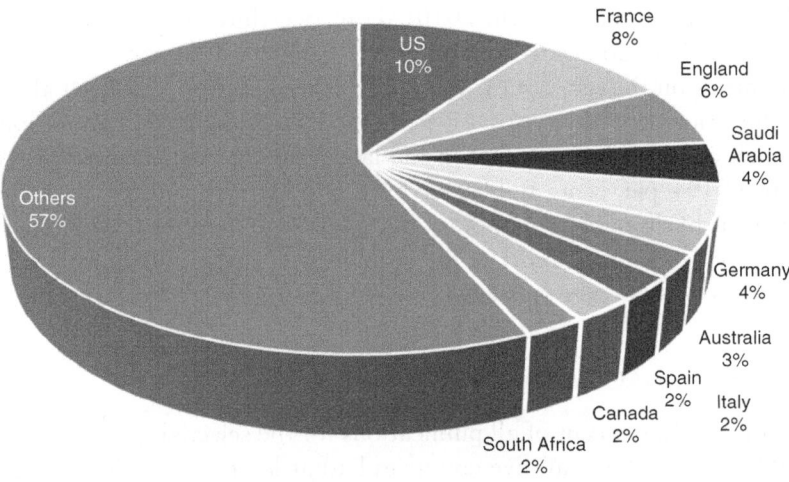

Figure 5.3 Ten top major contemporary international partners of Africa, 2011–15

   Tanzania had a good record of international collaboration from 1975, with 31 per cent of its publications having international participation. This reduced slightly in 1980 (24 per cent) and in 1985 (26 per cent) but increased later in 1990 (45 per cent), 1995 (59 per cent), 2000 (68 per cent), 2005 (86 per cent), 2010 (87 per cent) and in 2015 (89 per cent). Nigeria, which had the lowest percentage of internationally collaborated publications for the whole period of analysis, had begun to improve its position in recent years. In 2015, 58 per cent of its publications had international collaboration. Note that it had only 14 per cent of its publication in this category in 1975, which declined further until 1990. Since 1995, Nigeria strengthened its international collaboration, which was reflected in the increasing percentage of publications. South Africa, in 1975, reported the smallest share of publications with international collaboration (9 per cent). It began to increase the percentage of publications, particularly since 1995 (24 per cent). In 2015, it produced 31 per cent of publications under this category. Kenya in the beginning years produced 20–30 per cent internationally collaborated papers. In 1995, it had 48 per cent, which grew further and further. In 2015, it produced most of its papers (91 per cent) with international participation. Ethiopia also had a good beginning, which continued for some time. In 2000, it had a major shift in the production of internationally collaborated papers. By then it produced more than half of its publications with international participation and there was further increase of up to two-thirds in 2005 and later.

International partners, if not for historical and colonial reasons, tend to work with African countries that have made advancement in science and have earned a name for scientific productivity. These findings are mostly in line with other similar analyses and studies. Studies by Boshoff (2009a) and Nagtegaal and de Brun (1994) confirmed that the colonial legacy is a strong factor in the current international collaboration in Africa. Nagtegaal and de Brun (1994) found that ex-colonial countries in sub-Saharan Africa had stronger ties with their former colonisers than with other countries. Central African countries, as Boshoff (2009a) found in his study, continued to maintain their collaborative links with their former colonisers.

Regardless of colonial connection and contact, international partners have chosen their African partners carefully. The selection of partners is also influenced by the standing of the African countries, while most of the small countries have been omitted from any serious scientific association. Examples of this can be seen in the preference of international partners who were the colonisers of Africa. Invariably, countries such as Egypt, South Africa, Tunisia and to a certain extent Kenya, Nigeria and Uganda were among the

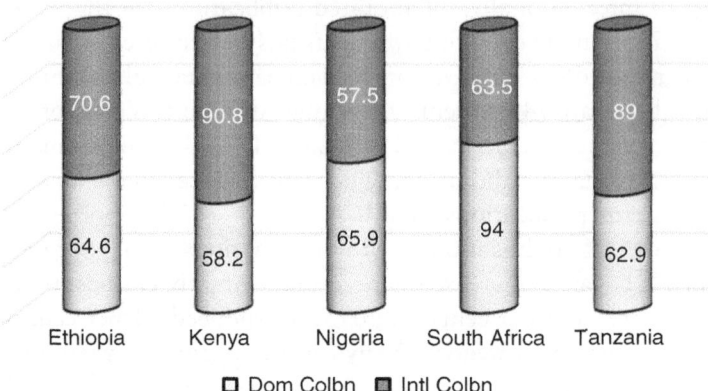

Figure 5.4 Domestic and international collaboration in major African countries (%), 2015

favoured countries of international partners for the joint production of science.

Regional collaboration (i.e., collaboration among countries within Africa) was not as significant as collaboration with countries outside Africa (Figure 5.4). Several studies have confirmed this trend in African science (Boshoff, 2009b; Onyancha, 2011; Owusu-Nimo and Boshoff, 2017). Investigation into the collaboration patterns among SADC countries shows that this leaves much to be desired. Boshoff (2009b) found that only 3 per cent of SADC papers produced during 2005–8 were joint papers by two or more SADC authors. On the other hand, 47 per cent of SADC papers were written with authors from high-income countries (Boshoff, 2009b). An exception to this was South Africa. South Africa is the centre for regional collaboration within Africa. Onyancha's (2011) study reported that South Africa collaborated with almost all countries on the continent (fifty-one countries in total) during the period of analysis (1986–2005). During 2005–8, South Africa produced 81 per cent of all SADC papers and 78 per cent of all intra-regional co-authored papers (Boshoff, 2009b). There is also evidence suggesting that collaboration is not always determined or facilitated by geographical, linguistic and economic affinities in the case of collaboration among advanced countries (Choi, 2012). These findings are generally in agreement with the data analysed in this chapter.

International collaboration in Nigeria turns out to be different from other African countries. Nigeria has an increased level of collaboration with China (Jonathan et al., 2010). This is true in the case of our data as well. During 2011–15 Nigeria published 6 per cent of its papers with China.

The findings of Adams et al. (2013) for the period of 2000–12 summarise the key features of Africa's research output clearly and briefly. These are generally in line with the discussion presented in this chapter and the data analysed in the previous chapter.

- The number of publications (papers and reviews) authored wholly within Africa but which do not have any collaborative author from outside the region have doubled since 2000;
- The autonomous research output of Africa clearly grew, which is sign of self-reliance;
- African science is dominated by three nations, namely, Egypt, Nigeria and South Africa;
- The countries collaborating most in Africa are the USA, France, the United Kingdom, Germany and Canada, and Saudi Arabia (mostly with North African countries and Egypt in particular);
- Marked interaction between countries in North Africa which share both culture and language;
- Collaboration between Kenya and its geographical neighbours in East Africa and with Nigeria, Ghana and Gambia can be explained by a common language of English or having a strong Anglophone influence; and
- The SADC does not emerge as a collaboration network.

Collaboration in Developing and Developed Countries

Several developed countries recognised the importance of collaboration with African countries. Such efforts were mutually beneficial and enabled the emergence of national scientific communities in the African countries which were integrated into the international scientific community (Gaillard, 1992a).

Cooperation between the developed and developing countries in S&T became a matter of interest soon after the UN conferences on S&T held in Geneva in 1963 and in Vienna in 1979 (Gaillard, 1992a). Scholars have however found structural weaknesses in research collaboration in Africa (Toivanen and Ponomariov, 2011). Concerns were raised about the asymmetry of the collaboration and the dominance of collaborators in the North (Gaillard, 1994). The leading science producers in the world, like the USA, tend to produce fewer international co-publications than do smaller countries (Krishna et al., 1998). In 1995, North America had only 10 per cent international co-publications as against 30 per cent for Latin America (Krishna et al., 1998). The USA, as the analysis of publication data for 1996–2008 presents, had 29 per cent publications that were internationally collaborative, and international collaboration involving the USA accounted for 17 per cent of all internationally collaborative publications (The Royal Society, 2011). This is because the larger the national scientific community, the greater its self-sufficiency, and the weaker the tendency to co-author publications with international partners (Krishna et al., 1998). The UNESCO declaration on science and use of scientific knowledge insists that cooperation between the developed and developing world should be implemented in conformity with the principles of full and open access to information, equity and mutual benefit (UNESCO, 2000).

Collaboration tends to occur within the periphery at an increasing level. Scientists in the developing world are poised to increase collaborations with their peers in other developing countries (Ynalvez and Shrum, 2011). As seen earlier, a few countries in Africa have established themselves as major science-producing countries. Africa looks up to them for scientific development in the continent. They have a larger role to play in scientific collaboration. Regional collaboration in Africa has a long way to go. Leading countries in Africa, such as South Africa, Egypt, Tunisia, Morocco, Nigeria and Kenya should take the initiative to establish strong regional cooperation within Africa, between the weak and the strong, and between the weak and the modest. They can play a leading role in regional scientific cooperation, as Japan did in Asia (Matsuura, 2000).

The ties from the peripheries to the centres have stimulated research in the peripheries but enhanced the dominance of the centres over and above the real research performance of the centres (Scott, 1998). The new rapidly growing countries (China, Turkey, Taiwan, India, South Korea and Brazil, for example)[10] are collaborating less than most of the

---

[10] These countries produced more than 70 per cent of their publications from national researchers alone (The Royal Society, 2011).

developed countries (The Royal Society, 2011), whereas smaller and less developed countries collaborate at a much higher rate with other countries (The Royal Society, 2011).

Internationalisation is inevitable in science, and no society can be aloof from the world scientific community for long. However, it brings both benefits and challenges to Africa. Africa can benefit in terms of staff development, networks that will strengthen their research capacity, development of information and communication infrastructure at centres of research, and a new generation of African scientists. This internationalisation, as Tijssen (2015) warns, on the other hand, can tempt African scholars to leave the continent for better opportunities abroad and Africa can lose out on this human resource and research capacity. Apart from this brain drain, commodification and commercialisation of research outputs and unfair collaborative practices dominated by the hegemony of international partners do not always work for Africa (Tijssen, 2015).

If collaboration is accepted as an essential ingredient for science development, then there should be efforts to support it. Collaboration is encouraged when equal credit is given to both single-authored and co-authored papers. In both South Africa and Tanzania this is not the case. In the Tanzanian university system, single-authored papers gain a higher score than co-authored publications (Freudenthal, 2014). In South African universities where productivity units are counted and funded, a higher unit for single-authored publications is allocated when the units are divided by the number of authors for joint publications.

Collaboration is shaped by political decisions. The effect of changes in the political alignment of countries on science might be shown in the collaborative patterns of authors (Lancaster and Abdullah, 1992). Good international relations can foster research collaboration through research agreements and inter-country alliances in common areas of interests and benefits. While this is often the case, it is not always so. Scientists often take the initiative for joint research endeavours that might emanate from their contacts, networks and interest in working in the same specific area or subject. These initiatives of research collaboration occur regardless of international relations between two countries. This is why researchers work with their peers in other countries. Indian scholars work with their counterparts in Pakistan or in China despite having political tensions and issues between them. Cuban scientists produce research papers with American scientists. American scholars work with their Iranian peers. Scientific collaboration between the USA and Iran that was evident in publication outputs has increased four-fold between the two time periods of 1996–2002 and 2004–8 (The Royal Society, 2011). This occurred while political tensions remained high between these two countries. They have no problem working together as part of multinational research teams. Several similar cases can be gleaned from the publication records stored in bibliometric databases. The details of authors of several joint publications show this kind of collaboration goes beyond political allegiance of countries.

An often-raised concern in international collaborations involving both developed and developing countries is the part played by the local researchers and their contributions to joint collaborative research. African researchers who benefit from international collaboration are concerned about their role in such research which is often limited to the supply of data (Waast, 2002). The bibliometric records of joint publications of African scholars examined provided the details of the order of authors as well. In the majority of the cases, scholars from Africa in most of the countries whose data was analysed were not among the first few authors if there were more than two authors: South Africa is an exception. Only in a very small number of publications did authors from the originating African countries appear as the first few authors. It is not clear whether this order of authors is a reflection of the magnitude of the contribution of authors or the funding that authors bring in.

Based on a large number of publications from the *Current Contents* database for 1999–2000 and combining it with survey data, Dahdouh-Guebas et al. (2003) demonstrate how international collaborations work. They revealed that publications produced in developing countries do not have co-authorship of local research institutes in 70 per cent of cases; there is unjustifiable under-representation of co-authorship with research institutes from the least developed countries; corresponding authors from developed countries failed to acknowledge their collaborators from developing countries as authors in about 70 per cent of the papers; and co-authorship of developing countries are deliberately and systematically excluded by the authors from developed countries. The study also confirmed that scientists from advanced countries view developing countries as a source of data gathering and field research and not as real partners in their scientific collaboration, which is evident in the neglect of the scholars from developed countries in peer-reviewed publishing.

As scholars have cautioned (Marton-Lefèvre, 2000), the direct and indirect contributions of local and indigenous scientific communities must not only be respected and recognised but also be given due weight. African collaboration is not driven by local researchers searching for collaborators. Rather, it is driven by the availability of resources and interests that come from outside Africa (Pouris and Ho, 2014). The interests of the local researchers differ from those of the international researchers (Onyancha, 2011).

The intention of international partners in collaborative research with scientists in Africa has also been in question. Invaluable data from the continent is crucial and indispensable for advanced scientific research in several fields, for instance, general and clinical medicine. The lack of resources and infrastructure to conduct research on their own is a great handicap and constraint for African scientists. International partners have resources, funding, technology and modern equipment that are an attraction for African researchers. Partnerships are not often equal, and somebody has an upper hand in the control of the research process and its outcomes. While international collaboration is interpreted as a positive aspect of knowledge production, a high incidence of it clearly denotes dependency of African science on world science (AOSTI, 2014). Scientists in African countries should also bear in mind that collaboration with their international counterparts also materialises due to the fact that they have the knowledge and expertise that prompts international partners to approach them. The US-based scientists in a study agreed that they were motivated to engage in collaboration with scientists in developing countries because the scientists in developing countries had expertise in the subject area (Wagner et al., 2001).

For international partners, Africa is a gold mine of resources for knowledge, which is a major attraction. New scientific discoveries and advancements are made in science with the data gathered from Africa. In research partnership, this kind of division of work is prominent. The question is whether the partnership is equal and benefits are shared equally or not. African scholars collect data from the field, which is otherwise hard for international partners to obtain, and the necessary resources are provided by international partners. This has been the case in several African countries.[11] Collaboration, whether it is international, regional or local, is not a recipe for science development in Africa. There are many other conditions that are required for the development and advancement of S&T in Africa. International and regional efforts to produce more relevant science for Africa falter because of the continuously weak nature of the existing expertise and facilities in Africa (Gruhn, 1984).

Science production is also associated with the science, technology and/or innovation policies countries develop, implement and monitor. They are to be aligned with the national development policies for meaningful outcomes, and growth and development. How far this has happened in Africa is reviewed in the next chapter.

---

[11]   As shown in a recent study on collaboration among Ghanaian scientists, Owusu-Nimo and Boshoff (2017) found that Ghanaian collaborators were largely involved in fieldwork and data collection while their international partners provided funding.

6 | Policy Matters in Science and Development

Science and science development rely on policies that are well conceived, implemented, monitored and evaluated. Policy matters most when it comes to scientific development, advancement and progress. Recent changes in Africa, as Juma and Clark (1995) argued, have brought policy reforms into focus in creating a suitable environment for change. Countries that have a substantial amount of scientific activity will have some kind of empirical policy in place, which may not necessarily be a comprehensive one (Shils, 1968). But something to begin with and to develop at a later stage is essential.

The basic concepts of science policy have been changed in the open knowledge systems (Hackmann and Boulton, 2015). As science has advanced, the structures to administer science have had to be adjusted and amended accordingly. Necessarily, this is reflected in any policy initiative that is taken to manage science for its own development and for the development of society. Science policy itself has become a separate field, fundamentally different from other policy areas (Barke, 1998). Now called science, technology and innovation (STI) policy,[1] it merits examination to comprehend how science is progressing in countries, Africa in particular.

Social policy has several components that constitute its overall purpose, strategies and actions. The most important of them are the mechanisms by which science and technology (S&T) contribute to find solutions to the problems of society and economy, and the mechanisms for strengthening the infrastructure of S&T (Fusfeld, 1979). In addition to these, there are a few more. Science constants, as Long (1981) called these, are the essential parts of science policy. These constants comprise science education (education of scientists

---

[1] In the 1960s, it was known as science policy or research policy and in the later years (i.e., in the 1970s and 1980s), it became more common to use the term science, technology and innovation policy (Martin, 2012).

and education about science); improved output of basic research in science (supporting scientists and facilities, establishing priorities and identifying quality); increased efficiency in support services in institutions for research; productive relations between university, industry and government; and better mechanisms for collaborations in developing countries for science education, research and national development.

Policy formulation occurs at many levels and stages. At the macro level, international organisations take the lead for policies meant for the world or for regions. UNESCO or the Organisation for Economic Cooperation and Development (OECD), for example, occasionally provide such leadership. At the micro level, the process begins at the national level where individual countries are engaged in the development of policies that will meet their national requirements in the coming years. Between these two levels, lies the meso level of policy formulation initiated at the regional level, aiming for specific regional development, taking into account regional needs. The African Union (AU), Southern African Development Community (SADC) and New Partnership for Africa's Development (NEPAD) engage in regional policy development for Africa.

Any policy is a blueprint for action that should be pursued and executed. Policy allows efficient management and organisation of resources to achieve intended and planned objectives. A general consensus is discernible that for any country, developed or developing, to advance S&T there should be a plan and a policy (Boon, 1979). STI policy broadly integrates the policies for science, technology and innovation. Science policy is a set of policies that influence the growth of knowledge and its rate and relative growth in different fields of science, and it should provide the best institutional conditions for scientific progress (Radnitzky, 1983). Technology policy, on the other hand, influences the rate of the transformation of scientific knowledge, which has practical implications for the production of goods and services (Radnitzky, 1983). The last part of STI policy is innovation policy that deals primarily with the promotion and support of innovation.

Policies emerge out of decisions. Science policy is a deliberate and coherent basis for national decisions that influence investment, institutional structure and the utilisation of scientific research for development (Hetman, 1979). In this process, anticipatory decisions are made about the development of S&T that are to be incorporated into the development plans (Sagasti, 1979).

Theoretical Basis

Science policy also has theoretical underpinnings. The national system theory approach which the United Nations Conference on Trade and Development (UNCTAD) adopted in many of its studies is based on the perception that innovation and technology development are the outcome of a complex set of relationships among all actors, namely, enterprises, universities and government research institutes, which are part of the system (UNCTAD, 2008). This underlines the role stakeholders have to play for the benefit of science and society. For policy development, participation and inputs of all partners of science have a contribution to make. Policies are the concrete versions of decisions made at different levels through processes to meet the objectives of a nation. The objectives are ultimately meant for growth and development and all who are affected by it should be part of the process of policy formulation. Scientists, academics, engineers, administrators, politicians, policymakers, professionals and representatives from civil society organisations, professional associations, industry and the public are to be involved in the process for their input but without losing the broader picture and implications for society. The input from potential users, such as the export commodity producers whose needs should reflect the research agenda, is often not taken into account in the formulation of policies (Eisemon and Davis, 1992), but they should also be part of this process leading to policy.

Progress and development in science can be accelerated through rational organisation and management of resources (Schott, 1993). This is where the relevance of policy comes in. There is now an increased awareness about the strategic nature of science, technology and policy as a driving force for economic growth and development (UNCTAD, 2011). A framework for S&T planning, as Sagasti (1979) observed, should consider a number of ideals that underlie the decisions. They are the long-term ideals that can be achieved to create a desired future image of the scientific system; those related to the institutional infrastructure of the scientific system; those regarding the pattern of interaction with related systems; those referring to the scope and nature of the activities to be performed; and those related to the allocation of resources for the implementation of the decisions in the policy (Sagasti, 1979).

As regards policy framework for development of science, there are two areas to be taken into account. The policy that directly relates to S&T is one, and the other consists of areas of government policy that include economic policy. The first refers to the strategies, plans and measures concerned with S&T. The latter may appear to have little to do with S&T but will have a major impact on the paths of scientific and technological development (Bell, 1988).

According to Moravcsik (1984), planning in science has three meanings. Firstly, is the weakest one that denotes considerations of actions that should be taken if and when a contingency arises. This is the response to events that may occur in future. Secondly, is the meaning that denotes making provisions for the inputs of the needed resources and for the procedure to be used. This is to make a blueprint for working towards the objectives of developing science. The third meaning is to set up targets for the output of the activity planned, which is not easy in science. In other words, planning science has to take into account all these elements for the successful completion of both planning and desired outcomes.

There is a difference between policy with and policy within science. Moravcsik (1975) makes the distinction between 'policy within science' and 'policy with science'. The former (policy with science) depends on the particular country and time. Policy with science changes all the time, subjected to the changes in cultural traditions, political systems and national goals. Policy within science, for Moravcsik, involves the measures taken to assure creative and productive scientific activity.

A science policy designed for the advancement and progress of the country has a range of functions. Hetman (1979) has a list of such functions: advisory and conceptualisation, strategic planning and decision-making, coordination, operational responsibilities, promotion, performance and legislative control. A policy should be derived from the development objectives of the country, which will create opportunities for the development process, and at the same time it should not be subordinate to development planning (Sagasti, 1979). Countries that are in the beginning stages of the evolution of local scientific infrastructure, as Moravcsik (1984) proposed, should have a bottom-up approach and not top-heavy policy. This approach heeds the problems of local scientists and people who make use of the results of science.

Science Policies in Africa

In the early years (i.e., up and until the 1960s), science policy was not an independent one but regarded as part of the educational policy (Hetman, 1979). The OECD in 1963 called for the establishment of mechanisms in each country to review the state of S&T, the allocation of resources for it, and the relationship of S&T to national issues of importance (Hetman, 1979). This integration was missing in several African countries where science had a strong presence. It was missing on both sides, namely, the government and the scientific community. In Nigeria, before the 1970s, the scientific community was hardly governed or influenced by the socio-economic objectives of the nation and was totally disconnected from those development needs (Lebeau, 2003). Angola's strategy to combat poverty, which was launched in 2004, did not have any explicit reference to the role of S&T in achieving the objectives (UNCTAD, 2008). However, this was realised and addressed in later years. It was different in Algeria where scientific research had direct association with economic and social development, and which was commonly shared by scientists, academics and policy-makers (Khelfaoui, 2004).

Science policy received serious attention in African countries only after their accession to independence (UNESCO, 1974a). In the early period of independence, several African countries engaged themselves to formulate national S&T policies to direct their scientific development. At the time of independence, many of them had neither a separate ministry for S&T nor a concrete S&T policy. As of 1973, only five countries (Algeria, Egypt, Guinea, Ivory Coast and Tunisia) had a ministry of science or a ministerial science policy committee. Cameroon, Ghana, Liberia, Mali, Niger, Nigeria, Senegal, Sudan, Tanzania, Uganda, Zaire and Zambia had a general science planning body. The Gambia, later to become one of the leading countries in science, firmly placed S&T as an integral part of its development since its first development plan drafted in the 1960s (Teng-Zeng, n.d.). In order to achieve the objective of becoming a middle-income country, The Gambia developed a vision that integrates STI to the national development strategies. However, there is a view that research and science in The Gambia is still in a rudimentary state, the S&T policy is not clearly articulated and it lacks capacity for research (Ozor, 2014). It has also been reported that The Gambia suffers from other inadequacies such as the lack of legal and strategic frameworks for research, lack of coordination of research activities, lack of demand for research, problems with policy and implementation, and insufficient financial and human capacity (Ozor, 2014). In the analysis of publications carried out in Chapter 3, the production figures for The Gambia does not even put the country among the middle-range countries. During 1945–2015, it made a contribution of only 0.33 per cent to the total publications in Africa. The percentage was 0.32 during 2010–15.

As far as Africa is concerned, policy development in S&T is still an underdeveloped or a developing domain. In many African countries, since the beginning of the 1960s, science policies were shaped by concern for shortages of trained scientific and technical manpower (Eisemon and Davis, 1992). This was the consequence of the colonial past. Many of them did not have a clearly defined scientific policy, nor were they in the process of developing one.

It was first recognised at the Lagos conference in 1964 that African countries were handicapped by the absence of national science policies and appropriate machinery to coordinate and prepare national science policies in these countries (UNESCO, 1974a). For the first time, the Lagos conference made it known that there was an inadequacy of national science policies in Africa and no effective machinery to coordinate and implement policies (Forje, 1992). The Lagos Plan of Action encouraged the African member countries to formulate their own national S&T policies. The consequent Yaoundé symposium in Cameroon acknowledged the developments in African countries that reinforced their national agencies for science planning and development. The Yaoundé symposium on science policy and research administration in Africa evaluated the situation of science policies and research organisation in relation to geographical, political and economic realities (Forje, 1992). By the time the Yaoundé symposium was held in 1967 in Cameroon, many African states had established their own national agencies for science planning, coordination and decision-making (UNESCO, 1974a).

In 1970, another UNESCO-initiated regional symposium was held in Addis Ababa, Ethiopia. This symposium also produced some recommendations for policy-making in Africa. The Vienna Programme of Action in 1979, adopted at the UN conference on S&T for development, called for the formulation of individual national policy for S&T

in all developing countries (UNESCO, 1986). It was made clear in the plan of action, adopted in April 1980, that each member state of the African Union was to establish a national centre for S&T for development and for formulating an explicit national S&T policy that translated into a national policy for socio-economic development.

The UNESCO-led conferences of the Ministers of African States Responsible for the Application of Science and Technology for Development, CASTAFRICA, were to make much headway in Africa. The Dakar (in Senegal) Conference (CASTAFRICA I) of the ministers of the African member states for the application of S&T to development in 1974 offered the forum to exchange information on national science policies. It was also the intention of the Conference to improve the application of national S&T policies for research activities, and to encourage scientific and technological research in member countries (UNESCO, 1974b). At the time of CASTAFRICA I, only a few African countries had national S&T policies and structures for their implementation. Since then, the number of countries on the continent that established organs for the implementation of S&T have increased considerably (Forje, 1992).

The CASTAFRICA II conference of 1987, held in Arusha in Tanzania, was meant to assess the situation of S&T in Africa after its first CASTAFRICA conference was held in 1974. It also deliberated on S&T policies for development, the basis for the development of S&T and scientific and technological cooperation in Africa, which are keys to advancement of science and for development. Effects of these were evident in the institutionalisation of S&T in Africa. These conferences and symposia were instrumental in reinforcing the need for solid science policies for Africa and served purposes for science development in distinguished phases. At the time of the Addis Ababa symposium, most of the African countries had put central machineries for science policy-making into place (Forje, 1992).

Science policies in Africa evolved from a concern that the allocation of resources for support of research is the recognition that S&T are key components in development (UNESCO, 1974a). There has however been a surge since the 1970s in the formulation of policies and formation of appropriate structures for it in Africa. By the early 1970s, there has been remarkable growth in planning and policy-making bodies for science in African countries (UNESCO, 1974a). In general, they were meant for augmenting the scientific and technological capacities of countries for social and economic development (Bell, 1988). But in most of these cases (for instance, Nigeria, Ghana and Kenya), the policies were sectoral, focusing on specific strategic industries (ECA, 2016; FRON, 2011; Ndemo, 2015). The national priorities as enshrined in the science policy of Nigeria are the fields of agriculture, water resources, biotechnology, health research and innovation, energy, environmental S&T, mines and material development, ferrous and non-ferrous materials and chemical technologies, ICT (Information and Communication Technologies), space research, industrial research and development, new and emerging technologies, transport, raw materials and manufacturing (ECA, 2016). During 1973–86, the number of policy-making organs relating to sectors such as agriculture, medicine, industry, environment and others increased from 69 to 197 and the number of ministries responsible for natural S&T increased from 5 to 27 (Thisen, 1993).

The environment of the newly independent countries in Africa has not been conducive to the success of their policies (Enos, 1995). The scientifically backward status of several African countries is not only related to the paucity of resources (Adeboye, 1998) but reflects the lack of awareness of the centrality of S&T to global competitiveness. There has also been a mismatch between the policies aimed at building technology and the measures needed to build technological capabilities in Africa (Hewitt and Albu, 1998). The countries that have been able to conceive their science and technological strategies in the form of explicit policy documents have not been successful in implementing those policies (NEPAD, 2005). Ghana, for instance, adopted a science policy in 2000 and prepared a working document in 2001, neither of which were implemented (Owusu-Nimo and Boshoff, 2017; ROG, 2010). A new policy was introduced only later in 2010. Until then, a definitive national science, technology and innovation policy was absent in Ghana (ROG, 2010).

The institutions responsible for the policy were mainly concerned with the allocation of resources to research and development (R&D) and the development of new research institutions and human resources for R&D (Bell, 1988). With a few exceptions, an organised S&T policy has not been the main driving force for the application of S&T for Africa's development (Lemma, 1988). Referring to a report of the National Council for S&T in Kenya, Eisemon and Davies (1997) observed that Kenya did not have a comprehensive and integrated policy on S&T. A UNESCO survey of the West African countries showed that while governments were aware of the significant role of S&T in development, very few countries had a clearly defined national S&T policy aimed at development (UNESCO, 1986). It has also been reported in a study (AOSTI, 2013) that in many African countries ministries and departments responsible for STI policy tended to operate in isolation from the rest of other policy agencies of government.

The efforts of the Eastern African Technology Policy Studies (EATPS), which began in 1982, and the West African Technology Policy Studies (WATPS), established in 1984, generated information for the formulation of technology policy in some countries in Africa (Vitta, 1990). Between 1974 and 1987, the S&T structures established in African countries increased from four to twenty-eight (Adeboye, 1998). The first NEPAD ministerial conference on S&T adopted an action plan to promote the development and application of S&T in Africa (UNESCO, 2007a). This high-level ministerial conference sought to develop policies and priorities on science, technology and innovation for Africa's development. This was later to become a Consolidated Plan of Action for Africa at the second African ministerial conference on S&T held at Dakar, Senegal, in September 2005. The Consolidated Action Plan, developed by the initiative of NEPAD, was meant to enable Africa to harness and apply science, technology and innovations to attain the developmental goals of Africa and to contribute to the global pool of scientific knowledge (NEPAD, 2005).[2]

The Cairo Declaration adopted at the extraordinary conference on African ministerial council on S&T in November 2006 contained ministerial commitments to the future of S&T in Africa, and the establishment of a fund for the implementation of the consolidated plan of action. An African Science and Technology Consolidated Plan of Action was adopted by the African Union (AU) in 2006, which consolidated the S&T programmes of AU-NEPAD for Africa (NEPAD, 2011). The plan was to improve policy and strengthen innovation, implementation, governance and funding in research and development. The programmes under the plan stressed African ownership

---

[2] The foci of this consolidated plan, which continues to be the basis for science policy development in Africa, include inter alia, building infrastructure and facilities for R&D, improving human skills, increasing the number of scientists, and building a strong political and civil society constituency of S&T (NEPAD, 2005).

and leadership and for the creation of networks of centres of excellence in African countries (NEPAD, 2011). This plan identified key areas in STI: biodiversity, biotechnology and indigenous knowledge, energy, water and desertification, material sciences, manufacturing, laser and post-harvest technologies, ICT, space science and technologies (ECA, 2016). The plan was soon to be replaced in 2014 by the Science, Technology and Innovation Strategy for Africa 2024 (STISA-24).

STISA-24 set the goal high for the use of S&T for the continent. It aims at providing an enabling environment for the development of science, technology and innovation in Africa. This development is to occur through improvement in human capital, technical competence, infrastructure and innovation (AU, 2014a). STISA-24 also has plans to implement policies and programmes in STI to address societal needs in a holistic and sustainable way. STISA-24 underlines the need for revamping STI structure in Africa, enhancing technical and professional competencies, achieving critical mass of human capital, building a strong culture, taking steps to curb brain drain, and encouraging collaboration within and between countries in Africa (AU, 2014a). Undoubtedly, these are crucial steps for Africa to take and achieve in a specified period of time. This can only be assured with strong political determination to support, sustain, implement, monitor and evaluate science policy.

It should however be noted that the impact of the policies was not impressive. The STI policies were not able to improve science, technology and innovation in Africa (ECA, 2016). Africa continues to perform poorly in tertiary education, innovativeness, productivity and competitiveness (ECA, 2016).

Review of Science Policies in Africa

Not all countries in Africa have a well-formulated science policy and there is shortage of information as well. Some countries in Africa, as noted earlier, have drafted detailed science, technology or innovation policies, which is not the case with many others. In some cases an S&T policy has been part of the national development policy. The policy-making process in Africa should be viewed in the diverse regional, cultural and economic contexts (Juma and Clark, 1995).

South Africa is one of the few countries in Africa that has seriously undertaken a science, technology and innovation policy (STI). Since its transition from apartheid to democracy in 1994, a series of policy documents in STI has been produced. The White Paper on Science and Technology brought out in 1996 established a policy framework for S&T based on the concept of a national system of innovation (NSI) (DST, 2002). The NSI was conceived as a set of institutions, organisations and policies for the pursuit of social and economic goals. The NSI is more integrated and aligned to national development objectives (Kaplan, 2004, 2008). The National Research and Development Strategy (NRDS), which was approved by the government in 2002, is the basis for its National Innovation System (NIS), which has identified three key areas of enhanced innovation, providing science, engineering and technology with human resources and transformation, and creating an effective government S&T system (GCIS, 2015). NRDS was able to identify major deficiencies ranging from the lower percentage of the GDP spent on R&D to declining human resources in science and engineering.

South Africa's Ten-Year Innovation Plan (TYIP), targeted to be achieved by 2018, is to drive towards a knowledge-based economy and economic growth is to be propelled by the production and dissemination of knowledge (DST, n.d.). More vehemently, the plan states that it is to help solve the deep and pressing socio-economic challenges of the society. A series of policy documents preceded this ten-year innovation plan. The ten-year plan focuses on the great challenges that the country is facing. Through this plan, the country aspires to become one of the three top countries in the world in the pharmaceutical industry, deploying satellites to provide scientific, security and specialised services, being the world leader in climate science and meeting the 2014 Millennium Development Goals (MDG) to halve poverty by 2014 (DST, n.d.). The plan is centred around the key principles of strategic decision (choices to be made to convert ideas into economic growth), competitive advantage (to invest in areas where there is high socio-economic return), critical mass (investment in key research areas to be made at a critical mass), sustainable capacity (R&D to have appropriate absorptive capacity) and life-cycle planning (R&D infrastructure to be considered over the long-term taking into account depreciation, skills and running costs) (DST, n.d.). The plan also emphasises international cooperation. By the end of this plan in 2018, South Africa should have foreign funding of 15 per cent of the total GERD, become a preferred destination for S&T investment and a leading player in the implementation of the African Research Area, establish stronger relationships with regional entities such as ASEAN and the ACP and with leading scientific nations, and to work on a functioning strategy to draw South Africans into the diaspora.

STI in South Africa is driven by the Department of Science and Technology (DST), which derives its mandate from the 1996 White Paper on Science, and Technology (DST, 2010). In 2002, the National Research and Development Strategy was adopted. The vision of South Africa is to create a prosperous society that derives enduring and equitable benefits from S&T. It is committed to the mission of developing, coordinating and managing a national system of innovation for sustainable economic growth and improved quality of life for its people (DST, 2015a).

In 2004, the Strategic Management Model for the Public Science and Technology System was adopted in South Africa. This model recognised the roles of the DST and other relevant government departments in funding S&T activities and oversight to scientific research institutions in the country (DST, 2013). The country has established several organisations responsible for S&T. The Council for Scientific and Industrial Research (CSIR),[3] National Research Foundation (NRF),[4] Human Sciences Research Council (HSRC), Africa Institute of South Africa, Academy of Science of South Africa and several other agencies and organisations are part of the national scientific system.

The Department of Science and Technology (DST), in its strategic plan for 2011–16, identified five priority focus areas: human capital development, global and Africa collaboration, knowledge generation (R&D), knowledge exploitation (innovation), and infrastructure development (DST, 2015b). Renewed focus and emphasis of areas were outlined in the new strategic plan for 2015–20. The new plan is to boost human capital development for STI with a special focus on transformation, promote government–business–university investments in R&D, effectively use outcomes of research for new products and

---

[3] Established under the Scientific Research Council Act of 1988, the CSIR is meant to foster scientific and industrial development in the national interest, through multidisciplinary research and technological innovation (DST, 2010).

[4] Founded in 1988, the NRF is responsible for promoting and supporting research not only in the natural sciences and engineering but also in the humanities and social sciences. It supports research innovation and national research facilities (DST, 2010).

services for the economy, and contribute to capacity building in STI in Africa (DST, 2015b).

Apart from South Africa, which had a long history of science development, Egypt and Ghana were pioneers among the independent countries in Africa to incorporate S&T in their national development plans (Teng-Zeng, n.d.). Ghana now has a well-developed national STI policy (ROG, 2010). The policy is contextualised and integrated into its national development plan and strategies to deal with a host of national objectives. Specifically, its STI policy is meant to facilitate scientific and technological capabilities by critical mass, improve scientific and technological infrastructure, and promote a scientific culture in the society. The policy is structured according to short-, medium- and long-term objectives. The long-term objective which is to be achieved in the next ten years is to create necessary S&T capacities within the country appropriate to national needs, priorities and resources, and to create an S&T culture in which solutions to social and economic problems are sought within the domain of S&T. The policy also underlines its role in all sectors of the economy, including agriculture, health, education, environment, energy, communications and industry. A sector ministry has been constituted for the management and implementation of policy in Ghana.

In many Arab countries that are part of Africa, rational policies and supportive institutional frameworks for a knowledge society (UNDP, 2003) that needs the backing of science are insufficient. The problem in these countries is the belief that a knowledge society can be built through the import of scientific products without investing in the local production of knowledge, and by depending on cooperation with universities and centres of research in advanced countries (UNDP, 2003). While these are important for scientific advancement, no society can rely completely on them without developing a scientific culture (which is more applicable to Arab African countries) and infrastructure (including human capital) for its own scientific research. Any S&T policy should pertain to all research and experimental development operations and the process of transfer and innovation ensuring the effective use of discoveries and innovations (UNESCO, 1986).

Acknowledging the value of science and scientists, Morocco took steps to strengthen science and research in the country. It established a dedicated office at the government to support scientific research (Waast, 2002). With the introduction of the law on scientific research and technological development in 1996, changes were to follow in the Tunisian national R&D system. Tunisia went for the restructuring of its national R&D system, which led to the creation of more new research laboratories and research units (Madikizela, n.d.). The ministry responsible for scientific research took special care of the researchers in the country by providing essential support in terms of equipment and infrastructure. The expenditure on R&D has been increased steadily from 0.45 per cent in 2000 to 1.25 per cent of the GDP in 2009 (Madikizela, n.d.) and as a consequence, research output (publications) increased. As shown in the analysis (Chapter 4) of WoS data for 1945–2015, Tunisia performed well on publications in several important research fields.

Until 2009, when Uganda formally formulated its science, technology and innovation policy, there was not much R&D in the country (GOROU, 2009).[5] GERD in Uganda was in the region of just 0.3 per cent of the GDP, and the institutional capacity and research output were poor. By its own admission, underdevelopment in the country was due to the lack of a coherent and overarching national STI policy, overdependence on imported technology and inadequate social awareness about the role of STI in national development. The policy is meant to strengthen the capacity of the country to generate, transfer and apply technologies for the realisation of national development objectives (GOROU, 2009). Matters were soon to change for the better. Later, the focused Uganda Millennium Science Initiative (2007–13) brought in substantial progress in certain areas of its scientific system. The initiative was instrumental in increasing human capital in STI, establishing a competitive funding mechanism with the involvement of international scientists and developing the capacity of the Uganda National Council for Science and Technology for the production of national STI statistics (Blom et al., 2016). A new ministry of education, science, technology and sports was created to augment the efforts of the government in STEM (Ecuru and Kawooya, 2015). The national STI policy was welded to the national development objectives of poverty eradication, industrialisation and productivity. Uganda's Vision 2040 emphasises STI system as well. Science,

---

[5] Uganda was actually involved, in stages, in formulating a national policy on STI since 1994, and the draft policy was reviewed prior to the 2009 policy, in both 2001 and 2006 (GOROU, 2009, 2013).

technology, engineering and innovation (STEI) have been identified as the fundamentals for economic growth under Vision 2040. Uganda is aware that it has low levels of STEI (GOROU, 2013). Its transformation from a peasant to a modern society rests on its capacity to strengthen its fundamentals such as STEI (GOROU, 2013). The Vision 2040 is meant to reorient Uganda, adopting STEI as the main driver of economic growth, establishing a national science, technology and innovation system, allocating a minimum of 2 per cent of the GDP to the STEI sector, investing in the education system with a focus on STEI and R&D to produce competent human resources. As shown in Chapter 4, research publications in Uganda improved recently which strengthened its scientific standing. Uganda had a share of 2.5 per cent of all scientific publications in Africa during the period of 1945–2015. This increased to 3.6 per cent for the period of 2011–15. Uganda has also made strides in the publication outputs and in its share of publications in Africa in several research areas. Among them are agriculture, general and internal medicine, immunology, infectious diseases, parasitology, public, environmental and occupational health, science, technology and others, tropical medicine and virology.

Burundi has a 2025 vision document (MPCDU & UNDP, 2011). S&T is not a major part of this vision, nor has it been integrated into its key focus of economic growth. The country identified some key challenges which do not have a direct reference to science, technology or innovation. This is despite the situation that its technological development is in its infancy. The vision document however states that economic growth and competitiveness will depend on local capacities to adopt advanced technologies. It also wants to promote technologies through reforms in education and stressing research. No clear-cut plan is presented for the objectives that are to be achieved by 2025. Burundi during 1945–2015 produced only 0.08 per cent of the total publications in Africa. In recent years (2010–15) its contribution was 0.06 per cent.

Burkina Faso designed a strategy for accelerated growth and sustainable development for the period of 2011–15 (IMF, 2012). In order to consolidate its economic function, the state promotes the management and use of knowledge and information, and confirms that science, industrial, technological and medical production are preferred areas for the state. There is no S&T mentioned in the document that presents a strategy for economic growth for a five-year period. No strategy can be expected when there is not even a plan. Later in July 2016, it adopted a national plan for 2016–20 namely, the National Economic and Social Development Plan (PNDES). The plan is unequivocal about its overall objective to structurally transform the Burkina Faso economy to achieve strong and sustainable growth (GOBF, 2016). Under its plan to develop human capital in the country, PNDES intends to promote research for the transformation of the economy. Other than these, the national plan document does not say anything about science, technology or innovation. For the long period 1945–2015, the country had 0.65 per cent of the publications in Africa. The percentage for 2010–15 was 0.83.

Ethiopia, drawing inspiration from South East Asian countries, developed its STI policy in 2012 (FDRE, 2012; Mamo et al., 2014). This is part of its Growth and Transformation Plan for 2011–15. The objectives to be covered include building STI capacity, promoting research for technology learning and adaptation, promoting and commercialising indigenous knowledge and technologies and implementing STI activities in coordination with economic and social development plans. Ethiopia acknowledges the need for research to resolve its major social and economic problems, and the contribution of scientific research to achieve its national goals. The gap between research activities and national development goals has also been identified in the policy document (FDRE, 2012; see also Nordling, 2010b). These measures have paid off for Ethiopia if its publication productivity is taken into account. It is one of the major science-producing countries in sub-Sahara, contributing to about 2 per cent of the total publications in Africa and 3 per cent in sub-Sahara. Also, in recent years (2010–15) it improved output over previous years.

In the National Policy on Science and Technology 2006–11, Lesotho is set to adopt and use STI to transform its economy and to meet its developmental objectives. Its vision, like many others in Africa, is to use S&T for a prosperous and progressive economy. To realise this vision Lesotho is to create a stronger S&T human resource base, develop a culture of innovation, build a vibrant information society and promote its indigenous knowledge systems. Lesotho, compared to other Southern African countries, has the highest percentage of tertiary students in science and engineering (UNCTAD, 2010a). However, its low enrolment numbers in subjects like mathematics and science at higher education institutions is a cause for concern, as it is important to have sufficient graduates to drive STI in the country (UNCTAD, 2010a). The effect of the policy on science production in the country is yet to be seen in the WoS data.

S&T is a key priority area for Malawi, and it has been considered vital for national productivity and development (GOM, n.d.). Reforms have taken place in the S&T sector to enhance its contribution to economic development. Its key strategies include adoption and transfer of technologies, prioritised and focused multidisciplinary R&D and strengthened institutional and regulatory framework and capacity for R&D. Malawi has introduced S&T as a subject at the primary school level. The publication figures for Malawi for 2010–15 and 2005–9 show some improvement.

Mauritania has made some progress in some areas of S&T since its independence in 1960. It lacked a clear STI strategy and suffers from inadequacies. There exists no institutional mechanism for coordinating research activities, not enough public funding for R&D and low levels of awareness about science and innovation in the country (UNCTAD, 2010b). Collaboration with international organisations and research institutes has assisted the country to manage skills development and technology transfer (UNCTAD, 2010b).

Mozambique advanced a strategy for STI in 2006 that can be achieved in a span of ten years (ROM, 2006). The strategy, called Mozambique Science Technology and Innovation Strategy (MOSTIS), is to establish an enabling framework to harness STI and focus on exploring solutions to its problems including poverty, further economic growth and social well-being. It developed mechanisms such as the National Science and Technology System to implement the strategies outlined in MOSTIS. The policy also recognises the need to develop a culture of innovation that is founded on S&T.

The Vision 2030 of Namibia, which became an independent country in 1999, is centred around the strategy of transforming into a knowledge-based economy, through rapid development of a knowledge and innovation system (Nyiira, 2005). Obvious from the vision is that Namibia is to take advantage of the use and application of STI. Towards this end, Namibia passed the Research, Science and Technology Act in 2004. The government was keen to restructure and strengthen research S&T to serve the national developmental objectives (Nyiira, 2005). The national research S&T policy of Namibia sought to increase investment in S&T in both public and private sectors, promote and apply traditional technologies and improve conditions for advancement of S&T professions. As part of the national development plan, Namibia adopted a National Programme on Research, Science, Technology and Innovation (NPRSTI) that identified priority areas to address its social and economic challenges (RON, 2014). The focus of NPRSTI is to increase funding for S&T, build research capacities, strengthen necessary infrastructure, promote public understanding of S&T, and establish national and international collaboration. Certain priority research areas such as health, agriculture, fisheries, water, energy, geosciences, mining and indigenous knowledge were to receive special attention. Its target funding for R&D from 2014–15 to 2016–17 was to increase minimally from 0.2 per cent to 0.3 per cent of the GDP (RON, 2014). The WoS figures for Namibia show that its production capabilities have developed since 1985–89. From a share of 0.20 per cent of publications in Africa during 1985–89, it increased to 0.29 per cent during 1990–94. By the period of 2010–15, it grew to 0.34 per cent.

The Vision 2020 of Rwanda underlines the development of S&T to support the development of a prosperous knowledge-based, technology-led economy (ROR, 2006). The development of S&T in the country is to stimulate a steady growth in GDP, improve skills and knowledge among the population and integrate science and technical education with commerce, industry and the private sector (ROR, 2006). The key science and policy objective of the country is to build capacity in scientific research, develop initiatives for technological transfer and promote STI. Rwanda plans to cultivate and retain interest in S&T among the population and to train scientists (in various fields including medical and agricultural), engineers and technicians who can support the national objectives. In order to achieve the objectives in the policy a series of strategies has been designed: the national integrated innovation framework, governmental reforms (public sector reforms, S&T outreach, incentives for S&T implementation, public-private partnership and intellectual property). Perhaps as a consequence of these, Rwanda made a rapid jump in its scientific output during 2010–15, doubling its share in Africa from the previous years (0.17 per cent to 0.34 per cent).

Zambia acknowledges that its socio-economic deterioration since independence in 1964 was due to the neglect of S&T and its significance in national development and its heavy reliance on technological support from outside (ROZ, 1996). With the constitution of a ministry for S&T and the formulation of a national S&T policy in 1996, Zambia recognised the relevance of S&T for its development. The policy of Zambia, as emphasised in the policy document, aims to develop indigenous technological capacity while promoting and exploiting S&T for development. The policy objectives were to change the institutional structure, to conduct scientific research guided by national developmental goals, and to establish a mechanism for increased innovation, transfer, diffusion and commercialisation of technology with an emphasis on indigenous technology (ROZ, 1996). The policy stipulates that at least 3 per cent of the GDP should be allocated to scientific and technological activities. Notably, the Zambian policy finds the need to conduct sociocultural research to determine the priorities of the economically disadvantaged people that can be taken into account in scientific research.

The 2012 STI policy of Zimbabwe is a culmination of a number of policies and acts since its independence in 1980.[6] It had a department of S&T development (established in 2002), which became a fully fledged ministry of S&T in 2005. Zimbabwe, like many other African countries, wanted S&T to be an integral part of national development. Under this policy, which is not a well-formulated document in regard to the goals to be achieved, the government aims to strengthen its capacity in STI, use technology to accelerate development and popularise S&T (GOZ, 2012). The policy recommends that pupils in schools should devote at least 30 per cent of their study time to science subjects, and at least 60 per cent of university education should be in skills development in S&T. Branches such as biotechnology, nanotechnology, ICT and space sciences are to receive increased attention. The use and development of indigenous knowledge systems finds a special place in the policy document. Other prominent features of the policy are the budgetary allocation of 1 per cent of the GDP to R&D, incentives and recognition for achievements in S&T, fostering international collaboration and popularisation of S&T in society.

Systematic reviews of policies and programmes in S&T in some countries have been undertaken. For many, the great challenge is to

---

[6] Among them were the Research Act of 1986, the Science and Technology Policy of 2002 (the first national policy on S&T), the Biotechnology Policy of 2005, and the ICT Policy Framework of 2006 (GOZ, 2012).

convert the policies into implementable initiatives (UNCTAD, 2010a). Theoretically, a policy document on S&T may be a strong one. The objectives of the policy are to be translated into action. The correspondence between policy and implementation is to be maintained, and followed by monitoring and evaluation. Some countries like Mauritania need to strengthen their skills base in science, technology and engineering and improved STI capabilities (UNCTAD, 2010b). A review on the policy in Lesotho showed that there was a need to prioritise collaboration and coordination for enhanced efficiency in STI sectors (UNCTAD, 2010a). The national STI system in Namibia was fragmented and there has been a lack of proper coordination between R&D institutions in the country which was further aggravated by policy inconsistencies (Nyiira, 2005). In order to deal with these drawbacks, Namibia introduced a National Commission for Research, Science and Technology. Namibia had a number of challenges to overcome, mainly the impact of S&T on the economic growth and research capacity and infrastructure (Nyiira, 2005).

The absence of relevant national strategies and S&T policies affect the production of science. In West Africa, for instance, there is a lack of national research and innovation strategies and political commitment (Essegbey et al., 2015). Some small countries on the continent still do not have a strong policy or strategy to move forward in S&T. The Seychelles, an island nation, has not evolved a comprehensive S&T programme or strong research institutions (ROS, 2013). Somalia, as shown in its development plan for 2017–19, is establishing institutions that can promote and support S&T in the country (FGOS, n.d.). Recognising the drawbacks of its S&T policy that was driven by external pressures and ignoring its diverse knowledge systems, Mozambique now acknowledges national development strategies as a key driver for its S&T approach in the country (Turpin and Martinez-Fernandez, 2003). As distinguished from other countries that are in a similar state of development, S&T policy is known to have some distinctive features.

In a review, Turpin and Martinez-Fernandez (2003) find three notable characteristics in the S&T policy in Mozambique. Firstly, the focus of the policy is not on institutions only as organisations, but as the places which can guide ways of thinking about STI. Secondly, the policy has an exclusive approach to the development of knowledge infrastructure that runs parallel and complementary to the development of physical infrastructure in the country. Thirdly, the current policy process in the country is a totally governmental approach and not a set of science- or technology-driven initiatives. This helps build links with other sectors of the economy, namely, agriculture, education, industry, health and the environment. These are relevant for other African countries as well.

Some other countries had a multisectoral body for coordinating scientific research.[7] Angola set up its Ministry of Science and Technology in 1997, mandated with the responsibility of developing a national policy for S&T. Mauritius, a small group of islands in the Indian Ocean, has a joint ministry for education, science, research and technology which undertakes initiatives in R&D (ROM, n.d.).

One can find a proper integration of the objectives of S&T with the objectives that are meant for the general development of the country. Also evident is the way it is proposed to have a close association with other sectors of the economy as development in one sector is dependent on other sectors.

Policy Challenges and the Road Ahead

There are now national S&T policies in most of the countries that vary in their content and focus. As the 1974 UNESCO conference recommended, a policy should be dynamic and adaptable to changing circumstances. The success and effective implementation of science policy largely depends on the consensus within the society, and that it should be an outcome of a collective process (UNESCO, 1974a). This process can be an ongoing one, relating to developments and achievements made in the sphere of S&T. The policy should take care of its relevance specific to the society and be properly integrated into the national development objectives, plans and strategies.

There is a need for accurate statistical data relating to the current state of science in any society, before a science policy is developed. The approach should be scientific for the genesis of a sound policy on science. The collection of scientific information on R&D data, research personnel, allocation of funds for sectors, contribution from public and private sectors to R&D, infrastructure for scientific research, performance of sectors within the fields of science (medical science or

[7] Gleaned from the data presented in Table 1, UNESCO, 1974a, p. 104.

agriculture), and collaboration constitute grounds for formulation of a science policy. National science policies and strategies should value collaboration and international activities (The Royal Society, 2011) as integral components of S&T.

Building a database of information pertaining to all aspects of science, technology and innovation is to be taken seriously in African countries. This has been a great challenge that had a detrimental effect on policy, producing bad outcomes. Some countries conduct regular surveys on R&D that allow them to make informed decisions. As has been argued, more empirical information about the sector and all aspects of it are indispensable. This information should be reliable, accurate to the extent possible and accessible to the scientific community and the public alike. The availability of such information, accessible to all, will lead to creative recommendations for science and its development. Often creative inputs emerge from the public and civil societies that are more pro-people than policymakers and will influence the latter in decision-making.

It is to be noted how information is used by policymakers on matters relating to S&T. An empirical study of policymakers in Nigeria, regarding the use of information that feeds into decision-making processes, demonstrates that the most available sources were newspapers and colleagues/superiors (Nwagwu and Iheanetu, 2011). For them, inaccessible information sources were theses and dissertations, research reports and books of abstracts. Policy makers would like to have better availability and accessibility of information that is produced in research institutes and universities (Nwagwu and Iheanetu, 2011). Scientists and researchers, and policymakers in science, have to assume a mutually beneficial and responsible role. When the findings and recommendations of scientific research are not accessed and used for policy purposes, the loss for a country is substantial. The consequence of this will be reflected negatively in future plans, programmes and strategies. Both scientific research and policymaking are ongoing activities and one feeds into the other. In order for a policy to be strong and achievable, available scientific information and data are unavoidable. For future scientific research programmes, a well-designed policy is a pre-requisite.

The continent is vast and heterogeneous in its social, economic, political, cultural, demographic and developmental indicators. Some have advanced on the path of development while others are struggling to move on and are trying to find their own standing in STI. One aspect they have in common, except for the countries in North Africa and some sub-Saharan countries like South Africa, is that all of them are developing, facing challenges that are not different from each other. They all need to develop, whether there is a determined political will or leadership, and move out of the miserable situations of poverty, ill-health or conflicts.

Most of the countries in Africa now have adopted S&T as a key area for national development that cannot be neglected. They share some similar components in their S&T policy that are in varying phases of implementation, both in time and phases. Valuable lessons that are unique to the countries are also to be learned, as they have advanced in S&T. The analysis of scientific publications presented in Chapters 3 to 5 suggests the relationship between scientific outputs and policies and strategies adopted by these countries. These lessons and experiences from the formulation of policies to implementation and achievements might have specific contextual singularities. Many of these experiences can be shared with other countries in the continent for them to try, experiment and emulate.

As has been shown, the policy directives of countries have distinctive components that, if they were found to be effective and successful, should be adopted in other countries. For this, there should be a common forum under the leadership of AU, NEPAD or SADC for regular feedback and sharing. This kind of cooperation will be mutually beneficial and supportive. Along with this, as Forje (1979) recommended, the creation of a corps of African science through which scientists can share and contribute to national development for the whole of Africa can be conceived.

Studies of STI policies in other similar countries offer valuable lessons. A study of STI policies in some central American countries describes the elements of these STI policies (Padilla-Pérez and Gaudin, 2014). By and large, they underline the importance of innovation for socio-economic development, identify key economic sectors and efforts meant to focus on these sectors, and promote interaction among actors (Padilla-Pérez and Gaudin, 2014). They have mechanisms for coordinating STI policies with agencies and stakeholders which many African countries have not taken up seriously.

Iran made rapid progress in its scientific output between 1996 and 2008. This progress was mainly due to its twenty-year Comprehensive

Plan for Science that emphasised higher education in the country and the establishment of stronger links between industry and academia (Sawahel, 2009). This plan was to sustain its scientific progress and self-reliance on its domestic pool of talents. Along with this it has plans to increase the number of scientists to 3,000 per million inhabitants, R&D investment to 4 per cent of the GDP and higher education investment to 7 per cent of GDP by 2030.[8]

Malaysia is another case. In a study of differential scientific output of ethnic communities in Malaysia, Lewison et al. (2016) examined the effect of the National Economic Policy (NEP) on scientific output which was introduced in 1970.[9] NEP promoted scientific activities among the ethnic community, the Malays, and their participation in research has risen. This is one example of the effect of policy that is specific and targeted to enhance scientific activity and productivity. Malaysia, which has plans to become a developed country by 2020, has made substantial investments in R&D under its national plans that resulted in an increase in its scientific output (Lewison et al., 2016).

As an example of what is happening elsewhere, the new policy initiatives that Germany adopted have contributed to improved performance of its science system. The Pact for Research and Innovation (PFI), the Strategy for Internationalisation of Science and Research and the German Universities Excellence Initiative were among them. PFI, adopted in 2005, guaranteed increased annual funding for major non-university research organisations in the country.[10] The research policy of the country covered inter alia the dynamic development of the scientific system, the creation of networks within the science systems to improve performance, development of new strategies of international collaboration and the establishment of science–industry partnership (Aman, 2016).

The immediate priorities of African countries are a deterrent for the development and growth of science. Many of them have to make substantial investments in education, health, agriculture and in basic

---

[8] Its R&D investment in 2006 was at 0.59 per cent and investment in education was 5.49 per cent of the GDP in 2007 (The Royal Society, 2011).

[9] For details, see Jomo, 2004.

[10] The Strategy for Internationalisation of Science and Research (adopted in 2008) promotes an internationally coordinated research agenda, while PIF acts to improve its science system through increased international collaboration (Aman, 2016).

infrastructure, which are the more pressing needs of these countries, and undermine the attention science and scientific growth deserves.

Broad policies for science, technology and development are in place for Africa that have been developed after deliberations under the patronage of the AU. It is up to the countries to develop their own individual policies and strategies, taking into account their unique, potentially dissimilar, situations. The challenge again is multi-fold. The elements of broad policies designed for the whole of Africa may not be successfully implemented across the continent in the same way. Individual countries have their strengths and weaknesses that would influence strategies and their implications. These relate to their existing human resources, infrastructure, economic health, governance, leadership and internal social conditions (including conflict and strife). There are countries that would fail the test on these measures, some miserably. Even if countries were in a position to spend the 1 per cent of the GDP on GERD as stipulated in the STISA-24 plan, the actual amount of money spent would vary. Before the actual policies and strategies are converted into action, countries will have to prepare the ground for nurturing science and development in the country. Some countries might need to revamp their basic school education system, if that is the priority, particularly in science education. Even some of the scientifically strong countries do not have strong science education system at schools.[11] The higher education system, which produces future scientists, should also be at the centre of attention.

Political ideologies and interferences may not work in favour of science or science policy. A policy is often the outcome of a political decision. The argument for this is the belief that science is too powerful to be left to scientists and researchers. The issue however is how politics affects the design and implementation of science policy. Strong political will and determination will prepare the ground for science to grow and advance, with the backing of a stable science policy in place. Often it is the lack of political support for the implementation of STI policies that holds back the momentum of science and science development (Padilla-Pérez and Gaudin, 2014). There should be a firm national commitment to science development.

---

[11] A study conducted by the Human Sciences Research Council (HSRC) found South Africa among the five lowest-performing countries in the 2015 Trends in International Mathematics and Science Study (Wolhuter, 2016b). In contrast, South Africa is a leading country in science in Africa.

The involvement of scientists in the formulation of science policy at the macro and national levels contributes to a robust policy that will take care of the current needs of both science and society. Scientists would understand the link between science and policy better than others. They will have clear perceptions about both the directions the discipline should take in the immediate and in the distant future in their specific national context and the needs of the country to meet its developmental objectives. This knowledge serves both the discipline and the nation alike. Science is not for science itself. It has a function to perform in the society. Its development is connected to the development of the nation where it is conducted and applied. The growth of science is reliant upon the growth of the nation and vice versa. This is a kind of symbiotic relationship in which each needs the other. The participation of scientists in policymaking is often influenced by factions in the community of scientists or interest groups who compete and lobby for the protection of the interests of their disciplines. The rise of interest groups in developed countries has been a cause of concern as this can affect all other disciplines (Barke, 2003).

The pre-modern notion of science as an individual activity (Jalali, 2000) seems to be disappearing from the minds of policymakers. This is a helpful change for science. Science is more and more viewed as a public good for public use and well-being, and its role in the society is enlarging day by day. It is no longer an individual or a small-group activity but one that does not have any borders. The frontiers of science expand from local research stations to international laboratories. Science is both local and global, which influences the decisions of policymakers.

It is not easy to prescribe a general policy for S&T. Policy making is a very complex procedure and it should take into account all aspects that concern a country – its economy, resources, developmental priorities and so on. There is no theory to guide policy prescription (Enos, 1995). Instead, it is easier to establish a list of priorities, commencing with those that require the most attention and terminating with those that need the least attention (Enos, 1995). Priorities are subjected to shifting from those that need urgent and immediate attention to those that need moderate or low levels of attention. Regular assessments of these priorities to measure the achievements and impact are helpful in determining new priorities that are well knit and entrenched in the current S&T and developmental demands.

In order for a country to keep track of the progress and development in S&T, it should have clearly constructed indicators. The absence of relevant STI indicators is a problem for many developing countries (Siyanbola et al., 2016). The indicators are also useful in the implementation of policies. The incongruity between planning, decision-making, allocation of funds and resources, implementation and evaluation has to be stressed. They form the integral components of science planning and integration of these phases is therefore vital for a successful science policy. Often it is the case that there is no true mechanism to ensure the last two stages of implementation and evaluation. The importance of S&T policy as a tool for national development planning is now evident in the policy documents of several African countries. Many of them have a long plan or vision to achieve progress in S&T.[12]

Policy development in S&T is a two-way process. Policies produce information to make decisions that are useful for further development of policies. Sound and stable S&T policies at the national level give rise to best practices and are capable of evolving a scientific culture in the country. Policies are meant to support the national scientific system by way of strengthening and establishing new institutions (research and higher education) for the production of knowledge and building structures meant for its overall growth. When properly drafted, policies are implemented within a stipulated period and the outputs are to be assessed and evaluated at the end of a period. At this stage, a country will be able to see where it has reached so far in terms of its achievements, drawbacks and bottlenecks in S&T, and that knowledge will provide information for the paths the country has to choose in future. A new set of policies or amendments in the existing policies might be required after careful assessment of the policies that have been implemented. Based on this assessment and evaluation, new policies can be designed, planned, drafted and implemented. A country which is committed to its S&T will go through these iterative cycles many times during its course of development that will ultimately position it on a higher pedestal in the world of science.

---

[12] For instance, Ghana has drafted its Vision 2020 that underlines the need to apply S&T for development. As the discussion showed, many countries in Africa have this in practice.

A gap is often found between policy and reality. In a report that summaries the situation of science in fifty-five developing countries, Mouton and Waast (2009) noticed a gap between what is stated in the policy and the actual plans and commitment to achieve what is implied in the policy. They believe that science policies in many countries are vacuous documents which have only symbolic effect. Serious commitments to science policies are yet to be seen in Africa.

Countries are taking special care about their science and science policy for bringing in science development. Several OECD countries, for instance, have taken serious initiatives to reform their STI policies in order to increase the contribution of science to economic growth (OECD, 2000). Africa cannot lag behind in this journey.

S&T often has been applied without due regard for the needs and interests of the population (UNESCO, 1974a). The role of social scientists becomes indispensable on such occasions. Social scientists are often neglected and sidelined in identifying the priorities of scientific research a country sets forth. Natural scientists or policy-makers do not always appreciate the role of social scientists. The major drawback of this is that research is conducted without proper consideration of the social questions involved (UNESCO, 1974a). In developed countries, social sciences play an increasing and indispensable role in the formulation of policies (Dedijer, 1968). Social scientists, through their research, can feed information to natural scientists about what kind of research is timely and warranted at a particular time to contribute to national development. In under-developed countries, social sciences have a special task of holding a mirror to the face of the nation to show what it is (Dedijer, 1968). Social scientists can supply information that can influence science policy decisions and direct the foci of scientific research that a country needs. In other words, social scientists act as the link between scientific research and national developmental priorities.

It is evident from the review of science policies in specific countries and in the analysis of scientific publications that both are inseparable. One has an effect on the other or is its cause. The production figures of countries such as South Africa, Egypt, Tunisia and many others have been the outcomes of a strong science policy in these countries. Admittedly, as elaborated earlier, science policy alone cannot lead to advancement in science. Other facilitating conditions are also necessary. Science policy is often complemented and supported by educational, economic and trade policies.

Therefore, the S&T policies of countries cannot be viewed in isolation. Several other realms of policies are complementary to S&T policies. Advancement in S&T is not just the outcome or impact of relevant policies in the area but also the cumulative effect of developments in other areas such as the economy, investment, trade and education. Often development in S&T is the direct or indirect result and consequence of developments in other areas such as education, trade and industry. Policies in these areas have an impact on S&T policies and advancement in science and technology. S&T and the economics of trade, as Sebastian (2013) notes, are intertwined in many different ways. There are examples from around the world showing how trade policies influence and advance S&T in countries.

The trade agreement between the USA and the European Union and with Japan grounded in the Transatlantic Trade and Investment Partnership (TTIP) had fruitful interaction between trade and science policy (Sebastian, 2013). In Africa, some of the best-performing countries in science have made their trade and economic policies very effective, which has assisted their science and technology policies. Those trade and economic policies have made progress in boosting technology-based manufacturing (ECA, 2016). Illustrations can be drawn from the NAFTA (North American Free Trade Agreement). This agreement between the USA and Canada shows how trade and national S&T policy were mutually influenced (Sebastian, 2013). Supportive policies and their successful implementation can take S&T to a higher level of achievement. Often such policies are mutually influential and beneficial, one contributing to the other, and effectiveness in one area will have a corresponding effect on another.

In many African countries there is a strong perception that S&T policies should work closely with trade, industrial and educational policies (Wakhungu, 2001). The educational system, as Wakhungu (2001) argued, should be made more accessible to the practitioners of S&T. Some African countries have followed the path of adopting a blueprint approach to designing and implementing industrial policy instruments (tariff protection, tax rebates and R&D subsidies) for growth (ECA, 2016). Uganda's Vision 2040 aims at increasing its industrial share of the GDP by 5 per cent (ECA, 2016). The National Industrial Policy of Uganda is meant to build a modern, competitive and dynamic industrial sector which cannot be achieved without the support of a strong S&T policy. This is important, as Africa is keen on industrialisation and has advantages for the successful implementation of their S&T policies.

Science policies often contain sectoral policies that contribute to overall national development policies. These focus on improving the efficiency and competitiveness of existing sectors such as agriculture and manufacturing (UNECA, 2016). Under the S&T policies, sectors including education and industries are often covered. This indicates the relationship between STI policies and industrial and other policies. In many African countries, S&T policies are carried out by government sectoral ministries and departments (UNECA, 2014, 2016).

South Africa, as seen earlier, is a top producer of scientific knowledge. South Africa has a record of good trade performance. The trade performance of the country presents itself as one of those BRICS countries that have been expanding their relative exports much faster than the USA and even surpassed China's relative performance in certain years (Kowalski et al., 2009). It has developed a set of comprehensive and effective trade and economic policies. The South African trade and economic policies have enabled the country to attain significant socio-economic developments (Makgetlaneng, 2003). This was obvious in other sectors of economy. As shown in the previous chapters, South Africa performed well in its agricultural sector, which is also a reflection of its development and advancement in agricultural research. Apart from other factors, this development was caused by South Africa's participation in different trade agreements and labour policy reforms (Chitiga et al., 2008).

South Africa adopted policy reforms and liberalisation measures (OECD, 2006). The reforms made in trade were meant to promote and integrate the sectors of the economy into the global economy (OECD, 2006). South Africa also gained from reforms that were introduced in the economy since it became a democracy in 1994. A growth-oriented policy was instrumental for development in other non-agricultural sectors (OECD, 2006). South Africa's trade policy promotes its integration into the world economy to encourage greater access to markets, technology and capital, which had direct impact on the country's economic growth (Chitiga et al., 2008; OECD, 2006). A number of trade agreements that South Africa entered into with the world community have served the country in its growth and development.[13] The Trade Policy and Strategy Framework of the country identifies its major national development goals, such as economic growth and industrial development (Sandrey et al., 2011). South Africa is seen as an attractive preferential trade partner for African commerce (Kowalski et al., 2009).

As far as education policies are concerned, South Africa had a huge task of addressing its past inequities in access to education. Democratic South Africa had to rebuild the system by creating a single unified national system, increasing access to those who were disadvantaged in the past, decentralising governance, revamping the curriculum, and nationalising and reforming the higher education system (OECD, 2008). The National Education Policy Act, along with other policy frameworks such as the National Qualification Framework and the South African Higher Education Funding Framework,[14] were fundamental to the revamping of the education system. The National Policy for Higher Education System focused on transforming the system, and on increasing the number of graduates in the country. Measures such as the integrated quality management system were made to improve the quality of education. In other words, since the dawn of democracy, the South African education system has undergone fundamental transformation with several important policies and their implementation being made in various stages (OECD, 2008). These changes were made along with increased allocation of funds for education. In 2016, education spending was 5.9 per cent of the GDP, as against 5.2 per cent in 1987.[15]

Egypt is another major producer of science. On the economic front, Egypt has made several drastic reforms, particularly in its macroeconomic policy that covered foreign direct investment, monetary policy, taxation reform, trade liberalisation, international trade agreements and public sector reform (OECD and World Bank, 2010).

---

[13] South Africa is a founding member of the General Agreement on Tariffs and Trade (GATT) and later the World Trade Organisation (WTO), and has agreements with the European Union, the Southern African Customs Union (SACU), the Southern African Development Community (SADC), the Trade, Development, and Cooperation Agreement (TDCA) between South Africa and the EU, SACU (with Botswana, Lesotho, Namibia, South Africa), EFTA (European Free Trade Association), along with other preferential trade agreements (OECD, 2006).

[14] The funding framework is to achieve equitable access to education, and quality teaching and research (Essack et al., 2010).

[15] https://data.worldbank.org/indicator/SE.XPD.TOTL.GD.ZS?locations=ZA

Egypt embraced a market approach to economic development and policy reforms in the form of opening up for foreign capital and reducing governmental controls on the agricultural and industrial sectors which have led to some success (Morley and Perdikis, 2000). Egypt rationalised its economy through the formation of its Economic Organisation in the 1950s, which promoted the government's economic policies through wide ranges of economic planning (Ikram, 2006). The five-year plans programme the country adopted was meant to achieve economic growth and development. The five-year plans that began in 1960 made policymakers become more specific about the economic system to be developed for the country (Ikram, 2006). The reforms were continued in the later years. These were manifest in the liberalisation of trade, improvement in the tax system, foreign investment and many others.

During the 1990s, Egypt followed successful macroeconomic stabilisation policies to put the country on a high-growth path (Licari, 1997). Under the economic reforms that were initiated in 1991, Egypt relaxed price controls, reduced subsidies, controlled inflation, cut taxes and liberalised trade and investment (OECD and World Bank, 2010). Reforms were to be seen more clearly in the manufacturing sector with more private participation, and the deregulated agriculture sector (OECD and World Bank, 2010). Major reforms were to continue into the 2000s and the industrial sector was the focus. In 2008, Egypt was ranked the second largest country in attracting foreign investment in Western Asia (OECD and World Bank, 2010). The policy reforms also found Egypt establishing free trade agreements with other countries including the EU.

Egypt was under pressure, as the OECD and World Bank (2010) review showed, to improve its competitiveness in the knowledge-based economy, and to provide for a larger and more diversified student population. Efforts were made in the education sector, which also had an impact. The net enrolment rates showed a significant increase: 83.7 per cent in 1985 to 98.3 per cent in 2003 in primary education; 64.7 per cent to 87.1 per cent in secondary school; and 18.1 per cent to 32.6 per cent in tertiary education during the same period (World Bank, 2006). Egypt's output of graduates grew by more than one million (i.e., 116 per cent during 2005–6), and graduates supplied to health professions such as medicine and pharmacy rose considerably (OECD and World Bank, 2010). Egypt has undertaken a range of reform initiatives on the education front as well. At the national conference held in 2008 and meant to reform secondary education and formulate policies for university education, Egypt received recommendations which had an impact on its education system. These were instrumental in improving the quality of the higher education system by shaping the admission criteria to universities.

Egypt's higher education system grew after 1957, and later in the 1970s the government took measures to consolidate the higher education system (OECD and World Bank, 2010). Free higher education in Egypt was introduced in the early 1960s. It spent a modest of 4.1 per cent of its GDP on education and its higher education share of 28 per cent is higher than that of the average low and middle income countries and even some OECD counties (Assaad, 2010). The government is aware of the directions it needs to take in areas of the capabilities of graduates, graduate output to fit the labour market, supply of higher education opportunities to student bodies with varying needs, and to strengthen university research capacity and its links to innovation. Some of the initiatives that have been undertaken are worth noting. They are measures to improve the quality of basic and secondary education including the quality of teachers and teaching, doubling the funding for higher education under its five-year plans, the formation of the Higher Council for Science and Technology, the S&T development fund, technology transfer centres of networks, as well as competitive funding for performance improvement (OECD and World Bank, 2010).

Until recently, the research, development and innovation system in Egypt functioned under a highly centralised model. This was revamped in 2007. The new system became a predominantly public-funded system consisting of various public sector ministries (OECD and World Bank, 2010). It secured funding from the EU, and was intended to enhance the overall performance of Egypt in its research, development and innovation and to facilitate Egypt's participation in the programmes of European research areas (OECD and World Bank, 2010). As a result, Egypt could develop and advance some of the research areas analysed in the previous chapter. The establishment of the Higher Council for Science and Technology was another major feat for Egypt. This enabled Egypt to align its research, development and innovation system to the national development goals and strategies (OECD and World Bank, 2010).

Nigeria is another major force to reckon with in science production. Since 1999, Nigeria began to introduce reforms in its economy. As a consequence of these, Nigeria transformed into a strong economy in Africa. In its early post-independence years, Nigeria opted for an import-substitution economic policy along with mid-term planning (OECD, 2015). The oil boom boosted its economy in the beginning but was hit later when the oil prices were slashed. In the late 1980s and early 1990s, Nigeria went for a large-scale privatisation programme under which trade was liberalised and government expenditure was controlled (OECD, 2015). Reforms were continued when democracy returned to Nigeria in 1999. The 2003 National Economic Empowerment and Development Strategy emphasised the continuance of economic reforms for growth and development (OECD, 2015). Nigeria made a high-level investment, about 15 per cent of the GDP, in 2003 (OECD, 2015).

Nigeria developed its first trade policy in 1991. It was revised later in 2002. The policy identified its priority areas of export with incentives, outlined the objectives for the most notable sectors and developed an institutional framework (OECD, 2015). As a federal state, the Federal Ministry of Industry, Trade and Investment is responsible for formulating and implementing the country's trade policy. The new National Trade Policy aims to broaden the national product base, bring in more fiscal restrictions, encourage non-oil exports, promotes competitiveness and commit to substantial tariff liberalisation (OECD, 2015). Nigeria has ratified the WTO (World Trade Organisation) agreement and has market-expanding bilateral trade agreements and memoranda of understanding on trade relations and economic cooperation with other countries. The frameworks for these agreements are to optimise benefits of trade and include provision for other areas such as investment, services, intellectual property and competition policy that made an impact on investment and trade (OECD, 2015). The Nigerian Investment Promotion Commission supplied the framework to promote investment in the country. The measures taken in this regard have improved the business environment in the country, attracting foreign investment to different sectors of the economy. As a result, Nigeria became a preferred country in Africa for investment.

Along with the political changes that occurred in Nigeria since its independence, changes have been obvious in other realms including education. There was a departure from the colonial times when the

New Education Policy on Education was formulated in 1977. This national policy is dynamic and subject to changes and amendments to make it more appropriate for meeting the needs of Nigerian society (Imam, 2012). The changes were evident in the planning and administration of universities in Nigeria (Arikewuyo, 2004). The educational objectives that are to be achieved in the country are well stated in its constitution. Policy reforms that occurred in Nigeria are based on these constitutional objectives and imperatives.[16] Accepting education as the tool for socio-economic development, leaders focused on the provision of free, universal primary education and on the creation of a rich stock of talented and skilled manpower (Nwagwu, 1998). The National Policy of Education, first published in 1977 and revised in 1981, 1998 and in 2004, deals with aspects of education such as the philosophy, levels and structure, finance, types of education, administration and planning of education (Sofolahan, 1998). The policy considers education as an instrument par excellence for effecting national development (FRN, 2004). It brought about drastic changes in the curriculum that were aligned to the sociocultural and political economy of the country (Ivowi, 1998).

As regards tertiary education in Nigeria, the policy states that a greater proportion of expenditure on university education will be dedicated to S&T and that no less than 60 per cent of places will be allocated to science and science-related courses in conventional universities and 80 per cent in universities of technology. As a result of the introduction of this policy, there has been a substantial increase in the number of tertiary institutions and in the number of graduates. This has had major implications for the production of scientists and scientific knowledge in the country. The policy emphasised the philosophy and objectives to include the need for the acculturation and use of S&T, and the need for technology transfer and acquisition through mass science education (Ivowi, 1998; UNESCO, 1998).

Section 7 of the policy deals with science education in Nigeria. At all levels of education, S&T education is emphasised to equip students to live effectively in the modern age of S&T (Ivowi, 1998). Under the national education policy, science is to be taught in schools at all levels

---

[16] It stipulates that the government shall direct its policy towards equal and adequate educational opportunities at all levels, promote science and technology, and provide free, compulsory, universal primary and secondary education.

of education, is compulsory in primary and secondary schools, and it is part of general studies at the tertiary level (Ivowi, 1998). The uniform educational standards prescribed in the national policy on education refer to thirty-six different educational prescriptions. They include industrial, technological and manufacturing standards for national development (Akpofure and N'dupu, 1998). The policy also insists on limits of enrolment at various levels of education.

Under its policy, the goals are to produce scientists for national development, cultivate inquiring, knowing and rational minds, make special provisions for the study of science at all levels of the education system, and popularise the study of the sciences. Technology education is also emphasised in the policy. Polytechnics are to provide courses in engineering, technologies and applied science for the production of trained manpower for Nigeria, and to provide technical knowledge and skills required for agricultural, industrial, commercial and economic development of the country. It is also part of the responsibility of polytechnics to offer training for the production of technicians, technologists and other skilled personnel.

7 | Science and Development
Challenges and Prospects for Africa

The study presented in this book began with several research questions that have been examined in the data presented and analysed in the previous chapters. With the support of data, the connection between science and development, science production and economic determinants such as GDP and GERD, prominent scientific research areas, emerging research areas in African countries, dimensions of international partnerships in the production of science in Africa, and policy and its interrelationship with development in science and national development have been examined in detail. In this chapter, the foci are on science from the developmental perspective and its prospects and challenges for contemporary Africa.

The challenges Africa had in past decades have not been completely removed. Africa has a poor record of building institutions for the development and advancement of S&T, and most of the national and regional mechanisms that have been established were only to wither away in time (Gruhn, 1984). Most parts of Africa, the sub-Saharan region specifically, have lost several precious years through wars, genocide, ethnic violence and conflicts, natural calamities and epidemics and due to poor political leadership. Many of them began their lives as stable nations only recently. They now think rather seriously about the development of their people and the use and application of science, technology and innovation to accomplish the priorities and objectives of national development.

The challenges that Africa is currently facing in science development are not limited to the continent. They are common to many other developing countries. Some of these, as Moravcsik (1984) listed, remain valid and relevant for Africa. His view is that the systemic and intricate nature of the infrastructure of science in developing countries is not well appreciated by the decision-makers, which gives rise to consequences for science. One major consequence of this lack of appreciation is that scientific research suffers due to the faulty provisions made for science and that there are an insufficient number of people who know how to make provisions for productive science. Secondly, science education is poor and dysfunctional, as science teachers do not have the personal understanding and contact with science. The third consequence is the weak link between science and technology (S&T) due to the lack of tradition and institutional opportunities to build this crucial link. A sub-consequence of this is that the productive forces in these countries fail to make use of the fruits of local scientific knowledge. The building of linkages between the parts of the system and the elements of science is often neglected, which has major consequences for science development (Moravcsik, 1984).

In this context, education is a key factor. Literacy levels, enrolment figures at various levels of education, and reduced dropout rates to keep the students at school and to feed into tertiary levels of education where scientists, engineers and technologists are formed are important. As seen in the data analysed in the previous chapters, Africa, sub-Saharan Africa in particular, has to move faster in all these measures to catch up with the world. Unless the foundation is strong (school and higher education systems) a strong scientific development is hard to build on. It is obvious that countries that had higher tertiary enrolment ratios either performed well or have a better success in science development.

The declining interest of students in science courses is not new. It is a perennial issue (Balaram, 2009) for countries, specifically in the developing world. Elsewhere it is a different issue. For countries like the United Kingdom, it is not students studying science that has changed but the type of science courses studied (Smith, 2010). Increased enrolment rates in science and engineering subjects and graduates in these courses can assist science production in Africa. As found in the data, in most of the African countries there has been a growing interest in these subjects and an increase in the production of graduates in these between 2011 and 2015. The enrolment ratios for some of the countries are growing encouragingly. Countries need to strengthen their education systems at both school and tertiary levels, through improved enrolment rates and reduced dropout rates. At the same time, the production of graduates in science and engineering subjects is subject to enrolment, advanced facilities for learning, and job opportunities that graduates find attractive.

For Africa to move on the path of scientific research and advancement, a host of issues deserve attention. The modification and orientation of international scientific and technological activities, generation of local scientific and technological capabilities, and the incorporation of S&T within the broader scope of development planning (Sagasti, 1973) are unavoidable. They need to be addressed for the revival and rejuvenation of science in Africa. There are some facilitating conditions, such as the structural and operating conditions, training and qualified research personnel, and the social recognition and status of scientists, that assist in the development of science in society (Gaillard, 1991). Let us consider some of the pertinent issues for Africa that relate directly to both the development of science and society.

Scientific Dependency

In the globalised world, no country is completely independent or dependent. In S&T, every country in the world is dependent on the other (Moravcsik, 1989). Absolute independence or self-reliance in S&T is elusive. Countries can reach a level of independence provided they have made efforts towards independence and minimised their level of dependency. The level of dependency cannot be measured easily. The percentage share of a country to the world production of science (Moravcsik, 1989) is not an accurate indicator either. Perhaps another useful indicator is the per capita contribution to the output of new science (Moravcsik, 1989). This can provide a fair idea about the level of dependency and self-reliance. But the measurement remains a complex one to arrive at an accurate figure. It will have to take into account several indicators: the share of world science, per capita contribution to the output of science, indebtedness in terms of financial and trade balance, and reduced indigenous decision-making power about the courses of action in S&T (Moravcsik, 1989). The dependence of Africa on the developed world for its S&T requirements raises questions about self-reliance and independence in the near future. Dependency is to an extent inevitable but it is not an excuse for failure to develop their own systems that can meet the national objectives of African countries.

In reality, moving from dependency to self-reliance is a gradual and slow process, and demands strategies to reach the destination of a higher level of independence. There are countries that have adopted successful models to achieve this. Japan is a strong case. It achieved technological independence. Its policies were designed to create a complementary relationship between the import of foreign technologies and skills, and learning by doing local technologists (Forje, 1979). Many newly industrialised countries in East Asia have the potential to both obtain technology transfer (that was made possible through flexible industrial structures suitable for rapid adaptation of technology) and to promote their endogenous technological development (Castells and Tyson, 1989).

The transfer of technology is as important as the development of technology and innovation. Not all countries, regardless of their being developed and developing, are self-sufficient in the production of technology and innovation. Complete self-sufficiency is not an achievable goal for any country. Some technologies and innovations are cheaper to buy than to produce. Transfer of technology is increasingly becoming the basis for growth and development (Osabutey and Debrah, 2012). In developing countries, and Africa in particular, technology is still evolving, and adoption and transfer of technology is critical for building capacity (Appiah-Adu, 2016).

The series of studies by the United Nations on the transfer of technology in Africa provide knowledge about its contribution to innovation, entrepreneurship, investment, efficiency, productivity and export performance of African countries. The transfer of technology, as these studies underline, is a means of assisting African countries to meet their various needs in health, nutrition, sanitation, energy and communication. The success of the transfer of technology is possible only if appropriate support mechanisms that encourage technology transfer and its diffusion are in place (UNECA, 2013).

Technology transfer assumes a dual role of building capabilities and improving business performance (Appiah-Adu, 2016). A recent study of this in selected sub-Saharan countries by Danquah (2018) found that the interaction of trade openness and human capital, and that of machinery imports and relative R&D, play a significant role in national efficiency and productivity. Exposure to foreign technology alone cannot bring about the needed impetus to national efficiency and productivity (Danquah, 2018).

There are challenges for African countries in the transfer of technology and its implementation. Some of these challenges are the lack of clear guidelines and institutional policies, lack of sufficient funding, and the low prioritisation of technology transfer as a core activity of

R&D (UNECA, 2013). The study of a sample of African countries by UNECA revealed that the most commonly used mechanisms to transfer R&D outputs are training and the provision of products and services (UNECA, 2013). The transfer of technology in the form of licensing technology to third parties and developing start-ups and joint ventures were not very common in the sampled African countries.

Some contextual factors are also influential in the transfer of knowledge and technology in African countries. Osabutey and Jin (2016) showed in their analysis that factors such as less congestion of firms (foreign and local), policy incentives, effective intermediate industry institutions for the transfer of technology, and educational effectiveness are significant.

The ability of R&D institutions to transfer technology depends largely on the absorptive capacity of firms and communities (UNECA, 2013). Specific studies of technology transfer in Africa suggest that it does not necessarily lead to enhanced capability. In Ghana, as the study by Appiah-Adu (2016) indicated, the relationship between transfer or outsourcing of technology and enhanced capability is significant but weak. In some eighteen sub-Saharan countries, as the analysis by Danquah et al. (2018) showed, the absorptive capacity proxied by R&D has an amplifying effect between technology transfer and efficiency.

Technology transfer is not always the only option for Africa. This occurs even in areas where Africa has the competency for developing local S&T. Countries have to find the right balance between technology transfer and the technology that is developed locally. Local, indigenous and traditional knowledge systems have the potential to contribute to human understanding (IAP, 2016). Any attempt to neglect a planned and vigorous development of indigenous research will endanger the entire process of national development (Dedijer, 1968). Many developing countries imitate their developed counterparts in pure and applied research with very little progress in technological innovation (Wionczek, 1979). It is also important to have local S&T capacity and local knowledge for making a valuable contribution to world knowledge (IAC, 2004). As has been argued, there should be a balance between indigenous technological development and imported technology in Africa, which should be emphasised in S&T policies (Thisen, 1993). No society can rely on technology transfer without doing the groundwork for the production of their own sustainable scientific system and essential scientific personnel. Danquah (2018) argued that for technological advancement to be achieved, the availability of highly trained manpower in science and technology (S&T) and an improvement in relative R&D are required. Similarly, as a study reported from Ghana (Osabutey and Debrah, 2012), technology transfer is likely to be enhanced if the science and technology base is high enough. In other words, a country should achieve a given minimum human capital threshold to enhance the potential of technology transfer (Xu, 2000, cited in Osabutey and Debrah, 2012: 453). The importance of human capital in facilitating the adoption of technology implies that sub-Saharan Africa should promote education policy and enhance the quality of its human capital (Ahmed and Suardi, 2007). It also depends on the success of attracting direct foreign investment with which technology transfer occurs. Support for government policies to attract direct foreign investment, adaptive capabilities, and R&D are instrumental in transfer of technology.

The relevance of local research in a globalised world is a real challenge for Africa. The forces of globalisation, as Krishna et al. (2000) argue, can undermine the relevance of local research if it is not geared to high levels of professional standards. What is required is not just the competency of scientists but also an adequate support system to conduct high-level scientific research.

An important component of technology transfer is the knowledge and skills necessary to install and operate the new technology (Hewitt and Albu, 1998). The effective use and mastery of technology requires knowledge of procedures and understanding these procedures and having skills in putting them to use (Hewitt and Albu, 1998). Quoting a technology transfer expert, Yriart (2000) notes that the higher the level and the more technology one imports, the more one has to know about the basics of the technology to make rational and responsible decisions about it and to make use of it. Countries cannot rely always on skills from overseas that are required to man the imported technology. For the success of any technology transfer, countries need a spectrum of skills, knowledge and capacities for searching, selecting, assimilating and adapting techniques (Hewitt and Albu, 1998). While a country depends partly on technology transfer, it can consciously make efforts to build the supportive systems to absorb the technology first and to create an environment in which it can develop its own technology where it has the advantage and skills. Several African countries have now begun to accept the importance of local, indigenous and traditional knowledge that can be tapped into national developmental strategies. Some countries specifically include this in their science, technology and innovation (STI) plans and strategies (ROK, n.d., 2008; UNCTAD, 2010a). Overdependence also affects the development of indigenous technologies and the capacity to generate new technologies that suit the local conditions in Africa (Wakhungu, 2001). S&T of the Western kind can only perpetuate the deteriorating situation in Africa and therefore Africa should initiate, control and direct its own pattern of development (Forje, 1979). We must also note that, as Sagasti (1973) warned, the S&T in the developed countries are not, in the main, the kind required in underdeveloped countries.

There are more reasons for building Africa's indigenous capabilities (Gruhn, 1984). The diffusion of S&T is based upon the ability of local researchers to modify it to suit local situations. As technology is science-based, it requires doing science first to adapt to local needs. Since the lead-time between a scientific discovery and its technological application is getting shorter, the capacity to participate in the process requires indigenous resources and expertise. If the national capacities in S&T are not developed, Africa will continue to be handicapped in making public choices about S&T (Gruhn, 1984).

There is a trade-off between developing own technologies and complete reliance on S&T transfer. No country can fully bank on buying technology all the time and for all of its developmental needs. It will have to produce its own for its own use.

Note that the institutionalisation of science is the net result of complex social processes and not just the transfer of knowledge (or technology) between centres and peripheries (Saldaña, 2000). Africa stays with its dependency on outside S&T. This comes with a cost. The cost of technology transfer is quite heavy for several African countries. Strong currencies and balance of payments, appropriate structures, skills and resources to absorb imported technology form the basis of successful technology transfer and are necessary to reap its benefits fully. Cost aside, technologies bought from abroad are often outdated and not economically sustainable (Wakhungu, 2001).

Africa needs a local scientific community, as Moravcsik and Ziman (1975) have underlined for developing countries, which is essential for science to have an impact on society. No importation of S&T from overseas can serve the purpose without having local scientific infrastructure and manpower. Even if a country is heavily reliant on adoption of technology produced elsewhere, no country can sustain that adoption. It still needs the supply of scientists, engineers and technicians to run the adopted technology. The skilled manpower required to handle adopted technology has to be produced within the country and cannot be dependent on the importation of human skills. Africa needs to build local human resources to the highest level (Okeke, 2010) so that they become leaders in their own fields of research.

The world scientific community, which is dominated by scientists from advanced industrial countries, so far contributed very little to the fostering of indigenous science in less-developed countries (Moravcsik, 1966; Moravcsik et al., 1975). Developing countries, by and large, are not in a position to generate their own technology or in a bargaining position due to the monopolies of the advanced countries (Brockway, 2011). In the UNESCO Declaration on Science and the Use of Scientific Knowledge, traditional and indigenous knowledge has a firm focus. It states that the traditional and local knowledge systems have made and can make a valuable contribution to S&T and they have to be preserved, protected and promoted (UNESCO, 2000).

Science depends on funding, from both national and international sources (Midgley, 2005). Dependence on foreign funding is quite substantial in Africa. Africa relies heavily on external funding through international collaborative enterprises (Blom et al., 2016; Maassen, 2012). Tom Egwang, an Ugandan immunologist and founding director of Uganda's Med Biotech Laboratories was honest in saying that 'If you look at any of the researchers who carry out any significant research in Africa, 99.9 per cent of their funding comes from outside. I really think that all these programmes are killing African science' (Nordling, 2010a: 1).

Foreign dependence of R&D is up to 60–70 per cent in several African countries (Krishna et al., 2000; Sanyal and Varghese, 2006). This overreliance has its implications on the priorities, research concerns and science development in Africa. African scientists may not be in a state to pursue research in areas that require urgent and immediate local and national attention as external funders may not be interested in those areas. The funding priorities of external donors are not always aligned with local and national needs. It has been, for example, reported to be a matter of concern in Tanzania (Gaillard, 2003b). Tanzanian researchers depend hugely on foreign colleagues in different stages of their research work, varying from collecting literature, analysis of data and publishing the outcomes (Gaillard, 2003b).

While there are some positive changes due to international funding for research in Africa, there are also other effects on local science and development. It is also an issue when international funding agencies pursue their research agenda through those receiving funds, which is not meant for the sustainability of local science (Mouton, 2008). Here again, the issue is the kind of research supported by the research wings of multinational corporations. Such supported research may not necessarily advance local scientific interests but those that have use and value in the international market.

One cannot but agree with Ferenc Glatz here. While making the opening address at the UNESCO world conference on science, Glatz, who was then the president of the Hungarian Academy of Sciences, was emphatic. He said that it will never be the duty of foreign capital to support local science education and innovation as capital is always driven by profit (Glatz, 2000). It is the duty and responsibility of the developing country to find its own resources to develop local science, technology and innovation. It will become the responsibility of the state, as Glatz argues, to take care of science education and the training and supply of manpower, and to ensure that at least basic provisions are made to achieve these.

There have also been serious concerns about the dominance of foreign scientists on the development of science in Africa. The concerns extend from the beginning where a research problem is identified and formulated to the conduct of the research in Africa and to the final publications. This dominance is quite striking in the publications of scientific papers which were analysed in the previous chapters. In the large majority of joint publications in which foreign countries were involved, the first authors were Americans, Europeans or Scandinavians. Scholars (for instance, Okeke, 2010) feel that foreign dominance has persisted for generations in Africa, which presents a picture that research in Africa is externally imposed, which is being perpetuated.

The over-reliance on international collaboration for research output is yet another issue. It affects internal research capacity, the critical mass of scientists to produce quality research, and building networks of collaboration within Africa (Blom et al., 2016). African scientists feel that dependence on international projects affects them in various ways.

Their projects are disrupted when funding runs out. They have to work for the research agendas of the funders and not for African science and policymakers, making it difficult to synchronise with national development priorities (Nordling, 2010a).

Indigenous Knowledge in Africa

Indigenous knowledge, also known as local or traditional knowledge, refers to 'the understandings, skills and philosophies developed by societies with long histories of interaction with their natural surroundings'.[1] This is the body of knowledge that has been developed, preserved and transmitted outside the formal structures of knowledge and educational systems. It is integral to a cultural complex encompassing language, systems of classification, resource-use practices, ritual and spirituality.[2] Indigenous knowledge is experiential-based and depends on the subjective experiences of people to generate social interpretations, meanings and explanations (Dei, 2000) for their lives and livelihoods. It consists of systems of knowledge developed by communities and localities over lengthy periods of time and transmitted from generation to generation. These systems comprise the knowledge of communities in areas which include culture, food, security, health, land, medicine, education, social justice and the management of resources. More generally, they are about the relationship between nature, people and communities. Indigenous knowledge is characterised by a worldview that is based on wholeness, community and harmony embedded in culture and cultural values (Owusu-Ansah and Mji, 2013). Its application in the day-to-day life activities of people and communities has existed for a long time, but has either been neglected or not given the importance it deserved in the knowledge domain.

As indigenous knowledge has been developed by generations of people over a period of time and in a close relationship to nature, it has a system of classification, or sets of empirical observations about the environment which are familiar to the people, and form systems of self-management for their use and application (Kargbo, 2005). In these

---

[1] www.unesco.org/new/en/natural-sciences/priority-areas/links/related-information/what-is-local-and-indigenous-knowledge/. Accessed 30 May 2018.
[2] Ibid.

classifications, the enlarging nature of the knowledge is emphasised. The knowledge is both cumulative and dynamic, building upon the experiences of generations and adapting to changing technological and socio-economic conditions (Kargbo, 2005). Science advances through research while indigenous knowledge develops and builds not through research but through the process of adding new knowledge. Compared to western knowledge, it is typically practical and interpersonal as against cognitive academic knowledge (Mpofu, 2002, cited in Owusu-Ansah and Mji, 2013: 2).

The current debate and discourse concerning indigenous knowledge has earned it a position in a knowledge economy. There is a growing realisation that it cannot be neglected but should be treated as part of knowledge that can be developed and used for beneficial purposes. The modern world is now recognising the value, significance and relevance of indigenous knowledge. Evidence to this effect is to be seen in parts of the world such as Africa, Asia, Australia and Latin America where centres of indigenous knowledge have been established (Hunter, 2005). Such efforts, as Hunter (2005) argued, can lead to the collection, capturing and preservation of indigenous knowledge to increase community-based involvement in planning and development.

Africa is known for its rich resource of indigenous knowledge. Like rich mineral resources, Africa has an expansive reserve of such knowledge scattered far and wide. Elements of this knowledge have been part and parcel of the life of Africans and been vital for their survival. Historically and currently, the value of indigenous and traditional knowledge has been an inseparable part of African life. Even in modern times, the value of this knowledge has not been diminished nor dismissed. Despite advances in science, extensive use of indigenous knowledge prevails on the continent.

Indigenous knowledge in Africa is applied by the majority of communities in areas of health, agriculture and education (Moahi, 2012). The information about it in Africa is recorded through the memory of people and it is embedded in various forms. They are captured in the form of folklore, songs, stories, dances, proverbs, rituals, language, myths, beliefs, cultural values, agricultural knowledge, knowledge of species in their usefulness for medicinal purposes, religious knowledge, art and craft, architectural knowledge, musical knowledge and knowledge about the earth, the stars and water systems (Kargbo, 2005; Raseroka, 2008). Several African countries, namely, Burkina Faso,

Cameroon, Ghana, Kenya, Madagascar, Nigeria, Sierra Leone, South Africa and Tanzania have established indigenous knowledge resource centres (Damme and Neluvhalani, 2004; Moahi, 2012).

Modern science and technology has been experimenting with indigenous knowledge to make contributions to the development of science and technology. However, this knowledge has not been effectively tapped into for the development and well-being of the population; nor integrated with modern science for effective benefits.

The role of information and communication technology (ICT) in the capture, management and dissemination of indigenous knowledge is being investigated (Hunter, 2005). The use of ICT to preserve it has not been very extensive in Africa. As a study done in Tanzania indicated (Lwoga et al., 2010) very few respondents used ICT to preserve indigenous knowledge in agriculture. ICT is an important tool for the capture of the predominantly oral-based indigenous knowledge to facilitate its preservation and access (Raseroka, 2008). There is great potential for ICT to record, manage, disseminate and preserve such knowledge (Hunter, 2005). For instance, as Akinwale (2013) reported, digitisation of African indigenous knowledge can assist in resources management. In Tanzania, a good number of farmers used ICT, varying from cell phones to email, to acquire indigenous knowledge in agriculture (Lwoga et al., 2010). The indigenous knowledge management software system has been designed to enable indigenous communities to manage their own digital collections. This software system has tools for the users to describe digital objects and an interface to search, browse and retrieve collected objects (Hunter, 2005). Digitisation of indigenous knowledge can promote cultures and enhance indigenous capacities in fields such as environmental sustainability and effective management (Akinwale, 2013). This digitisation is underway in many African countries, namely, Kenya, South Africa, Mozambique, Nigeria and Egypt (Akinwale, 2013).

How indigenous knowledge is relevant for Africa is evident from several examples, as shown here.

Indigenous knowledge is being used in many areas of Africa. Traditional medicine is one such area. It has been estimated that one traditional healer interacts with every 500 individuals in sub-Saharan Africa (Maunganidze, 2016). In several parts of Africa, traditional medicine is used in the treatment of some of the continent's common diseases, including HIV/AIDS and cancer. In Uganda, indigenous knowledge is extensively used for the cure of diseases (Kibuka-Sebitosi, 2006). The health system in Uganda incorporates it with modern medicine in the prevention, treatment and control of diseases such as malaria, tuberculosis and HIV/AIDS (Kibuka-Sebitosi, 2006). In Ghana, the use of indigenous knowledge is extensive in the treatment of HIV/AIDS (Tharakan, 2015). In Ethiopia, the promotion of public health through traditional medicines has been validated (Tharakan, 2015). There have also been efforts to integrate the traditional with modern medicine in some African countries (Manyaka, 2006). Indigenous knowledge in treating diseases such as *trypanosomiasis* in animals in Nigeria has been successfully implemented (Atawodi et al., 2002). Morocco, which has a long history of traditional medicine and knowledge of medicinal plants, has done studies on the traditional uses of medicinal plants and indigenous knowledge for the general health care of the population (Bouyahya et al., 2017). The engagement of elders and traditional health providers has led to the reduction of female genital mutilation in Eritrea (Tharakan, 2015).

Agriculture is an area that has widely adopted indigenous knowledge in Africa. Malawi has resisted the use of scientific ideas in agriculture and preferred locally evolved practices (Briggs and Moyo, 2012). The importance and potential of indigenous knowledge in fisheries management in Malawi has been examined (Donda and Manyungwa-Pasani, 2018). Its application in agriculture, food processing and water management in Sudan has contributed to sustainable development (Tharakan, 2015). The use of indigenous practices in food processing in Kenya has been well recognised (Wane, 2014). In Ghana, the influence of indigenous knowledge is evident in agriculture and organic farming (Tharakan, 2015), while in Mali it enjoys the support of agricultural unions for local empowerment (Tharakan, 2015). In both Uganda and Tanzania, the indigenous knowledge systems were effective in the development of vegetable farming that resulted in increased household food security (Tharakan, 2015). In Tanzania, farmers applied indigenous knowledge for crop husbandry, new techniques and varieties and animal husbandry (Lwoga et al., 2010).

Indigenous knowledge has proved vital for the protection and preservation of ecosystems that are sources of food, water, timber, firewood and health in Burkina Faso (Ouédraogo et al., 2014). Environmental management issues in South Africa are receiving the benefits of traditional and indigenous knowledge (Damme and

Neluvhalani, 2004). Borana pastoralists in Ethiopia use indigenous knowledge for sustainable resource utilisation. This information and the related ecological interactions have been documented (Gemedo-Dalle et al., 2006). In Nigeria, studies have been undertaken to understand how traditional knowledge, beliefs and cultural practices strengthen and promote the well-being of children and families (Olaore and Drolet, 2017). In Uganda, the local knowledge system facilitated the engagement of development, implementation and acceptance of a telecentre for information and communications (Tharakan, 2015). The adoption of indigenous knowledge in environmental management has been found to be effective in southern Ethiopia, enabling pastoralists to adapt to the challenging environmental changes in the region (Liao et al., 2016). Community-based mapping projects are found to be beneficial for the preservation of indigenous knowledge and such projects are being undertaken in a few African countries such as Mozambique.

Countries have begun to realise the need to make use of indigenous knowledge for their development-oriented needs and objects. They are beginning to adopt it along with modern knowledge for their varied and relevant development objectives. Both modern knowledge and indigenous knowledge seem crucial for African countries and, in some African countries, policies are in place that deal with this. For instance, South Africa adopted an indigenous knowledge policy in 2004 which recognises, promotes and protects indigenous knowledge systems (Dyll, 2018). It has contributed to the recognition of this knowledge in the country. Since the introduction of this policy, indigenous knowledge has become part of the discourse in science, education, medicine and law (Green, 2012). Namibia stresses in its policy documents that the role of indigenous knowledge is a key engine to solving many of its social challenges and to boosting its economic development (Hooli and Jauhiainen, 2018). This is a contemporary trend in most southern African countries (Hooli and Jauhiainen, 2018).

Universities in Africa are offering accredited teaching programmes in indigenous knowledge systems.[3] At the University of Botswana, a Centre for Scientific Research, Indigenous Knowledge and

---

[3] For instance, North-West University in South Africa has an accredited teaching programme at undergraduate and post-graduate levels (Mubangizi and Kaya, 2015).

Innovation has been established (Moahi, 2012). At the University of Limpopo in South Africa, a Centre of Excellence of Indigenous Knowledge System Curriculum Development exists (Moahi, 2012). The University of KwaZulu-Natal in South Africa has a Research Chair for the development of a bachelor of Indigenous Knowledge Systems degree (Moahi, 2012). The Centre for African Studies at the University of the Free State in South Africa has indigenous knowledge systems as a specialist field (Moahi, 2012).

While the value of indigenous knowledge is recognised, it needs to undergo the same critical scrutiny as western knowledge (Dyll, 2018). This means that no knowledge is accepted before its validity has been tested and approved by the scientific community. The recognition and validation of the legitimacy of indigenous knowledge is hence unavoidable (Moahi, 2012). One way to engage indigenous knowledge and modern science is to use science to evaluate the former and vice versa (Onwu and Mosimege, 2004). This will lead to the complementary existence of both types of knowledge.

There are issues regarding the sourcing of indigenous knowledge from communities. Africans need to make concerted efforts towards the documentation, preservation, protection and promotion of indigenous knowledge systems (Moahi, 2012). A number of actors come into play here. Higher education and research institutions need to have a greater role in documenting and promoting indigenous knowledge (Moahi, 2012). They can conduct and publish research on it (Moahi, 2012). Going further, the use of indigenous knowledge can be explored for innovations. Collaboration with government, universities and research institutions can lead to the use of indigenous knowledge for useful innovations (Moahi, 2012).

Unfortunately, there is a lack of understanding about the relationship between indigenous knowledge systems and other fields of knowledge such as S&T (Mubangizi and Kaya, 2015). But there are exceptions to this as reported from a few African countries. In Namibia, for instance, the role of indigenous knowledge is being recognised in strategies for innovations in university research programmes and in new institutions (Hooli and Jauhiainen, 2018). The institutions and strategies for an indigenous system in Namibia currently focus on the science–technology–innovation mode requiring high-level analytical knowledge and a well-functioning indigenous system (Hooli and Jauhiainen, 2018). Science and technology teachers integrate indigenous knowledge into their teaching in Nigeria (Singh-Pillay et al., 2017).

The importance of indigenous knowledge in policy development has been recognised in Africa. In the realm of science policy, there have been examples showing how local communities with their knowledge have been involved in science policy negotiations (Diver, 2017). In many sub-Saharan countries, indigenous knowledge is a strategic direction of the innovation policy (Jauhiainen and Hooli, 2017). The innovation system, as studies show, can be enhanced by the inclusion of indigenous knowledge (Jauhiainen and Hooli, 2017) from the discovery of appropriate indigenous knowledge to innovation. Once identified, indigenous knowledge can be enhanced through R&D and finally become an innovation which can be introduced to the market (Jauhiainen and Hooli, 2017). We find the meeting point of indigenous knowledge and science here when a piece of indigenous knowledge is extracted and explored for making use of a technological innovation through scientific research. Jauhiainen and Hooli (2017) conclude that, in the development of innovation policies in many countries, the role of indigenous knowledge is significant.

Indigenous knowledge has yet to play an active role in higher education and the knowledge economy. This can be achieved by the promotion and support of the indigenous knowledge system at various levels. Beginning from its validation, indigenous knowledge can be used for the advancement of modern scientific knowledge through its integration with it. Innovations in science and technology are also possible through documentation, preservation, dissemination, validation and integration.

The call for indigenous knowledge to feature in the curriculum (Matemba and Lilemba, 2015) is becoming stronger. This makes for its entry into formal education systems that will pave the way for its further advancement through research. South Africa has included indigenous knowledge systems as one of the principles underpinning the South African curricula (Taylor and Cameron, 2016). The country's new curriculum policy and the curriculum assessment policy encourage the inclusion of indigenous knowledge systems. The science curriculum development takes into account indigenous knowledge systems in its early years (Diwu and Ogunniyi, 2012; Onwu and Mosimege, 2004). Its inclusion in the curriculum policy is regarded as a positive step and it opens up opportunities for a debate on the interaction between indigenous and Western worldviews (Grange, 2007).

The integration of traditional knowledge with modern knowledge is to recognise that different knowledge systems can coexist and complement each other (Dei, 2000). The call for the integration of two systems of knowledge has met with mixed responses as it has proved to be an elusive goal (Diwu and Ogunniyi, 2012). For a genuine synthesis of knowledge, a collective and collaborative enterprise is required (Dei, 2000). The challenge however is to validate the legitimacy of indigenous knowledge and to integrate it as a pedagogic and instructional tool in education (Dei, 2000). In order to realise this, indigenous people should be allowed to produce and control knowledge about themselves and they must own their past, culture and traditions (Dei, 2000). A few examples can be cited.

In countries like South Africa, indigenous knowledge systems as a research, teaching, learning and community engagement programme is a new research and teaching field (Mubangizi and Kaya, 2015). In Namibia, under a traditional life skills project, the traditional skills and knowledge were integrated into the extracurricular activities of schools in the Karas region, which has had positive outcomes (Klein, 2011). In Mali, the indigenous knowledge system is integrated into the implementation of new technologies and practices, enhancing sustainable development (Tharakan, 2015). In Burkina Faso, indigenous knowledge in rainfall forecasting by traditional farmers who follow indigenous practices is adapted into modern scientific rainfall forecasting (Tharakan, 2015). The integration of mainstream indigenous knowledge into scientific knowledge systems has been called for in areas such as sustainable development (Okoti et al., 2006).

As no knowledge is complete or incomplete, comprehensive or incomprehensive, superior or inferior, perfect or imperfect, the use of it for common benefit is what is needed. Indigenous knowledge in Africa is able to contribute a great deal to the understanding of many phenomena and problems varying from agriculture to diseases. Africans cannot afford to neglect or sideline the value of this knowledge. Indigenous knowledge can be used along with modern science. In other words, integration of both indigenous and modern science can make a difference to the development of the body of knowledge.

Funding Allocation

Funding for science has many objectives to achieve. It enables young people to pursue a career in S&T and it allows society to make better decisions for itself (Besley, 2016). While funding for specific subjects or focus areas is unavoidable, a country cannot neglect the overall benefit expected to accrue from it. Funding is invariably limited for science and no amount of funding is enough. Judicious allocation and use of funds is therefore a matter of national interest, and to be aligned with national priorities, strengths, weaknesses and targets. This is why funding agencies, including governments, identify key areas they find it necessary to support. But this should take into consideration both the short-term and long-term objectives outlined in the policy document. Individually funded projects in any of the focus areas should have an underlying thread that connects to all those to achieve a common aim and purpose that the country aims to achieve. The outcomes of those several individual projects are to contribute either to develop capacity of the country in a specific field or to find solutions to contemporary problems. In the long run, the country should have benefitted from the funding.

A funding mechanism at the national level has been found to be effective in many countries. South Africa is a case in point. It has a strong national organisation, the National Research Foundation (NRF), that distributes millions of rands for research every year. The NRF, from time to time, identifies focus areas of research taking into account national developmental needs. The scientific community now largely accepts a system of rating scientists, adopted by the institution. Funding is linked to one's rating obtained through a peer-review process. Substantial funding in South Africa is associated with enhanced research performance (Fedderke and Goldschmidt, 2015).

Funding for research should be open to all and should be through a transparent, competitive and peer-reviewed process, which is the 'gold standard' for research funding (Blom et al., 2016). The approach of funding agencies towards supporting research has also shifted, accepting them as instruments of funding for attaining national, technological and social priorities (Gibbons, 1999). In some countries, this has to be aligned with the national policy to accommodate some groups of scientists. Research funding in South Africa takes into account the transformation agenda of the nation. This is to encourage scientists who were disadvantaged during the apartheid period which officially ended in 1994. As part of transformation, black African scientists are supported more than other racial groups.

It is not unusual to find that some institutions are preferred to others when it comes to research funding, either by the government or the private sector. Well-run and efficiently managed institutions receive continued support as they perform better than others. It is easier for them to source funding as they already have an impressive track record. They have earned a name for their competence and transparency which weigh in their favour in competitive funding applications. Funders do not hesitate to put their money into them as there is a guarantee that the money will be well spent. Such institutions are careful to maintain their reputations. They have proper mechanisms in place to ensure that the research conducted is up to standard and will be maintained or improved. Robert Merton calls it the 'Matthew effect' which applies not only to individual scientists but also to institutions. Skewed distributions of resources and productivity are well known among scientific institutions (Merton, 1988). The centres of historically demonstrated accomplishments attract larger resources of every kind, human and material, than other organisations which have yet to make their mark (Merton, 1988). For instance, universities with great resources and prestige attract disproportionate shares of the most promising students and are able to identify and retain prime movers in contemporary science (Merton, 1988). One cannot find fault with this current practice. The domain of science is very competitive, and scientific research has to be path breaking and cutting-edge, for which competent scientists and adequate infrastructure are unavoidable. Funders want to see concrete outputs and deliverables for the money they are ready to spare for research. New and young institutions can also grow and become recognised in due course.

Brain Drain and Gain

Since the term 'brain drain' was first used in 1963 when the United Kingdom was losing their best talents and brains to the USA (The Royal Society, 2011), it has become a serious and contentious issue in science. The brain drain has become an important area of concern as highly educated workers generate positive externalities for society while they are lost to their countries of origin (Özden and Schiff, 2006). The positive externalities, according to Özden and Schiff (2006), that are lost for countries of origin include the positive effects on the productivity of colleagues, employees and workers, and the provision of key public services with positive externalities such as education and health. The extreme size of the brain drain makes it relevant to study the issue (Weinberg, 2011). The movement of individuals with expertise and skills that are in demand is progressively increasing in the academic profession (Ben-David, 2008). In 2017, the UN estimated that 258 million people are living in a country other than that of their birth, which is an increase of 49 per cent over the 2000 figures.[4] About 3.4 per cent of the world's inhabitants are now international migrants.[5] The out-migration of workers with a tertiary education from low-income countries to middle-income countries is in the region of 6.1 to 7.9 per cent.[6] According to Mazrui (2002), two interrelated forces are responsible for the brain drain: the uneven development between the global North and South, and the new levels of mobility fostered by globalisation. The disparities between countries in the North and South facilitate emigration and immigration.

The brain drain is a very complex phenomenon that has political, economic and psychological dimensions (Moravcsik and Ziman, 1975), and far reaching consequences for developing countries.[7] Developing countries from where the flight of skilled workers occurs suffer from lower levels of human capital and skilled workers (Özden and Schiff, 2006). The effect of the outflow is manifest not only in science, technology and innovation but also in the general development of the country. The study of the problem is essential for the understanding of the scientific performance, knowledge production and knowledge diffusion in developing countries (Weinberg, 2011). Countries struggle to deal with the situation. In China, it became a policy priority to bring back talent when 70 per cent of the

---

[4] www.un.org/development/desa/publications/international-migration-report-20 17.html. Accessed 30 May 2018.

[5] www.unesco.org/new/en/natural-sciences/priority-areas/links/related-information/what-is-local-and-indigenous-knowledge/. Accessed 30 May 2018.

[6] Docquier and Marfouk (2006) cited in Weinberg, 2011: 95.

[7] Studies (Weinberg, 2011) show that large countries with higher per capita income and higher education levels in comparison to other countries, experience relatively smaller brain drains. As Weinberg (2011) noted, large countries produce more important scientists than smaller countries, and education is also related to the number of important scientists born in a country.

1.06 million Chinese who studied abroad during 1978–2006 did not return (The Royal Society, 2011). They introduced the Thousand Talents programme to bring back their people who lived elsewhere. India established a separate ministry to make the return of the Indian diaspora easier. Malaysia introduced a Talent Corporation to address the issue effectively.

The haemorrhage of talent from Africa (Jonathan et al., 2010) is serious.[8] The brain drain has hit Africa very severely as it is still the poorest region in the world (Campbell, 2007). One estimate (Kaba, 2011) shows that at least 16 million African immigrants live outside the continent, with a very high number in developed countries. In terms of the proportion of the total educated force, the highest migration rates are from Africa, the Caribbean and Central America (Özden and Schiff, 2006). Since 1980, international migration of skilled human personnel from and within sub-Saharan Africa in particular has surpassed previous figures (Campbell, 2007). This period was characterised by political instability, ethnic conflict, poor governance, the decline of economies (Campbell, 2007), corruption, human rights violations, natural disasters and epidemics. Emigration from Africa and within Africa began to peak from the 1980s onwards. The USA was the main winner in this. In 2002, 400,000 African workers 16 years of age and over arrived in the USA, and 36.5 per cent of them were in the managerial and professional ranks (Mazrui, 2002).

Africa needs to keep its human resources but arguably offers the least to attract them back (The Royal Society, 2011). The brain drain results in the loss of skilled human resources in countries which had invested hugely in them. The loss of physicians from African countries, as the study of Kalipeni et al. (2012) showed, is troubling. They are lost to Europe and North America. Nearly 86 per cent of the Africans practising in the USA are from Nigeria, South Africa and Ghana (Hagopian et al., 2004). This has implications for science and development in African countries. For instance,

---

[8] Some figures are in order here. According to the Research and Development Forum for Science-Led Development in Africa, up to 30 per cent of African scientists are lost to the brain drain; the Economic Commission for Africa and International Organisation for Migration estimated that 27,000 skilled Africans left their home countries for industrialised countries between 1960 and 1975; and since 1990 at least 20,000 qualified Africans left the continent every year (cited in Mouton, 2008: 49–50).

very few of the bright students who go to developed countries for higher education return to Africa (Jonathan et al., 2010). In the case of Nigeria, the outflow of qualified scientists destabilised many of its research institutes and universities (Lebeau, 2003). Likewise, Ethiopia continues to lose its talent to other countries in Africa and elsewhere. The highly qualified people look for positions in international and multinational corporations, not always for a better pay but for pursuing a successful career (Ayalew, 2012). Ghana admits in its national STI policy that it is affected by the brain drain (ROG, 2010).

Both economic and non-economic factors are part of the push and pull. Poor working conditions, low salaries, political instability, conflicts and proximity to OECD countries are some of the pull factors for this phenomenon (Confraria and Godinho, 2015).

Beine et al. (2008) found factors such as GDP per capita, geographical distance between countries of emigration and immigration, colonial links between the host countries and their former colonies that minimise cultural distance, linguistic proximity, sociopolitical environment of the country of emigration and the size of the country of origin. Gaillard (2003a) reported that over 90 per cent of scientists in sub-Saharan Africa are not satisfied with their salaries. South Africa, Botswana and Namibia may be an exception.

Empirical studies on the brain drain shine a light on this phenomenon: maintaining a strong scientific community is important for development; developing countries should produce and retain important scientists; while developing countries produce a sizeable number of important scientists, they experience tremendous brain drain; education levels, population and per capita GDP are positively related to the number of important scientists born in and staying in a country (Weinberg, 2011).

Campbell's (2007) study of students in Botswana identified factors that motivated them to consider migrating. Prominent among them are cost of living, ability to find a job, prospects for professional advancement, HIV/AIDS in the home country, job security, income, good schooling for children, medical services, personal and family safety, quality of public amenities and better customer service. Issues of low wages, crime, armed conflicts, political repression and poor educational systems are also among the push factors from Africa (El-Khawas, 2004).

The migration of people endowed with high levels of human capital is undoubtedly detrimental to the growth of the country (Beine et al., 2001). Small states are the main losers because of the negative impact of country size and because of the skilled migrants' greater sensitivity to push factors (Beine et al., 2001). The study by Campbell (2007) highlights the impact of the potential migration and brain drain on the socio-economic development of Botswana. Campbell makes it clear that, given the massive financial and human investment that the government has made over the last three and a half decades, the prospects of the brain drain within the next decades have policy implications. The inability of the government to acknowledge the relationship between migration and development is partly responsible for its unpreparedness (Campbell, 2007).

Political stability provides a facilitating condition for scientists to stay and work for their country. When day-to-day life is disrupted due to conflicts, wars, ethnic struggles and political fluidity, the professional lives of scientists are affected. As the African Union (AU) reported, conflicts, violence and civil wars are negatively affecting universities and centres of research (AU, 2014b). Scientific research suffers in such an environment. Scientists who are committed to their discipline and profession look for opportunities where they can conduct research peacefully, efficiently and successfully. Scientists move countries, not always for better remuneration. They move when the existing environment is not conducive for their research aims, plans and progress. True scientists who are passionate about science and maintaining the spirit to undertake rigorous scientific research consider basic conditions for research above anything else. Salary and benefits appear to be secondary to basic conditions where there are advanced cutting-edge facilities for research (Castells, 1993). While it is not possible to stop educated and skilled people from migrating, there are other options. The theoretical model developed by Beine et al. (2001) suggests that subsidies to education are likely to be inefficient if the probability of migration is high. But no subsidy is required to foster human capital formation (Beine et al., 2001).

Africa in general has not been very successful in bringing back their talent who are working elsewhere in the world. Although those emigrant scientists wish to see their home countries developing, no compelling reasons attract them to return. The pull factors should be more forceful than the factors that force them to stay in the countries they work for. They should feel that they are wanted in their own countries and that the society values their knowledge and experience. The basic condition to attract the diaspora is to provide the necessary infrastructure for scientific research, and the commitment of the country and political leadership for scientific growth. Moravcsik (1986b) suggested that the best remedial measures to offset the brain drain are to provide a congenial research atmosphere, absence of bureaucracy, availability of auxiliary services for research and recognition. Iran, under its comprehensive plan for science, prevents emigration of scientists by boosting employment opportunities in higher education, providing them with special facilities and preserving their social status in the society (Sawahel, 2009).

Scientists move between countries within Africa as well. The pull is often from countries such as South Africa, Egypt and Tunisia where opportunities and facilities for scientists are tempting. Scientists from Kenya and Tanzania work in Botswana and Nambia (Campbell, 2007). Some countries, such as South Africa, attract scientists and students from other African countries. A good number of Masters, PhD and post-doctoral students seek admission to South African universities that have a higher international ranking. Many of these graduates do not return to their home countries but find employment in the higher education sector or in the fields of S&T. Countries will have to design strategies to bring them back to their home countries for their contributions.

African students, mainly from the sub-Saharan region, are the most mobile students in the world (UNESCO, 2006). According to 2013 figures, one out of every nine people who was born in Africa and who holds a tertiary diploma lived in OECD countries (Kigotho, 2013; OECD-UNDESA, 2013). The brain drain, as the OECD-UNDESA (2013) report confirms, is acute for small countries in Africa.

While it is not possible to stop the educated and the skilled from moving and migrating, attempts can be made to utilise this phenomenon to the advantage of countries that are affected by it. This is the brain gain. In order to address this effectively and turn it around, it should be approached from multiple fronts.[9] Establishing contacts

---

[9] There were several initiatives aimed at containing the brain drain. The UNESCO and UNDP joint initiative in 1996 with the University of Mali, the Transfer of Knowledge through Expatriate Nationals (TOKTEN), to encourage overseas academics and experts to take up short-term contracts in their home countries,

with the scientists in the diaspora, networking with them, joint meetings, conferences and workshops, scientific visits and exchange and graduate supervision (Badran, 2000) are some strategies for brain gain. The Inter Academy Council (IAC, 2004), in its well-prepared report, recommends that governments should consider providing special working conditions for the best talent in S&T at least for a temporary period of time, including income supplements and research support. Governments should also establish ties with their expatriate scientists and engineers who are currently working in the developed world (IAC, 2004). Cameroon is now offering competitive salaries and part-time positions for its scientists working overseas. The part-time opportunity allows Cameroonian experts in other countries to spend time in Cameroon while keeping their positions in developed countries (Oukem-Boyer et al., 2009).

If scientists who migrated to developed countries return to their home countries there will be a great impact on science and development in their own countries. This occurred in East Asia in the mid-1980s. Krishna et al. (2000) record that this brain gain helped East Asia in different ways. One, it helped increase the scientific output of publications. Two, it assisted in strengthening research institutions in the home country. Three, it contributed to the development of new and high-technology industries. The same would occur in Africa only if conducive conditions are created for African migrants to return to their home countries.

Science Education and Awareness

A group of young scientists from fifty-seven countries met at the International Forum of Young Scientists in 1999 to make recommendations that were submitted to the World Conference on Science. The recommendations of these young scientific minds are worth considering. According to them, science education must be strengthened at all levels. This is important to allow scientists to work with educators, scientists to assume increased responsibility of development programmes, young scientists to participate in decisions made about

the UNESCO joint project with Hewlett Packard in 2006, the Piloting Solutions for Reversing Brain Drain into Brain Gain, aimed at establishing links with people who left their native countries, and the Academics Across Border (AAB) are among them (Teng-Zeng, n.d.; UNESCO, 2006).

science, and scientists to increase their responsibility to inform the public about research and its wider implications (UNESCO, 2000).

Many reasons contribute to the lack or low levels of awareness of the importance of science in society. The low priority given to science in development and the lack of cultivation of scientific potential to produce science are among them. As has been identified in the Delphi survey conducted by the UNESCO (1974a) among the African and international communities, the lack of basic education of the masses and lack of awareness of the potential benefits are a barrier to scientific advancement in Africa.

There are challenges for science education at both school and university levels (Osborne, 1971, 1999). A functional basic education system is essential for the development of science in Africa (Allotey, 2000). In order to have good scientists, there should be good teachers; and to have good teachers, there must be good scientists (Moravcsik, 1986a). The lack of qualified and committed teachers and the lack of science facilities (Wandiga, 2000) have been persistent obstacles in science education. Any sustained efforts to deal with those challenges will take time to bear fruit. They are not easy to overcome for developing countries like those in Africa.

The backwardness in science, technology, engineering and mathematics (STEM) relates to the low levels of quality basic education in science and mathematics in sub-Saharan Africa (Blom et al., 2016). Even South Africa, the leading producer of science, cannot take credit for having a strong science education system at school level. In a frank statement, the minister of basic education admitted that there is a serious problem in teaching mathematics in South African schools (Wolhuter, 2016a, 2016b). The comments of the minister were made against the background of the recent Trends in International Mathematics and Science Study (TIMSS) in 2015, which ranked South Africa among the five lowest-performing countries. Empirical studies (Wiseman, 2012, for instance) have revealed that student poverty is a stable and strong predictor of science teaching and learning and student performance in science in South Africa. Students perform consistently poorly on the TIMSS (Wiseman, 2012).

Lack of interest in science education is mainly due to the way science is taught. Teaching and learning of STEM subjects generally occurs in a passive and inactive manner, inhibiting the creativity and participation of the learners in the classroom (Power, 2000). An understanding of the culture and logic of a society (Verran, 2001) for imparting science education cannot be undermined. Teaching science or learning to think scientifically is a long and complex process which cannot be speeded up by teaching scientific facts and theories by rote (Ziman, 1978).

A school system that supports science education is a prerequisite for budding new scientists and researchers. For children, science can become a passionate subject that challenges their inquisitive minds. Interest in science early in children's lives can influence their decision to pursue a career in science later (Xu et al., 2012). The beginning step is to ignite that passion and nurture it at the school level. The experiences of accomplished scientists who became interested in science subjects when they were young at school point to the way science was taught by their teachers. Interesting moments with science and spending time in laboratories watching the colourful and sometimes noisy chemical reactions, or touching and feeling the bones of primates, or curiously looking at the animals and plants preserved in glass jars and examining the viscera of small animals on the dissection table are pleasing environments for curious students. Science flourishes and attracts a new generation of children to science in such an inspiring environment. Better classroom practices, positive experiences in learning science, and providing information about the future use of science in life have also been suggested as ways to interest children in science (Singh et al., 2002).

There are success stories; for instance, the Asian tigers. These countries were once poor. They developed a good education system at all levels and government, the private sector and private universities invested in the quality and quantity of education in these East Asian countries (Castells, 2009). Finland is another example that is credited with an outstanding basic education system on which their higher education system is built (MacGregor, 2009). A reform in higher education may be warranted to have positive implications for science (Nour, 2011).

Kenya, recognising it as a critical area of concern, included STEM training and an education programme that begins at the school level in its Vision 2030. Uganda, under its Vision 2040, plans to run a special science and mathematics awareness programme from pre-primary to higher education levels for the younger generation to pursue careers in

S&T (GOROU, 2013). South Africa allocated funds to implement a coordinated approach to science education, science awareness and science communication through the South African Agency for Science and Technology (GCIS, 2015). It also promotes development in the public interest of discoveries, inventions and innovations, which is specified in the Technology Innovation Act, 2008. Ghana introduced strategies such as an incentive system to attract science teachers and students as part of its national developmental priorities (Teng-Zeng, n.d.). Under the new S&T policy, Rwanda is to introduce new incentive schemes such as packages to reward technical achievers, rewards for individuals and organisations for outstanding achievements in S&T and special benefits for those with skills that are in high demand in the country (ROR, 2006).

School curricula might require revision to provide the added focus on science. This is to produce a population who is receptive to the scientific approach and ideas, and that is willing to integrate science into their life and culture. Along with these, extra-curricular programmes in the form of science clubs, quizzes, exhibitions, discussion forums and competitions can supplement the promotion of science in schools. Children are to be attracted to science and told of the advantages of pursuing a career in science. Identifying talents and nurturing them should begin at this stage. They are also influenced by the importance science receives in the country. They become aware of the benefits of choosing science; not just monetary privileges, but the contribution science can make to the development of their own country. Knowing the role of schools and curricula, countries have begun to turn their attention to them. Many of them have integrated defined objectives as part of their S&T policy.

Malcolm Gladwell (2009) in his book, *Outliers: The Story of Success*, narrates the story of an experimental public school called KIPP Academy.

This school was opened in the mid-1990s in one of the poorest neighbourhoods (South Bronx) in New York City. At this KIPP Academy, which is a middle school, classes were large, and students were admitted not through entrance exams but through a lottery method. They were either African American or Hispanic, three-quarters of them from single-parent families and 90 per cent of them qualified for a free or reduced lunch. Inside the school, a different atmosphere persists. In ten years, it has become one of the most desirable public schools in New York City. In 2008, hundreds of families across the Bronx participated in the lottery for admission to KIPP. Why?

KIPP was most famous for mathematics. By the end of fifth grade at KIPP, most of the students call math their favourite subject. In the seventh grade, they start high school algebra and by the end of the eighth grade, 84 per cent of students were performing at or above their grade level. Now there are over 50 KIPP schools across the USA and its educational philosophies have become the most promising in the country.

Promotion of science awareness programmes and campaigns involving organisations such as non-governmental organisations (NGOs) have been successfully employed in many countries (Zachariah and Sooryamoorthy, 1994). A few central American countries have used methods such as fairs, academic contests and the inclusion of specific topics in the curriculum to promote scientific and innovation culture (Padilla-Pérez and Gaudin, 2014). These measures meant for the popularisation of science have a direct bearing on progress and prosperity, and the influence of science on society is largely determined by the development of science and the extent of public understanding of science (Xun, 2000). They can bring science closer to the public (Gaudin, 2000).

Public perceptions of science are also a combination of scientific and technological optimism, views of social responsibility and doubts about its social implications (Prpić, 2011). But the reality is that scientists do not show as much interest in the popularisation of their own science, discoveries and inventions as in their own research. In order to regain trust and support from the public, scientists have been urged to renew their commitment to public engagement (Rogers, 2005). It is the responsibility government and policymakers to ensure that science is administered in such a way that society wins the greatest return for the investment and efforts (Madox, 1968). In order for any society to affirm its faith and confidence in science, the returns science brings back to it, in whatever forms (technological advancement, for instance) are of great value. The support of society to S&T emanates from such circumstances.

As Piltrim Sorokin considered, the theories of science and the rate of scientific advancement are dependent upon the underlying cultural premises, and the forms of knowledge are dependent upon the type of sociocultural system (Merton with Barber, 1973). The cultural context that values scientific and technological innovation therefore has a direct influence (Merton, 1978). Any type of science, from the production point of view, cannot ignore the cultural premises in which the scientific system is embedded. The value of science and scientific culture is ultimately provided by society. The support science receives from society will not be unconditional but rather depend on some external criteria such as technological merit, scientific merit and social merit (Weinberg, 1968a). These are however not easy to assess and measure from time to time but scientists need to consider these in their scientific research so that these foci are not lost sight of in their scientific endeavours. It is also crucial that conflicts between science and the general public should be avoided so as not to imperil science (Polanyi, 1968). NEPAD (2005) acknowledged some time ago that the bond between science and society is weak in Africa and that there is an absence of a strong culture of science to promote science.

There are ways to strengthen science in society and to earn the support of society to scientific endeavours. Moravcsik (1984) thinks that 'linking science' should be one of the main objectives of science. According to him, if science remains a purely internal game within a group of scientists, the justification of science and support of society falter. Therefore, linking science with other entities outside its realm is the key. This link can be established and maintained in many forms; the link between S&T, for that matter. Science transmits the results of scientific problems to technologists in a form they can digest. Only then is the link complete and scientific knowledge transformed into technology that is useful for society. The ineffectiveness of scientific establishments in many developing countries is due to the lack of links between various parts of the establishment and the fragmented state of scientific and technological infrastructure (Moravcsik, 1982).

In the modern world and modern science new relationships are emerging. The interrelationship between science and society is closer and more intense than it was a few decades ago. Expectations from both sides have changed and transformed. Science spoke to society and now society is able to speak back to science, and reciprocity is required in which the public understands how science works and science understands how its publics work (Gibbons, 1999; Lane, 2000). The growing interest in the science–society relationship has led to more creative engagement and participation in science (ICSU, 2005). The relationship is set to influence both the directions and practices in science, and in this relationship scientists need to be responsive to the changing needs of society and society needs to understand and support science (ICSU, 2005). Public understanding of S&T should be fostered so that people can take advantage of the products of science and new knowledge (Lane, 2000). In the new scenario, this relationship has become a more mutual, bilateral and beneficial one. The role of science and scientists is being emphasised and society seems to accept the changing and enlarging role of science more than before.

The call for a new social contract in science (Agazzi, 2000; Alagh, 2000; Foray, 2000; Gibbons, 1999; Lubchenco, 2000; Vavakova, 1998) is likely to cement the bond between science and society for a more symbiotic and beneficial existence.

New forms of communication which are different from the conventional ones are emerging and are being experimented with. One of these is the form of science communication that is based on the principle of constructive dialogue between people and scientists. In this form, citizen juries, deliberate opinion polls and consensus conferences are undertaken (Durant, 2000). These bring together groups of citizens for careful deliberations on matters relating to S&T, and feed back these deliberations into the national policymaking process.[10]

How can the role of S&T be enhanced in the development of Africa? Forje (1979) believes that a more meaningful role for S&T requires extensive changes in the expansion of education and training along more socially constructive lines that are related to the needs of the people, resources and productivity. Efforts towards strengthening education and awareness will be useful for African societies given their character and nature. Such endeavours will contribute towards building a scientific culture and incorporating it into the values of society. Since independence, several countries in Africa have made efforts to concentrate on their science education 'through the development of curricula' (Ajeyalemi, 1990).

In the era of new media and communication technologies, dissemination of scientific knowledge and scientific ideas spread much faster and wider than previously. Accessing scientific knowledge in the past

---

[10] In Denmark, the consensus conference was successfully piloted to involve people in science and technology policy issues. This model began in the 1980s and has been adopted by many in later years (Durant, 2000).

was limited to manual searches in libraries resulting in restrictions of the expansion and spread of scientific knowledge. Information is now available at our fingertips on the worldwide web. Science can make use of this facilitating platform more effectively to attract young and curious minds, and can create a scientific culture in the society. Creative minds can now develop their creativity as information about ideas, principles and methods are in the open domain. However, one cannot overlook the existing unequal distribution of communication technologies in Africa (Sooryamoorthy, 2017b) that prevents the easy flow of information, its easy access and the unleashing of the potential for science development. This remains a challenge for Africa.

Media has an indispensable role in the promotion and support of science in society. Not only can media present news about break-throughs in science but it also has a responsibility in keeping up public interest in matters concerning science. Regular interesting features of science and how science is influencing and changing the lives of the people will be interest to the public. Recent research (Appiah et al., 2015) in Ghana reported the importance of science reporting of science stories and news in the media. One key finding of this study is that scientific institutions should make the contact information of research-ers and their research findings accessible to science reporters. This would help science and interest in science. In Uganda, under its Millennium Science Initiative, financial provisions have been made to improve public understanding and appreciation of science in society (Blom et al., 2016).

Increased productivity arises from exposure to information and communication technologies (ICT) (Ngwainmbi, 2000). The supply and use of ICT is rather limited in Africa and this limited presence in both public and private sectors is disadvantageous (Ngwainmbi, 2000), and has negative consequences for research in Africa. Scientists need fast and reliable ICT that is available to them at all times; not just in their offices and laboratories but also in their homes. Scientific research is not confined to work places or limited to office hours. It transcends from workplace to home and home to workplace. Communication therefore has to be incessant and uninterrupted. Modern communication technologies including Voice over Internet Protocol (VoIP) and other internet-based communication media come in handy for scientists in their daily routine research activities. They have become indispensable tools in scientific research. Electronic communication, including audio and video calls over the Internet, has become cheaper than before, but Internet connectivity has not. All free online communication, voice and video, is possible only when there is Internet connectivity with sufficient bandwidth. Access to the Internet has not yet permeated to Africa as it could, which frustrates scientists.

Research Capacity

At the plenary of the World Conference on Science held in 1999 at Budapest, Hungary, representatives from 155 countries and NGOs projected a new commitment for the future of science. One of the key priorities of this commitment covered the areas of capacity building in developing countries.

Research capacity in Africa, more specifically in some specific fields such as medical science, has improved substantially. There is a shortage of faculty and research leaders, inadequate research facilities and infra-structure, and few career opportunities for young researchers to advance (Davies and Mullan, 2016). According to the S&T capacity index, only one country from Africa, namely, South Africa, is among the scientifically proficient countries in the world.[11] A few are in the group of scientifically developing countries (such as Benin and Egypt) and most others are in the group of scientifically lagging countries.

Building research capacity should focus on all essential stages of research. In this process, building existing capacity, expertise and institutions are unavoidable (Jones et al., 2007). Research capacity cannot be achieved without making strenuous efforts on several fronts. National infrastructure of communication and transportation systems; legal and regulatory structures for science development; the pool of scientists, engineers and skilled personnel; research facilities such as laboratories and libraries; and academic institutions are part of this (Wagner et al., 2001). A country should be prepared to work on all these prerequisites, ranging from upgrading infrastructure for scientific research and promoting science through clearly defined policies,

---

[11] Wagner et al. (2001) have developed an index to measure S&T capacity in countries. The components of this index are the per capita national product (GNP), the number of scientists and engineers per million population, the number of S&T journal articles and patents produced by the population, the percentage of GDP spent on R&D, the number of universities and research institutions per million population and the number of patents registered.

allocation of resources and availability of manpower. Capacity building takes time and persistent effort on the part of governments for a lengthy period of time.

Initiatives to strengthen research capacity, research management and research performance in Africa are ongoing, and some of these are extensive (*Lancet*, 2009; Ubogu and Van den Heeve, 2013). The Initiative to Strengthen Health Research Capacity in Africa (ISHReCA) recommends that supporting individuals and institutions are the key requirements to strengthening research capacity. As shown in specific case studies, research capacity can be improved at the university level (Vellho, 2004).

Research capacity should be a primary focus of the sustained production of scientific knowledge. Knowledge production is not confined to a single centre. Increasingly, more non-academic institutions produce scientific knowledge (Hicks and Katz, 1996). Universities nevertheless continue to be the centre of knowledge production while other related sectors, such as hospitals, government laboratories and industries, are also part of it (Godin and Gingras, 2000). Universities occupy a privileged global institutional position as the focus of knowledge is legitimate and essential (Gabler and Frank, 2005). In sub-Saharan countries universities are the most important scientific institutions that account for a substantial proportion of the national expenditures, a larger share of scientists engaged in research, and the bulk of national production of influential scientific research (Eisemon and Davis, 1992). Their contribution however, as Cloete et al. (2011) noted, depends largely on the nature, strength, size and quality of the academics, and the pact between universities, political leadership and society. Unfortunately, the centres of research productivity, namely universities, in Africa are not well equipped to drive scientific production the way Africa needs at this time.

Although universities in Africa have the staff complement of competent academics to conduct scientific research most academics' time is not spent on research. There are no coherent research policies in several universities in Africa (Ahmed, 2004). While universities remain the centre of production of scientific research, in Africa, sub-Saharan Africa specifically, expectations and performance do not always correspond. Universities can take the lead in responding to the sustainable needs of society, making relevant changes in the science curriculum and in the organisation of scientific research, and establishing links with other relevant sectors in society (Malcom et al., 2002).

Universities in Africa face multiple challenges: lack of basic infrastructural facilities, inadequate research budget, poor research management, and exponential growth in student enrolment resulting in overburdened loads, shifting focus on higher education to basic education are only a few of them (Atuahene, 2011; ICIPE, 1988; Kwapong, 1988; Nour, 2012; Oanda and Sall, 2016; Teferra, 2013). A commensurate proportion of funding and resources has not accompanied the increasing student enrolment. In Tunisia, for instance, the number of students enrolled exceeded the projected figures. The researchers in these universities are finding it hard to balance their research, teaching and administrative activities (Siino, 2003). In Algeria, due to the massive intake of students, university faculties have collapsed, upsetting the balance between teaching and conducting scientific research (Khelfaoui, 2004). Mozambique found its higher education system expanding with an increase of institutions and student enrolment of about five times during 2004–6. Funding has been cut substantially in many African countries (Mouton and Waast, 2009; Teferra, 2013). The allocation of funds and resources for higher education in Africa continues to be one of the lowest in the world. In 2012, Africa had spent only 0.78 per cent of its GDP on higher education, with variations across the continent (AU, 2014b). Declining state support to African universities began back in the 1980s following the economic crisis and deterioration at several levels ensued (Sawyer, 2004).

African universities have been in perpetual decline (Kagame, 2009). Most of the universities in sub-Saharan Africa are struggling to improve their scientific productivity (Musiige and Maassen, 2015). A great deal of effort will have to go into the higher education systems to make them effective, efficient and productive to perform their functions. This is where scientists find their place to advance their knowledge and work for the production of new knowledge. They necessarily need updated library (subscriptions to journals and online databases) and information systems; data management software programmes, equipment, maintenance and upgrades from time to time; fast and reliable communication infrastructure; laboratories equipped for advanced research; opportunities for training to gain new skills; access to funds for research and professional activities; and above all a conducive working environment that supports and manages research well.

Researchers at the key centres of scientific production, mainly universities, are overburdened by teaching. The incentive to conduct research is not uniformly available in countries across the continent. Adequate opportunities for funded research or for career advancement are still issues in African universities. Generally, academics at universities have not taken research as seriously as teaching. There are exceptions to this general trend but they are rare. Research has not been fully integrated into the core functions of a university in several parts of sub-Saharan Africa. Higher learning institutions need to emphasise the importance of research for their own growth, reputation, quality of teaching and standing. This is about developing a research culture that is ingrained in the mission and vision of the institution. The research output of institutions also reflects the ability of the institution to provide quality teaching and training. A research-intensive university is expected to attract top academics and researchers, which will have an impact on the reputation and ranking of the university. On the other hand, an active and live research culture at universities can give rise to a better education system. Qualified and research-minded academics will be in a position to train the new generation with the advanced knowledge and skills they have acquired while engaging in research. They will be in touch with recent developments in their own fields of specialisation and can impart their new knowledge and skills to the learners. The standards of education with those teachers who are knowledgeable and skilled will improve and the quality of the outputs of the new graduates will also be better. Scientific research is the foundation of high-quality university education and if the quality of research declines it will drag down the education system (Yriart, 2000).

The situation for African scientists has not changed much. In the 1980s, an average African scientist was an isolated person, without adequate access to information, with very little support for research, meagre research facilities and infrastructure, low salary and low morale (Lemma, 1988). The primary resource for creating an environment for scientific research is infrastructure and trained researchers. Laboratories, equipment, libraries holding updated resources and access to journals, and communication infrastructure (high speed broadband and online access to journals, periodicals and databases) are indispensable parts of research infrastructure. Trained and committed researchers can be attracted and retained if there is adequate research support and prospects for career advancement. There should be ample opportunities for them to upgrade their skills and knowledge (through participation in international conferences, engaging in exchange programmes and allowing for sabbaticals). An efficient research management system can boost the research output of universities. However, science education precedes all these. Before the focus falls on higher learning institutions and research institutes, attention should be drawn to the basic education system.

Building strong institutions of higher learning is indispensable for scientific growth and development. The present higher education scenario in most developing countries is a cause of concern. The situation, as Moravcsik (1975) reported in the 1970s – that the scientific communities in less developed countries have been lethargic in reforming their educational systems – has not changed to an appreciable level. The current standards of both teaching and research are generally low in developing countries and room for a suitable institutional framework to improve them is lacking (Patel, 1993). Renowned universities in Africa that once enjoyed an international reputation have become institutions of ordinary stature. A system which emphasises quality education and qualified, well paid and respected teachers is what is required (Castells, 2009).

The higher education system in Africa as a whole is suffering from resource constraints, poor quality of teaching and research, politicisation of students that causes frequent disruptions of academic work, and poor throughput of graduates with necessary skills (Ransom et al., 1993). The changing priorities of the sector and investments have paved the way for a decline of higher education and research conducted in these institutions (Sanyal and Varghese, 2006). In a review of the implementation of the Plan of Action, the AU reported that the research environment in which universities and other institutions operate is replete with challenges.

The Higher Education Research and Advocacy Network in Africa (HERANA) study of the eight selected flagship universities in Africa (Cloete et al., 2015a) revealed some major strengths, weaknesses and challenges for the centres of science production.[12] In most of the

---

[12] These are the universities of Botswana (Botswana), Cape Town (South Africa), Dar es Salaam (Tanzania), Eduardo Mondlane (Mozambique), Ghana (Ghana), Mauritius (Mauritius), Makerere (Uganda) and Nairobi (Kenya).

universities studied, fewer than 50 per cent of their permanent staff had PhDs; there was a higher proportion of junior permanent staff, and the groups of staff who were expected to be active in research were largely unproductive. Notably, there has been an increase in both research outputs and doctoral graduates at some of these universities (Bunting et al., 2015; Tijssen, 2015). There are also issues related to management, the culture of research consultancy, and the lack of an incentive system to promote research (Cloete et al., 2011).

Universities in Africa are predominantly teaching institutions rather than research institutions without having strong graduate programmes, which limits their research capabilities (AU, 2014b). This does not augur well for African knowledge production. A shift in focus from teaching to research, without undermining the core function of teaching, can make a difference in Africa. A given reason for the performance of South Africa in knowledge production is that many of its universities have assigned more or less equal importance to research. In South Africa, the percentage of teaching that an academic is expected to carry is 45 per cent against 40 per cent for research while the rest is meant for other functions of universities such as community engagement. At the institutional level, where a performance management system is in place, this ratio is maintained and monitored. This institutional direction contributes to scientific research and knowledge production. The government, under its Department of Higher Education and Training (DHET), prescribes certain standards for academics to publish their research (in terms of number of publications according to rank), which is one of the parameters for the government subsidy to universities. The universities in South Africa, research universities in particular, offer benefits such as flexible working hours, travel grants, leave for attendance at conferences, research infrastructure, research support and incentives for research publications (Schoole, 2012). These are supportive measures for research capacity that have paid off well. In Ethiopia, while staff are expected to teach and do research, the proportion of these two responsibilities depends on the type of employing institution. In a full-time teaching position, incumbents are to spend 75 per cent of their time on teaching and the remaining 25 per cent on research. In research institutions the percentage is reversed (Ayalew, 2012).

Castells (2009) suggests ways to strengthen the research functions of a university without undermining other functions. When it is not possible for every university to be a research university in the true sense of the term, as he recommends, it should have access to research centres that exist in the university system for specific purposes. Many universities in sub-Saharan Africa, including top-ranked ones, do not have an integrated policy at the university level that outlines a research agenda and foci.[13] Castells (2009) thinks that universities can develop a nucleus of research that is linked to the needs of the society and economy and connected to networks of research that can be constructed in the entire university system. He argues for a differentiated higher education system for Africa, where the focus has been on building research universities to the neglect of other institutions (MacGregor, 2009). Castells supports this differentiation as all the experiences in university systems have differentiation between different levels of education (MacGregor, 2009). The universities, which Castells (2009) called entrepreneurial universities, that focus on innovation, and connection between S&T and the business world, are to grow in number. Africa needs such universities which are focused on S&T. In view of the feeble contact and association with industry, entrepreneurial universities can move ahead with innovation.

Some time ago, Ziman (1971) suggested that in a developing country research at universities should lead to the establishment of a self-sustaining scientific community, which trains its members and stands on an equal footing with other scientific communities in other countries. This scientific community should have the highest standards of scientific judgment and the members should be competent to earn international recognition in their fields of specialisation. African universities can aim at, and achieve this in the near future. An active scientific community in the country can contribute a great deal to scientific development and the scientific future.

As centres of science production, Africa requires world-class institutions. Universities have tilted their basic orientation towards research and towards more problem-oriented research (Gibbons et al., 1994). Africa needs more research-intensive universities that can give leadership in scientific research. These research-intensive universities, as Cloete et al. (2015b) prescribe, are to be committed to the production

---

[13] Makerere university, the largest public university in Uganda, for instance, does not have internal policies that encourage research among its academics (Ecuru and Kawooya, 2015).

and dissemination of knowledge in a range of scientific areas and are to be equipped with facilities and infrastructure to conduct high levels of research. The key components of these universities are scientific personnel, both scientists and administrators, who are competent, experienced and have an impeccable track record of conducting research in their own specific areas. Attracting them to institutions and retaining them might pose challenges. As institutions cannot be top-class in all areas of science, they will do better if certain areas are identified as their focus. Many institutions have made salutary contributions to scientific knowledge when they focused on their strengths and channelled their resources and energy to those specific focus areas. They thus become leaders and are known in those specific scientific domains. The quality of the products that come out of the university system determines the strengths and weaknesses of the national science system. The strength of a national science system depends on the quality of scientists who are trained and are capable of conducting world-class research.

To improve upon the research capacity, countries should have centres of excellence where competent scientists are located, advance facilities and equipment are provided, and autonomy in the management of these centres is granted. These centres can function as key nodal points of S&T and attract the best talent in the country. They should have collaboration with the international community of scientists and be focused on developing local science for resolving local, regional and national problems. The centres should also have sustainable financial support, knowledgeable and capable leadership, focused research agendas, and mechanisms for nurturing a new generation of talent in S&T (IAC, 2004). Except in South Africa, Zambia, Mauritius, Tanzania, Zimbabwe and in Mozambique, not all African countries have academies of science. Those currently existing are operating with an inadequate budget (Michelson, 2006).

Advancement in science occurs best when individual and independent issues are encouraged. Polanyi (1962), in his thought-provoking idea of the republic of science, states that the pursuit of science by independent initiatives ensures the most efficient possible organisation of scientific progress. This is often not the case in many African countries where independent self-initiated ideas do not receive the attention they deserve. This can happen only in an environment where freedom is granted to scientists, as long as the research problems they choose to study are relevant and appropriate to meet developmental needs.

It is highly likely that enrolment at universities in Africa will increase in the coming decades, which will put pressure on the already-stretched higher education system. In line with enlarging enrolment, an increase has not occurred, or will occur in future, in other resources which include academics. This is an international phenomenon which is applied to developed countries as well. Teaching loads will increase further, shrinking the research activities of academics at universities. How can this high enrolment at universities be turned in favour of science? Some countries in Africa, as specified in their science policies, are targeting increased numbers of students registered in science subjects. If this is achieved with the required support for universities and science education, it will go a long way towards growing their scientific communities. In a survey of selected universities in Ghana, Nigeria, Tanzania and Uganda it was reported that the majority of the universities prioritised training the younger generation and strengthening capacity for conducting scientific research (Kirkland and Ajai-Ajagbe, 2013). This priority, if transcended to implementation level, will have beneficial effects for Africa and for science in Africa.

Although there has been an increase in the number of students who are enrolling at universities in Africa, the increase in the proportion of students who opt for science and engineering courses has not been great. Increasingly, students at universities are not registering for science, technology, engineering and mathematics subjects. This feature stands out in African universities. It will be beneficial for both science and Africa if the increased enrolment in universities is translated into a proportionate increase in the size of the scientific community. It depends on how many new graduates coming out of universities turn to scientific professions. The outgoing chairperson of the African Union Commission of the AU, in her last 'state of the AU' report stated that although there has been a proliferation in the number of students (2.7 million students in 1991 to 11 million students in 2015) entering universities in Africa, most of them are graduating in the social sciences. Africa has the highest share of social science and humanities graduates at university level of any world region but the lowest share of engineers and only 2 per cent study agriculture (AfDB et al., 2012). This is when the continent needs science graduates (Hans, 2016).

Enrolment in science and technology subjects is falling in Africa, not only at the university level but also at all levels (AU, 2014b). The UNDP (2003) reported that only a small number of students have opted to study basic sciences, engineering, medicine and other scientific subjects in Arab countries. There are many Arab countries that are part of Africa. This low rate of science and engineering graduates, as the report concludes, undercuts efforts to build human capacity for S&T. University students being drawn towards the social sciences and humanities rather than to science and engineering subjects, as reported from Mozambique, is due to the poor quality of science education in schools and the lack of demand for engineers and scientists (ROM, 2006).

The massification of the higher education system has resulted in an abnormal increase in the number of graduates who are trained and skilled in research methods and who now possess specialised knowledge (Gibbons et al., 1994). Whether this massification of education is helping Africa is debatable. Are these trained and skilled new graduates able to find appropriate openings to fruitfully use their skills and talents? As long as there are not enough employment avenues to tap into these new talents, massification is not of much help. At the same time, Africa needs to grow its own resources. There should be a sufficient supply of graduates, Masters and PhDs in the continent to form the ground for science to grow and develop. They are the ones who become scientists one day, and will be expected to be active in scientific research. While everything that promotes and facilitates development of science in society is implemented, it is also important that these are accompanied by structures that are in place for capacity building. In order for S&T to achieve sustainable goals of development, societies need to engage creatively in science education and scientific research, and develop capacities for the application of science and technologies (Malcom et al., 2002).

All sectors and stakeholders have a role to play in capacity building for the successful engagement of S&T (Gaillard, 2003a). There are certain other preconditions that will make the process of capacity building effective and successful. The persistent development of science, as Merton (1938) noted, occurs only in societies that have a certain order. There should be, as Gaillard (2003a) confirms, certain levels of economic, social and political stability along with peace, equity and justice for the capacity in S&T to grow. Political uncertainties take a toll on science development, and can retard growth and development. In Algeria, scientific production (publications) stayed on hold during the years of civil war between 1991 and 1998 (Waast and

Rossi, 2010).[14] The national scientific system in Argentina suffered hugely during its periods of political instability and military dictatorship (Yriart, 2000).

Political instability has implications for science, in both the short and long term. Continuity and commitment to science policy are compromised in situations of political instability. Internal political disturbances disrupt effective implementation of policy and take the country back years. According to the Global Peace Index (IEP, 2016), the position of sub-Saharan Africa has declined slightly in 2016. The majority of the world's positive peace deficit countries are in sub-Saharan Africa while in some countries the score has been improved (Chad, Mauritania and Niger). There has also been an improvement in political stability in Nigeria, Guinea and the Central African Republic. Countries such as Djibouti, Burundi and Burkina Faso were on the side of poor performance on the index. The overall index for Tanzania, Ethiopia and Kenya has improved in 2016.

Scientists enjoy working in peaceful conditions where external forces do not affect their research. Instabilities in the political and economic domains, strife, conflict, violence, poor political leadership, corruption and wasteful use of resources operate as deterrent forces. These frustrate scientists. The dormant migratory habits of scientists become dominant in these circumstances. They look for opportunities within or outside the continent where they can carry on with their work. These also refer to their working conditions. Country-specific studies on developing countries have shown that remuneration of scientists has deteriorated since the 1980s (Mouton and Waast, 2009). A reasonable salary and benefits allows them to dedicate themselves to their profession, rather than looking for additional jobs and incomes that distract their attention and energies from scientific research. Along with adequate financial remuneration, decent social status is also important for scientists (Quartey, 1971) to sustain their interest in the field of science and become productive in their areas of expertise.

In industrialised countries, the contribution of the private sector, government agencies and the non-governmental sector are not insignificant (The Royal Society, 2011; Zymelman, 1990). Waast and Krishna (2003a) hint at concerns about the huge investments that are

---

[14] The cases in Lebanon, Jordan, Kuwait and in several other countries were similar (Waast and Rossi, 2010).

being made in R&D in the developed world by both public and private sectors. At the same time, this is not happening in the developing world where the public sector is withdrawing from scientific research and the private sector is not an active player in R&D. In developing countries, the participation and share of the private sector in R&D activities is noticeably poor and, in many instances, minimal. Private firms, as Enos (1995) found in his study of Ghana, Kenya, Tanzania and Uganda, have very little money for R&D due to uncertain returns. Parallel to this, a scenario exists in developed countries where the private sector is a major contributor to R&D. The contribution of the private sector to R&D in Africa has not been substantial, which is far below the contributions the governments are making (Nordling, 2010a; Nour, 2012). Africa is a source of wealth for private enterprises. There is wealth in the form of oil, minerals and precious metals. The profit generated within Africa by the private sector is invariably not ploughed back into the countries for their developmental needs. A fraction of these profits should have been channelled back to centres of research, namely, universities, laboratories and research institutes. Governments can encourage this by providing tax incentives for contributions the private sector is making to R&D. Governments can consider giving preferential treatment in the award or renewal of contracts to those companies that have made significant contributions to scientific research in the country.

However, increased participation and contribution by the private business sector in R&D is occurring in Africa (Siyanbola et al., 2016). The participation of industry also contributes to the support research derives from society. In the changing scenario in which the traditional boundaries between the university and industrial science are disappearing (Gibbons, 1999), the participation of industry is all the more important. Historically, there has been a symbiotic relationship between universities and industry in advancing S&T and both sectors have benefitted from such relationships (Dasgupta and David, 1994).

The triple helix model of university (and research centres), industry and government collaboration has not been adopted successfully in many African countries. Some countries, like South Africa, have recognised the usefulness of this model for S&T. Structures are in place which should be supported by policies and funding. Industries will be keen to associate with centres where knowledge is produced as long as that knowledge is useful for the production and development of goods and services. Initiatives should come from all the stakeholders and government can play the role of bringing them together for continued collaboration. This is already happening in the medical sector and can be extended to other sectors such as agriculture, engineering and ICT and to all countries in Africa.

Identifying key areas of research (engineering, medical science and ICT, for instance) that are more timely for the developmental challenges of Africa is necessary. Africa, as shown in studies like AOSTI (2014), has achieved critical mass (measured in terms of research efforts that are evident in the concentration of publications in specific areas) in general and internal medicine, tropical medicine, microbiology, virology, zoology, health policy and health care services and mathematical physics. In the previous chapters, we have seen the scientific research areas of Africa in general and countries in particular. In Africa, there are common grounds and common priorities. These priorities include education and human resources development, agriculture, energy, health, environment, transport and communications (ECA, 2016).

Priorities in science are commensurate with resources. When science grows, its demands on the resources of society also grow (Weinberg, 1968a). In this situation, choosing a scientific option that helps establish priorities among scientific fields is desirable (Weinberg, 1968a). Again, what should the basis for the choices be? It is a perplexing issue. What criteria should be used in the choice and allocation of resources to science? According to Weinberg (1968b), the criteria can be both internal and external.

An external environment that is created through sufficient funding and research infrastructure alone can make a notable difference or generate momentum in scientific research. Individual factors internal to the scientists who are supposed to make use of the provided favourable environment are not inconsequential. Specific case studies of researchers at African universities have acknowledged those individual factors such as personal academic interest and motivation, determination to engage in scholarly research and goal-driven activities (those meant for promotion, financial incentives and career advancement) (Musiige and Maassen, 2015).

Leading scientists are the ones who lead science in Africa. To have a supply of great and towering scientists to lead advanced scientific research (Salam, 1968) might help. The AOSTI (2014) highlighted the features of these top scientists and their impact on the science that is produced in Africa. They are the active and productive scientists in the continent who have a larger role than others to play in the development of science in Africa. One might see a kind of relationship between the number of such top scientists and the position of science in Africa. South Africa, according to the AOSTI study, has the largest number of leading scientists in Africa, and South Africa is a leading producer of science in Africa.

Scientific research, as Moravcsik (1975) rightly emphasised, is the essence of science and all other activities such as teaching, organisation and communication are only auxiliary to the *raison d'etre* of a scientist. Keeping this in mind, scientists should be provided with what is necessary for them to conduct rigorous scientific research that would produce new knowledge, and the material for new technological innovations. This is to assist them in their contribution to the development of science and society. Able scientists are therefore best left alone to continue as such, and they should not be pulled in different directions to become research managers or administrators or even full-time teaching professors. Their contribution is more valuable in the field of research than in other administrative and management positions. They can still teach and serve as advisors of policymakers which should not put undue pressure on their key task as scientists.

Research management is the key to building capacity. An efficient research management system can serve as a catalyst for the efficient use of resources (Kirkland and Ajai-Ajagbe, 2013). Advances in science management and science governance have been occurring in parts of the world. The emerging governance structure for science, as Macnaghten and Chilvers (2014) argued for a more inclusive form, has to have a new focus. The focus should be on public engagement for a broader appreciation of the governance system in which public engagement is a specific part. African leadership in scientific research and technology, as in many other areas, remains fractured (Odhiambo, 1993). As for research management, scientific structures with the necessary autonomy to contribute to scientific output and best research practices are indispensable (Ozor, 2014). Ozor's empirical study conducted in the Ghana presents us with the relevance of autonomous scientific committees to sustain research behaviour, practices and competencies of researchers as well.

Research consultancy has affected scientific research in Africa, which cannot be separated from the existing research scenario that prevails in many African countries. It is not uncommon for African academics to go after research consultancies to supplement their income (Benneh, 2002; Mkandawire, 1994). In some countries, such as Nigeria, consultancy services by academics in universities are encouraged (Bamiro, 2012) as a way to support their own salaries. More than 80 per cent of the staff in public institutions in Ethiopia are engaged in private work and consultancy research (Ayalew, 2012). The valuable skills and human resources spent on these short-term consultancy projects should have been otherwise utilised for conducting research aimed at resolving national issues.

Rewards, Incentives and Funding

The effect of an incentive system on the performance of researchers is not a new subject in science. The phenomenon of the Matthew effect of Robert Merton, after the New Testament verse in the gospel of Matthew (13:12 and 25:29), that 'for unto everyone that hath shall be given, and he shall have abundance. But from him that hath not shall be taken away even that which he hath,' speaks of a reward system that prevails in science (Merton, 1988). Before postulating his Matthew effect in science, he reviewed the reward and recognition system that influenced scientific productivity.[15] Merton noted that eminent scientists get disproportionately greater credit for their contributions to science while relatively unknown scientists are likely to receive disproportionately little credit for comparable contributions (Merton, 1968, 1973, 1988, 1995). The reward system in science consists of rewarding scientists with public recognition granted by their peers and all other rewards flow from it (Merton, 1995). It serves to heighten the visibility of contributions to science by scientists who already have a standing and reduce the visibility of scientists who are less well known.

The academic incentive system is a response to the competitive knowledge economy paradigm and it is part of a global trend in the political economy of knowledge production (Wangenge-Ouma et al.,

---

[15] His review of the contributions of other scholars shows that the highly productive scientists at major universities gained recognition more often than equally productive scientists in lesser universities, scientists who received recognition for their research done early in their careers are more productive than those who did not, and material rewards in science serve to reinforce a reward system (Merton, 1973).

2015). The reward system serves as a mechanism to reinforce certain patterns of scientific work. A reward system does not operate uniformly across scientific systems and a relationship between quantity and quality of work can be observed (Cole and Cole, 1967). The quality of research output is more significant than quantity for attracting recognition.

Often the pay of university researchers is linked to their publication record, citation count and successful funding (Frey and Neckermann, 2008). An incentive system to encourage scientific research has been experimented with in countries such as South Africa and Kenya. This has produced positive results in the quantity of production. The incentive system that prevails in South Africa works around circulating money in the research system. Scientists receive financial benefits for every publication they produce in indexed peer-reviewed journals and by reputable publishing houses. But this incentive is to be used for further research and not for personal use. As quantity is not always an index of high standards in research, a differentiated incentive system is to be put in place for high-quality research. Under its plan for STI, Kenya has also accepted the reward system as one of the ways to advance STI initiatives in the country. The National Council for Science and Technology in Kenya adopted a system of monetary incentives for publications (50–200 US dollars per publication) in international peer-reviewed journals (AOSTI, 2014; Wangenge-Ouma et al., 2015). Morocco had regulations affecting promotions at universities that took into account personal academic achievements, research and publications in the 1980s and 1990s (Waast and Rossi, 2010). At differing levels, forms and magnitudes, countries have experimented with a system that suits them.

The benefits of an incentive system are not negligible. South Africa surged ahead in scientific publications and became a leader in the continent (Blom et al., 2016; Jeenah and Pouris, 2008; Mouton and Gevers, 2010; Pouris, 2003; Sooryamoorthy, 2015). Research is valued, encouraged and recognised in South Africa. South Africa has been able to infuse a research culture into its institutions, and the value of research is uniformly subscribed. This also links to the status the society ascribes to its scientists and researchers. Science or scientists will not flourish in a society which does not value or respect them. South Africa acknowledges and rewards scientific research, discoveries and innovations that motivate scientists to do more and work hard.

Incentives are not the only means to encourage and motivate researchers to engage in rigorous research. They also need to be paid well so that their needs are met and money is not a distracting factor in dedicating themselves to research. Money acts as a threshold motivator in science. If researchers are not paid well, they will not be motivated and the outcomes of research will be poor (Pan et al., 2012).

Honouring scientists for their scientific accomplishments has an effect on their work and on maintaining their spirit in scientific research. Scientists would agree that scientific research is not always exciting. They have to work hard for several years in their laboratories and research stations before they make any breakthroughs. But they continue to sustain their efforts and find inspiration from several sources. Honouring, supporting and recognising their work are ways of keeping their interest alive. Institutions often celebrate the achievements of their scientists by organising annual research events. Scientists savour such events and are ready to come out of their workplaces to enjoy those moments where they are commended for their work. It is a way for society to manifest its interest in science and scientists and show the value attached to them. Although scientists are scientifically minded and have an objective approach to realities, they are also human beings with emotions and feelings.

Science and Development

The poor state of affairs in the domain of development in Africa is not due to only one factor or reason. A view which is often heard is that it is because of the lackadaisical attitude of African countries towards the application and utilisation of scientific knowledge for meeting economic objectives (Siyanbola et al., 2016).

The missing link between R&D activities and national development plans, as is the case in many instances, is a challenge. When Africa is keen to use STI mainly to find solutions to the problems that are specific to the continent, the link between R&D and national development plans needs to be built and strengthened. R&D activities should be tailored to meet the development goals of the country and the latter should precede the former. The strategies of development that are envisaged in development plans are achieved through the vehicle of R&D. The priorities and allocation of resources for R&D should take into account the core purposes of time-bound development plans for any country.

The portion of the GERD actually spent on science and scientific research has remained an insufficiently known area. A great deal of money allocated for R&D is being used for non-research purposes, including salaries, wages and maintenance of physical infrastructure. In some research institutes in Africa, the average expenditure on salaries and wages comes close to 95 per cent.[16] In most of the sub-Saharan African countries the disproportionate spending of funds had an effect on the advancement of S&T. One argument to balance the expenditure for research (equipment, supplies, material and others) and non-research activities (salaries, wages, travel, upkeep, etc.) is to reduce the number of scientists, engineers and technicians and apportion an increased amount per researcher. It is difficult to arrive at a formula for research and non-related expenditure, which are to be country- and research-specific. However, proper checks and balances need to be put in place to ensure that expenditures are allocated funds according to the budgeted amount. Wastage of expenditure must be curtailed and a proportionally high percentage should be devoted to research expenses.

Scientists are the centre of R&D. A regular supply of scientists is necessary for the production of science. Scientists are born when the education system, both at the basic and higher education levels, is capable of producing them. This means adequate investment should be ploughed into the education system, not just for R&D. The track record of Africa has not been very laudable in developing its science education. Low levels of investment are a reflection of the declining quality of science education in the continent (NEPAD, 2005).

In the analysis of science in Africa one issue should not go unnoticed. Does the development and advancement in S&T that is taking place in Africa actually assist in sustainable economic growth? Enos (1995) presents the scenario of what some countries (Ghana, Kenya, Tanzania and Uganda) had done. They chose areas of development based upon benefits for the developed countries. These countries in

[16] As the analysis of some African countries showed in the study of Enos (1995), Ghana specifically consumed 88–98 per cent of the budget allocated for its research institutes during 1974–81 for salaries and wages alone. Enos reported that a similar situation existed in countries such as Kenya, Tanzania and Uganda as well.

general spent proportionately huge sums of money on R&D for products and commodities that are in demand in developed countries to the detriment of their own long-term interests. The development of S&T and any investment in Africa should have a direct impact on its own sustainable growth. Science generated in Africa should primarily be for finding solutions to its unique problems and advancing its scientific and economic growth. This is also to achieve independence in areas where Africa can stand on its own feet.

Scientific research takes place mainly at public institutions (universities, research institutes and laboratories) in the developing world. The type of research that is being undertaken in these institutions is not the same in terms of focus and its relationship to national developmental priorities. Mouton and Waast (2009) find a divide between academic research and commissioned research, the former occurring at universities and the latter in research institutes. The divide they find is between research meant to advance the career aspirations of academics and that conducted at government institutes where the state has responsibilities in areas such as health and agriculture. This will be problematic for the future of both scientific research and development. Academics at universities pursue their own research interests, which they are passionate about. This often depends on the training they have received in the past and the skills and expertise they possess in specific scientific fields. It is not easy to find the common thread of national priorities and developmental needs of the country that unite the research interests of thousands of academics who work on thousands of research problems. Not all universities in any given country have a well-defined research programme that lists the priorities of the universities or the country and are assigned to academics to accomplish. Rather, academics work in their own areas, which may or may not be directly related to national priorities. Some, however, might have selected priority or focus areas for funding requirements. It is therefore important that the priorities and foci of the country percolate down to universities and research institutes that undertake research in those areas. This kind of coordination, which goes beyond the stipulations of national funding, is essential as researchers source funding not just from national agencies. This has to happen regularly and consistently by monitoring and evaluating priorities and achievements, and identifying new priorities and foci for future research for regional and national development.

The intimate connection between scientific research and development priorities needs to be emphasised in Africa. The argument raised is that there should be a balance between two research conditions; in other words, the social demand for science should correspond to a real capacity of research (Khelfaoui, 2004). The word 'real' is to make the distinction between true research and research in routine testing and consultancy (Khelfaoui, 2004). The social demand for science necessarily pertains to the science that is required to deal with developmental issues within the country. This is possible in countries where the policy is explicit in regard to doing science to address the problems the country is facing.

The backwardness in S&T and the production of scientific knowledge in Africa is also due to the more pressing problems many African countries have to address. When S&T is a second-order issue, to use the term of Rath (1990), that depends on many first-order problems, S&T is at the bottom of the priority list. There are more pressing issues for African countries to tackle which top the list of priorities. Conflict, strife, war, poverty, unemployment, diseases, food insecurity, crime and numerous others take precedence over S&T. These should not, however, undermine S&T or prevent its advancement, for many of the first-order issues can be tackled with the assistance of S&T.

All over the world, science and science development have gained unique characteristics that are transformative. Nowotny et al. (2003) speak about acceptable trends of some countries in this context that are applicable to others, such as the steering of research priorities of countries in scientific research, the commercialisation of research and the accountability of research. Increasingly, science is transforming through the steering of priorities that are identified to meet social and economic needs. Research programmes integrating these priorities are developed to meet both short- and long-term objectives. Commercialisation of science has become necessary in view of funding and resources for scientific research. In order to secure funding and resources, universities and institutions are encouraged to commercialise the production of knowledge.

However, as Moravcsik and Ziman (1975) cautioned, the over-ambitious promise that S&T can transform the economic and cultural situation in developing countries is a little overstated. Often, such promises cannot be fulfilled. S&T is not the only solution for the problems Africa is facing today, but it is a major means and instrument for development. Along with development in S&T, a country needs to have corresponding development in other domains, without which it cannot reap the full benefits of S&T. This further emphasises a two-way approach in strategies for S&T and national development that encompasses other important sectors of the economy. S&T cannot grow alone when other sectors (manufacturing, agriculture, education and health) remain in a state of underdevelopment. Overall growth, manifested in a steadily growing GDP part of which is going towards GERD, sets a facilitating environment for S&T to grow and develop, contributing towards national growth and development. An administrative climate characterised by improved efficiency and fewer bureaucratic bottlenecks supports the growth of not only science but also the country as a whole.

Africa can learn from other countries such as India, Brazil, China and Korea which have formed strategies and put structures in place. These countries have made sustainable investments in education, have long-term commitment to their country's S&T, built reasonable communication infrastructures for both to access knowledge and to interact with the international scientific community, and encourage excellence in science to compete with the international scientific community (Gaillard et al., 2005).

For African countries, as Krishna et al. (1998) rightly pointed out, there are two issues of great concern. One is that the social system of science has not been able to attract sufficient political legitimacy. Two, no amount of external aid and international cooperation can radically supplement or replace local and national support in harnessing S&T for development. Although there is an increased level of appreciation of S&T in the development of the continent by the political leadership (AOSTI, 2013), it has not been reflected uniformly in the allocation of funds for S&T. A clearer commitment to science by the government and political leadership is essential for any developing country (Badran, 2000). The GERD data of many of these countries is an indication of this lackadaisical approach on the part of leaders. Political stability is a facilitating condition for the development and application of S&T in the developing world. In order for them to enhance their capabilities in S&T, they require political stability; the commitment of political leaders, laws and administration; and good governance (IAC, 2004).

Backing of Policies

The ingredients for a national scientific system are many. Some are necessary, some are essential, while some others are desirable. When science becomes a means for national development, there are other things to note. STI policies must be part of the national development policy. These policies need to be well integrated with that of the country's other relevant policies. No scientific system can exist or thrive on its own. It requires the support of other systems which can contribute to its development and growth. For instance, advanced ICTs, without which communication between scientists, nationally and internationally, is not possible. Information cannot be stored in libraries and laboratories cannot be accessed efficiently without a backbone of ICTs. The policies relating to ICT are to be properly aligned with the objectives of STI policies. Essential infrastructure (transport, for instance), policies that build the link between partners of the triple-helix model (government–university–industry), and international scientific collaboration facilitates development of STI.

The reason for the poor rate at which scientific and technological capacity is accumulated in African countries is that the development policies of governments contain very little S&T content (UNESCO, 2007a). In most African countries, policies are made around poverty reduction or economic recovery strategies (UNESCO, 2007a). Policies in general have been either ineffective or unrealistic in African countries (Eisemon and Davies, 1997). A strong policy in S&T is necessarily an essential step towards planned activities for S&T. Planning for a science policy in developing countries is, therefore, as Moravcsik (1982) argues, not an ingredient for success in S&T without the implementation of policy.

African countries, by and large, developed S&T policies following their independence. But in many African countries, their STI policies are not properly linked or integrated into their developmental plans or strategies. Often efforts to ensure the relevance of scientific research to socioeconomic and developmental objectives remain one sided and government policies are not put into effect with reference to S&T (Hetman, 1979). STI policy must be attuned to the concerns and objectives expressed in the national development plans (Hetman, 1979). There is a glaring mismatch between R&D activities and national development plans and strategies in African countries

(NEPAD, 2005). The UNCTAD review showed that a close alignment of development and innovation policies is still elusive in some African countries (UNCTAD, 2015). While STI policies have a broader vision to build capacity, an innovation culture is still necessary to development (UNCTAD, 2015).

Building science and scientific systems in African countries is a hard task to accomplish in a few years or without the backing of supportive policies. Given the background of these countries, and what they have inherited from their colonial past, this can be a long-drawn process. With political will and determination that is also reflected in good governance and in a conflict-free environment, the future of science can be bright for Africa. A bright future for science also means a bright future for Africans, and their quality of life as well. A great deal of continued and sustained labour is called for to make this happen. Scientific communities, teachers, administrators, policymakers, politicians, interest groups and civil society must work hand in hand.

The future of science and the role of scientific knowledge in society are on the path of expansion. Africa is entering into a new phase of development that acknowledges and values the importance of S&T in development. African leadership, by and large, is beginning to see S&T as the key to progress and development (Irikefe et al., 2011). Countries on the African continent are prioritising S&T to achieve their national developmental objectives (ROZ, 1996; Webersik and Wilson, 2009). As Odhiambo (1993) called for, there should be a determined and long-standing commitment of African leadership to build and maintain Africa's capacity for science-led development.

It is now clear that considerable change will have to occur if Africa wants to improve its scientific systems and capabilities to serve the interests of the countries and the people on the continent. Improvements in scientific systems and all other related systems (basic education and university systems, for instance) that feed into this will have to be carefully catered to and nurtured in both the short and the long term. After more than half a century of independence it is time that Africa grows the science, technology and innovation that will ultimately take the continent to an advanced level of development.

References

Abramo, G., D'Angelo, C. A., & Costa, F. D. (2009). Research collaboration and productivity: Is there correlation? *Higher Education*, *57*, 155–71.

Abramo, G., D'Angelo, C. A., & Solazzi, M. (2011). The relationship between scientists' research performance and the degree of internationalization of their research. *Scientometrics*, *86*, 629–43.

Adams, J. (2012). The rise of research networks. *Nature*, *490*, 335–6.

Adams, J. (2013). The fourth age of research. *Nature*, *497*, 557–60.

Adams, J., Gurney, K., Hook, D., & Leydesdorff, L. (2013). *Collaboration in Africa: Networks or Clusters?* Paper presented at the 14th International Society of Scientometrics and Informetrics Conference, Vienna, 15–19 July.

Adams, J., Gurney, K., Hook, D., & Leydesdorff, L. (2014). International collaboration clusters in Africa. *Scientometrics*, *98*, 547–56.

Adams, J., King, C., & Hook, D. (2010). *Global Research Report, Africa 2010*. Leeds: Evidence Thomson Reuters.

Adamson, I. (1992). Access and retrieval of information as coordinates of scientific development and achievement in Nigeria. In R. Arvanitis & J. Gaillard (eds.), *Science Indicators for Developing Countries* (pp. 83–99). Paris: ORSTOM.

Adeboye, T. (1998). Africa. In *UNESCO, World Science Report 1998* (pp. 166–91). Paris: UNESCO.

Adedeji, A. (1984). The economic evolution of developing Africa. In M. Crowder (ed.), *The Cambridge History of Africa* (Vol. 8, pp. 192–250). Cambridge: Cambridge University Press.

AEO (African Economic Outlook). (2017). *African Economic Outlook 2017: Entrepreneurship and Industrialisation*. Paris: African Development Bank, Organisation for Economic Co-operation and Development and United Nations Development Program.

AfDB, OECD and UNDP. (2012). *African Economic Outlook 2012: Promoting Youth Employment*. Paris: African Development Bank, Organisation for Economic Co-operation and Development, United Nations Development Programme.

AfDB, OECD and UNDP. (2015). *African Economic Outlook 2015: Regional Development and Spatial Inclusion.* Paris: African Development Bank, Organisation for Economic Co-operation and Development, United Nations Development Programme.

Agazzi, E. (2000). What does it mean, a social contract for science? In *UNESCO, World Conference on Science: Science for the Twenty-First Century – a New Commitment* (pp. 349–51). Paris: UNESCO.

Aghion, P., David, P. A., & Foray, D. (2009). Science, technology and innovation for economic growth: Linking policy research and practice in 'STIG Systems'. *Research Policy, 38*, 681–93.

Ahmed, A. (2004). Making technology work for the poor: Strategies and policies for African sustainable development. *International Journal of Technology, Policy and Management, 4*, 1–17.

Ahmed, A. D., & Suardi, S. (2007). Sources of economic growth and technology transfer in sub-Saharan Africa. *South African Journal of Economics, 75*, 159–78.

Ajeyalemi, D. (1990). Science and technology education in Africa: a comparative analysis and future prospects. In D. Ajeyalemi (ed.), *Science and Technology Education in Africa: Focus on Seven Sub-Saharan Countries* (pp. 162–81). Lagos: University of Lagos.

Ajeyalemi, D., & Balyelo, T. D. (1990). Nigeria. In D. Ajeyalemi (ed.), *Science and Technology Education in Africa: Focus on Seven Sub-Saharan Countries* (pp. 59–95). Lagos: University of Lagos.

Akinwale, A. A. (2013). Digitisation of indigenous knowledge for natural resources management in Africa. *Africana, 6*, 1–32.

Akpofure, R. E. O., & N'dupu., B. L. (1998). National standards and quality control in Nigerian education. In UNESCO (ed.), *The State of Education in Nigeria* (pp. 119–31). Lagos: UNESCO.

Alagh, Y. K. (2000). The social contract with science in developing countries. In *UNESCO, World Conference on Science: Science for the Twenty-First Century – a New Commitment* (p. 351). Paris: UNESCO.

Allotey, F. K. A. (2000). A need for capacity-building in Africa. In *UNESCO, World Conference on Science: Science for the Twenty-First Century – a New Commitment* (pp. 352–4). Paris: UNESCO.

Aman, V. (2016). How collaboration impacts citation flows within the German science system. *Scientometrics, 109*, 2195–216.

Amankwah-Amoah, J. (2016). The evolution of science, technology and innovation policies: a review of the Ghanaian experience. *Technological Forecasting & Social Change, 110*, 134–42.

Ani, D. P., Biam, C. K., & Kantiok, M. (2014). Patterns and impact of public expenditure on agriculture: Empirical evidence from Benue State, Nigeria. *Journal of Agricultural & Food Information, 15*, 311–23.

Ani, O. E., & Biao, E. P. (2005). Globalization: Its impact on scientific research in Nigeria. *Journal of Librarianship and Information Science*, *37*, 153–60.

AOSTI (African Observatory of Science, Technology and Innovation) (2013). *Science, Technology and Innovation Policy-Making in Africa: an Assessment of Capacity Needs and Priorities*. Working Paper No. 2. Malabo, Equatorial Guinea: The African Observatory of Science, Technology and Innovation (AOSTI).

AOSTI (African Observatory of Science, Technology and Innovation). (2014). *Assessment of Scientific Production in the African* Union *2005–2010*. Malabo, Equatorial Guinea: The African Observatory of Science, Technology and Innovation (AOSTI).

Appiah, B., Gastel, B., Burdine, J. N., & Russell, L. H. (2015). Science reporting in Accra, Ghana: Sources, barriers and motivational factors. *Public Understanding of Science*, *24*, 23–37.

Appiah-Adu, K., Okpattah, B. K., & Djokoto, J. G. (2016). Technology transfer, outsourcing, capability and performance: a comparison of foreign and local firms in Ghana. *Technology in Society*, *47*, 31–9.

Archibugi, D., & Coco, A. (2004). A new indicator of technological capabilities for developed and developing countries (ArCo). *World Development*, *32*, 629–54.

Archibugi, D., & Coco, A. (2005). Measuring technological capabilities at the country level: a survey and a menu for choice. *Research Policy*, *34*, 175–94.

Arikewuyo, O. (2004). Democracy and university education in Nigeria: Some constitutional considerations. *Higher Education Management and Policy*, *16*, 121–33.

Arunachalam, S. (1992). Peripherality in science: What should be done to help peripheral science get assimilated into mainstream science. In R. Arvanitis & J. Gaillard (eds.), *Science Indicators for Developing Countries* (pp. 67–76). Paris: ORSTOM.

Arunachalam, S., & Garg, K. C. (1986). Science on the periphery: a scientometric analysis of science in the ASEAN countries. *Journal of Information Science*, *12*, 105–17.

Arvanitis, R., Waast, R., & Gaillard, J. (2000). Science in Africa: a bibliometric panorama using PASCAL database. *Scientometrics*, *47*, 457–73.

Assaad, R. (2010). *Equality for All? Egypt's free public higher education policy breeds inequality of opportunity*. Cairo: Policy Perspective No. 2, Economic Research Forum.

Atawodi, S. E., Ameh, D. A., Ibrahim, S., et al. (2002). Indigenous knowledge system for treatment of trypanosomiasis in Kaduna state of Nigeria. *Journal of Ethnopharmacology*, *79*, 279–82.

Atuahene, F. (2011). Re-thinking the missing mission of higher education: an anatomy of the research challenge of African universities. *Journal of Asian and African Studies*, *46*, 321–41.

AU (African Union). (2014a). *Science, Technology and Innovation Strategy for Africa 2024*. Addis Ababa: African Union Commission.

AU (African Union). (2014b). *AU Outlook on Education Report: Continental Report*. Tunis: Association for the Development of Education in Africa (ADEA), and African Development Bank (AfDB).

AU-NEPAD (African Union–New Partnership for Africa's Development) (2010). *African Innovation Outlook 2010*. Pretoria: AU–NEPAD.

AUC (African Union Commission). (2015). *Agenda 2063: Popular Version*. Addis Ababa: African Union Commission.

Ayalew, E. (2012). Salary and incentive structure in Ethiopian higher education. In P. G. Altbach, L. Reisberg, M. Yudkevich, G. Androushchak & I. F. Pacheco (eds.), *Paying the Professoriate: a Global Comparison of Compensation and Contracts* (pp. 125–35). New York, NY: Routledge.

Bachewe, F. N., Berhane, G., Minten, B., & Taffesse, A. S. (2018). Agricultural transformation in Africa? Assessing the evidence in Ethiopia. *World Development*, *105*, 286–98.

Badran, A. (2000). Building capacity and creativity in science for sustainable development in the South. In *UNESCO, World Conference on Science: Science for the Twenty-First Century – a New Commitment* (pp. 310–12). Paris: UNESCO.

Balaram, P. (2009). Science and engineering, theory and experiment. *Current Science*, *96*, 321–22.

Bamiro, O. A. (2012). Nigeria: Toward an open market. In P. G. Altbach, L. Reisberg, M. Yudkevich, G. Androushchak & I. F. Pacheco (eds.), *Paying the Professoriate: a Global Comparison of Compensation and Contracts* (pp. 245–54). New York, NY: Routledge.

Barke, R. P. (1998). Authority in science and technology policy. *Science Communication*, *20*, 116–23.

Barke, R. P. (2003). Politics and interests in the republic of science. *Minerva*, *41*, 305–25.

Barnard, H., Cowan, R., & Müller, M. (2012). Global excellence at the expense of local diffusion, or a bridge between two worlds? Research in science and technology in the developing world. *Research Policy*, *41*, 756–69.

Barnes, B. (1972). Introduction. In B. Barnes (ed.), *Sociology of Science: Selected Readings* (pp. 9–16). London: Penguin Books.

Barré, R. (1998). Indications of world science today. In *UNESCO, World Science Report 1998* (pp. 22–30). Paris: UNESCO.

Barré, R., & Papon, P. (1993). Global overview. In *UNESCO, World Science Report, 1993* (pp. 139–50). Paris: UNESCO.

Basalla, G. (1967). The spread of western science. *Science, 156*(3775), 611–32.

Beine, M., Docquier, F., & Rapoport, H. (2001). Brain drain and economic growth: Theory and evidence. *Journal of Development Economics, 64,* 275–89.

Beine, M., Docquier, F., & Schiff, M. (2008). *Brain Drain and Its Determinants: a Major Issue for Small States.* Bonn: IZA Discussion Paper No. 3398.

Bell, R. M. (1988). *The Development of Scientific and Technological Institutions in Africa: Some Past Patterns and Future Needs.* Paper presented at the Symposium on Scientific Institution Building in Africa, March 14–18. Bellagio, Italy.

Ben-David, D. (2008). *Brain Drained. Part 1: Soaring Minds.* London: Centre for Economic Policy Research (CEPR).

Benneh, G. (2002). Research management in Africa. *Higher Education Policy, 15,* 249–62.

Berg, P. (2000). Science: a legitimate path to understanding. In *UNESCO, World Conference on Science: Science for the Twenty-First Century – a New Commitment* (pp. 365–7). Paris: UNESCO.

Berlinguet, L. (1981). Science and technology for development. *Science,* New Series, *213,* 1073–6.

Besley, J. C. (2016). The National Science Foundation's science and technology survey and support for science funding, 2006–2014. *Public Understanding of Science, 27,* 94–109.

Blom, A., Lan, G., & Adil, M. (2016). *Sub-Saharan African Science, Technology, Engineering, and Mathematics Research: a Decade of Development.* Washington, DC: International Bank for Reconstruction and Development/The World Bank.

Bonneuil, C. (2000). Development as experiment: Science and state building in late colonial and postcolonial Africa, 1930–1970. In R. Macleod (ed.), *Nature and Empire: Science and the Colonial Enterprise* (Vol. 15, pp. 258–81). Chicago, IL: Chicago University Press.

Boon, G. (1979). Science and technology planning: Possibilities and limitations. In V. L. Urquidi (ed.), *Science and Technology in Development Planning: Science, Technology and Global Problems* (pp. 5–20). Oxford: Pergamon Press.

Borgatti, S. P., & Everett, M. G. (1999). Models of core/periphery structures. *Social Networks, 21,* 375–95.

Boshoff, N. (2009a). Neo-colonialism and research collaboration in Central Africa *Scientometrics, 81,* 413–34.

Boshoff, N. (2009b). South-South research collaboration of countries in the Southern African Development Community (SADC). *Scientometrics, 84,* 481–503.

Boshoff, N. (n.d.). *Mapping Research Systems in Developing Countries: Country Report–The Science and Technology System of Malawi.* Paris: UNESCO.

Boshoff, N., & Kleiche, M. (n.d.). *Mapping Research Systems in Developing Countries: Country report – The Science and Technology System of Morocco.* Paris: UNESCO.

Bouyahya, A., Abrinib, J., Et-Touysa, A., Bakria, Y., & Dakka, N. (2017). Indigenous knowledge of the use of medicinal plants in the North-West of Morocco and their biological activities. *European Journal of Integrative Medicine, 13,* 9–25.

Briggs, J., & Moyo, B. (2012). The resilience of indigenous knowledge in small-scale African agriculture: Key drivers. *Scottish Geographical Journal, 128,* 64–80.

Brockway, L. H. (2011). Science and colonial expansion: the role of the British Royal Botanic Gardens. In S. Harding (ed.), *The Postcolonial Science and Technology Studies Reader* (pp. 127–39). London: Duke University Press.

Bunting, I., Cloete, N., Wah, H. L. K., & Nakayiwa-Mayega, F. (2015). Assessing the performance of African flagship universities. In N. Cloete, P. Maassen & T. Bailey (eds.), *Knowledge Production and Contradictory Functions in African Higher Education* (pp. 32–60). Cape Town: African Minds.

Campbell, E. K. (2007). Brain drain potential in Botswana. *International Migration, 45,* 115–45.

Cano, V. (1992). Bibliographic control and international visibility of Latin American periodical publications. In R. Arvanitis & J. Gaillard (eds.), *Science Indicators for Developing Countries* (pp. 511–26). Paris: ORSTOM.

Carter, C. F. (1968). The distribution of scientific effort. In E. Shils (ed.), *Criteria for Scientific Development: Public Policy and National Goals* (pp. 34–43). Cambridge, MA: The MIT Press.

Carty, A. J. (2000). R&D, innovation and the knowledge-based economy: the Canadian experience. In *UNESCO, World Conference on Science: Science for the Twenty-First Century – a New Commitment* (pp. 370–2). Paris: UNESCO.

Castells, M. (1993). The university system: Engine of development in the new world economy. In A. Ransom, S.-M. Khoo & V. Selvaratnam (eds.), *Improving Higher Education in Developing Countries* (pp. 65–80). Washington, DC: The International Bank for Reconstruction and Development / The World Bank.

Castells, M. (2009). On Higher Education. Lecture delivered at the University of Western Cape, South Africa, 7 August.

Castells, M., & Tyson, L. D. A. (1989). High technology and the changing international division of production: Implications for the U.S. economy. In R. B. Purcell (ed.), *The Newly Industrializing Countries in the World Economy: Challenges for U.S. Policy* (pp. 13–50). Boulder and London: Lynne Rienner Publishers.

Chatelin, Y., Gaillard, J., & Keller, A. S. (1997). The Nigerian scientific community: the Colossus with feet of clay. In J. Gaillard, V. V. Krishna & R. Waast (eds.), *Scientific Communities in the Developing World* (pp. 129–54). New Delhi: Sage Publications.

Cherry, M. (2010). South African science: Black, white and grey. *Nature*, *463*, 726–8.

Chitiga, M., Kandiero, T., & Ngwenya, P. (2008). Agricultural trade policy reform in South Africa. *Agrekon*, *47*, 76–101.

Choi, S. (2012). Core-periphery, new clusters, or rising stars? International scientific collaboration among 'advanced' countries in the era of globalization. *Scientometrics*, *90*, 25–41.

Clapham, C. (2017). The Ethiopian developmental state. *Third World Quarterly*, *39*, 1151–65.

Cloete, N., Bailey, T., Pillay, P., Bunting, I., & Maassen, P. (2011). *Universities and Economic Development in Africa*. Wynberg, South Africa: Centre for Higher Education Transformation (CHET).

Cloete, N., Bunting, I., & Maassen, P. (2015a). Research universities in Africa: an empirical overview of eight flagship universities. In N. Cloete, P. Maassen & T. Bailey (eds.), *Knowledge Production and Contradictory Functions in African Higher Education* (pp. 18–31). Cape Town: African Minds.

Cloete, N., Maassen, P., Bunting, I., Bailey, T., Wangenge-Ouma, G., & Schalkwyk, F. v. (2015b). Managing contradictory functions and related policy issues. In N. Cloete, P. Maassen & T. Bailey (eds.), *Knowledge Production and Contradictory Functions in African Higher Education* (pp. 260–89). Cape Town: African Minds.

Coccia, M., & Bozeman, B. (2016). Allometric models to measure and analyze the evolution of international research collaboration. *Scientometrics*, *108*, 1065–84.

Coile, R. C. (1977). Lotka's frequency distribution of scientific productivity. *Journal of the Association for Information Science and Technology*, *28*, 366–70.

Cole, S., & Cole, J. R. (1967). Scientific output and recognition: a study of the operation of the reward system in science. *American Sociological Review*, *32*, 377–90.

Colglazier, W. (1981). Science and development. *Harvard International Review*, 3, 6–7.

Collyer, F. (2014). Sociology, sociologists and core–periphery reflections. *Journal of Sociology*, 50, 252–68.

Confraria, H., & Godinho, M. M. (2015). The impact of African science: a bibliometric analysis. *Scientometrics*, 102, 1241–68.

Cooper, C. (1994). *Science and Technology in Africa under Conditions of Economic Crisis and Structural Adjustment*. Working paper No. 4. Maastricht: The United Nations University.

Cooper, C. M., & Zammit, J. A. (1964). An aspect of the planning of science in developing countries. *The American Economist*, 8, 16–21.

COSTECH (Tanzania Commission for Science and Technology). (2015). *Building Systems for High Quality, Relevant Research in Tanzania*. Dar es Salaam: Tanzania Commission for Science and Technology.

Coughlan, S. (2017). How Canada Became an Education Superpower. www .bbc.com/news/business-40708421, 26 May 2018.

Court, D. A. (1988). *Institution Building in Africa: Reflections on the University Development Program of the Rockefeller Foundation*. Paper presented at the Symposium on Scientific Institution Building in Africa, March 14–18, Bellagio, Italy.

Dahdouh-Guebas, F., Ahimbisibwe, J., Moll, R. V., & Koedam, N. (2003). Neo-colonial science by the most industrialised upon the least developed countries in peer-reviewed publishing. *Scientometrics*, 56, 329–43.

Dahoun, A. M. (1999). Black Africa in the Science Citation Index *Scientometrics*, 46, 11–18.

Damme, L. S. M. V., & Neluvhalani, E. F. (2004). Indigenous knowledge in environmental education processes: Perspectives on a growing research arena. *Environmental Education Research*, 10(3), 353–70.

Danquah, M. (2018). Technology transfer, adoption of technology and the efficiency of nations: Empirical evidence from sub Saharan Africa. *Technological Forecasting & Social Change*, 131, 175–82.

Danquah, M., Ouattara, B., & Quartey, P. (2018). Technology transfer and national efficiency: Does absorptive capacity matter? *African Development Review*, 30, 162–74.

Dasgupta, P., & David, P. A. (1994). Toward a new economics of science. *Research Policy*, 23, 487–521.

Davies, J., & Mullan, Z. (2016). Research capacity in Africa: Will the sun rise again? *The Lancet Global Health*, 4, 287.

Davis, C. H. (1983). Institutional sectors of 'mainstream' science production in sub-Saharan Africa, 1970–1979: a quantitative analysis. *Scientometrics*, 5, 163–75.

Debru, C. (2000). How modern science was born and developed. In *UNESCO, World Conference on Science: Science for the Twenty-First Century – a New Commitment* (pp. 82–4). Paris: UNESCO.

Dedijer, S. (1968). Underdeveloped science in underdeveloped countries. In E. Shils (ed.), *Criteria for Scientific Development: Public Policy and National Goals* (pp. 143–63). Cambridge, MA: The MIT Press.

Dei, G. J. S. (2000). Rethinking the role of Indigenous knowledges in the academy. *International Journal of Inclusive Education*, 4, 111–32.

Dillon, W. S. (1966). Africa Science Board, National Academy of Sciences, U.S.A. *The Journal of Modern African Studies*, 4, 98–101.

Diver, S. (2017). Negotiating Indigenous knowledge at the science–policy interface: Insights from the Xáxli'p Community Forest. *Environmental Science & Policy*, 73, 1–11.

Diwu, C. T., & Ogunniyi, M. B. (2012). Dialogical argumentation instruction as a catalytic agent for the integration of school science with Indigenous Knowledge Systems. *African Journal of Research in Mathematics, Science and Technology Education*, 16, 333–47.

Donda, S., & Manyungwa-Pasani, C. L. (2018). Understanding indigenous knowledge: Its role and potential in fisheries resources management in Malawi. *Aquatic Ecosystem Health & Management*, 21, 176–84.

Dorosh, P., & Rashid, S. (2012). Introduction. In P. Dorosh & S. Rashid (eds.), *Food and Agriculture in Ethiopia: Progress and Policy Challenges* (pp. 1–20). Philadelphia, PA: Pennsylvania University Press.

Dow, M. M. (1988). *Issues in African Scientific Institution Building*. Paper presented at the Symposium on Scientific Institution Building in Africa, March 14–18, Bellagio, Italy.

DST (Department of Science and Technology). (2002). *South Africa's National Research and Development Strategy*. Pretoria: Department of Science and Technology, The Government of the Republic of South Africa.

DST (Department of Science and Technology). (2010). *Corporate Strategy, 2009/10*. Pretoria: Department of Science and Technology, the Government of the Republic of South Africa.

DST (Department of Science and Technology). (2013). *2012/13 Report on Public Funding for Scientific and Technological Activities*. Pretoria: Department of Science and Technology, Government of the Republic of South Africa.

DST (Department of Science and Technology). (2015a). *Annual Report 2014/15 Financial Year*. Pretoria: Department of Science and Technology, the Government of the Republic of South Africa.

DST (Department of Science and Technology). (2015b). *Strategic Plan for the Fiscal Years, 2015–2020*. Pretoria: Department of Science and Technology, the Government of the Republic of South Africa.

DST (Department of Science and Technology). (n.d.). *Ten-Year Innovation Plan.* Pretoria: Department of Science and Technology, the Government of the Republic of South Africa.

Dubow, S. (2000). Introduction. In S. Dubow (ed.), *Science and Society in Southern Africa* (pp. 1–10). Manchester: Manchester University Press.

Durant, J. (2000). Public perception of science: Between acceptance and rejection. In UNESCO, *World Conference on Science: Science for the Twenty-First Century – a New Commitment* (pp. 256–9). Paris: UNESCO.

Dyll, L. (2018). Indigenous environmental knowledge and challenging dualisms in development: Observations from the Kalahari. *Development in Practice, 28,* 332–44.

ECA (Economic Commission for Africa). (2016). *Assessing Regional Integration in Africa (ARIA VII): Innovation, Competitiveness and Regional Integration.* Addis Ababa: Economic Commission for Africa.

Ecuru, J., & Kawooya, D. (2015). Effective innovation policies for development: Uganda. In S. Dutta, B. Lanvin & S. Wunsch-Vincent (eds.), *The Global Innovation Index 2015: Effective Innovation Policies for Development* (pp. 147–52). Geneva: The World Intellectual Property Organization.

Ehikhamenor, F. A. (1988). Perceived state of science in Nigerian universities. *Scientometrics, 13,* 225–38.

Ehikhamenor, F. A. (1990). Productivity of physical scientists in Nigerian universities in relation to communication variables. *Scientometrics, 18,* 437–44.

Eisemon, T. O. (1979). The implantation of science in Nigeria and Kenya. *Minerva, 17,* 504–26.

Eisemon, T. O. (1980). African academics: a study of scientists at the universities of Ibadan and Nairobi. *The Annals of the American Academy of Political and Social Science, 448,* 126–38.

Eisemon, T. O. (1981). Scientific life in Indian and African universities: a comparative study of peripherality in science. *Comparative Education Review, 25,* 164–82.

Eisemon, T. O. (1986). Foreign training and foreign assistance for university development in Kenya: Too much of a good thing? *International Journal of Educational Development 6,* 1–13.

Eisemon, T. O., & Davis, C. H. (1991). Can the quality of scientific training and research in Africa be improved? *Minerva, 29,* 1–26.

Eisemon, T. O., & Davis, C. H. (1992). Universities and scientific research capacity. *Journal of Asian and African Studies, 27,* 68–93.

Eisemon, T. O., & Davis, C. H. (1997). Kenya: Crisis in the scientific community. In J. Gaillard, V. V. Krishna & R. Waast (eds.), *Scientific*

Communities in the Developing World (pp. 105–28). New Delhi: Sage Publications.

Elalami, J., Dore, J. C., & Miquel, J. F. (1992). International scientific collaboration in Arab countries. In R. Arvanitis & J. Gaillard (eds.), *Science Indicators for Developing Countries* (pp. 357–71). Paris: ORSTOM.

El-Khawas, M. A. (2004). Brain drain: Putting Africa between a rock and a hard place. *Mediterranean Quarterly, 15,* 37–56.

Elmslie, B., & Criss, A. J. (1999). Theories of convergence and growth in the classical period: the role of science, technology and trade. *Economica, 66,* 135–49.

Enos, E. L. (1995). *In Pursuit of Science and Technology in Sub-Saharan Africa: the Impact of Structural Adjustment Programmes.* London: Routledge.

Esau, S. (n.d.). *Mapping Research Systems in Developing Countries: Country Report – The Science and Technology System of Kenya.* Paris: UNESCO.

Esau, S., & Khelfaoui, H. (n.d.). *Mapping Research Systems in Developing Countries: Country Report – The Science and Technology System of Algeria.* Paris: UNESCO.

Eslami, H., Ebadi, A., & Schiffauerova, A. (2013). Effect of collaboration network structure on knowledge creation and technological performance: the case of biotechnology in Canada. *Scientometrics, 97,* 99–119.

Essack, S. Y., Naidoo, I., & Barnes, G. (2010). Government funding as leverage for quality teaching and learning: a South African perspective. *Higher Education Management and Policy, 22,* 1–13.

Essegbey, G., Diaby, N., & Konte, A. (2015). West Africa. In *UNESCO Science Report: Towards 2030* (pp. 471–97). Paris: UNESCO.

ESTA (Ethiopian Science and Technology Agency) (2006). *National Science, Technology and Innovation (STI) Policy of Ethiopia.* Addis Ababa: The Ethiopian Science and Technology Agency.

FDRE (The Federal Democratic Republic of Ethiopia). (2012). *Science, Technology and Innovation Policy.* Addis Ababa: The Federal Democratic Republic of Ethiopia.

Fedderke, J. W., & Goldschmidt, M. (2015). Does massive funding support of researchers work? Evaluating the impact of the South African research chair funding initiative. *Research Policy, 44,* 467–82.

FRN (Federal Republic of Nigeria). (2004). *National Policy on Education* (4th ed.). Lagos: Nigerian Educational Research and Development Council.

Ferreira, E. de. S. (1974). *Portuguese Colonialism in Africa: the End of an Era.* Paris: The UNESCO Press.

FGOS (Federal Government of Somalia). (n.d.). *National Development Plan, 2017–2019.* Mogadishu: Federal Government of Somalia.

Foray, D. (2000). Building a new social contract. In *UNESCO, World Conference on Science: Science for the Twenty-First Century – a New Commitment* (pp. 361–4). Paris: UNESCO.

Forje, J. W. (1979). Science and technology: the African search for a third way to development. *Alternatives, 4,* 355–69.

Forje, J. W. (1989). *Science and Technology in Africa.* Essex: Longman.

Forje, J. W. (1992). The role and effectiveness of national science and technology policy-making bodies in Africa. *Journal of Asian and African Studies, 28,* 12–30.

Frame, D. J., Narin, F., & Carpenter, M. P. (1977). The distribution of world science. *Social Studies of Science, 7,* 501–16.

Freudenthal, S. (2014). *Tracing Research Capacities in Tanzania: a Study of Tanzanian PhD Holders Trained within the Tanzania–Sweden Research Cooperation.* Stockholm: Swedish International Development Cooperation Agency.

Frey, B. S., & Neckermann, S. (2008). *Academics Appreciate Rewards: a New Aspect of Incentives in Research.* Working Paper No. 400. Zurich: Institute for Empirical Research in Economics, University of Zurich.

FRON (Federal Republic of Nigeria). (2011). *Science, Technology and Innovation (STI) Policy.* Abuja: Government of the Federal Republic of Nigeria.

Fusfeld, H. I. (1979). Overview of science and technology policy 1979. *Annals of the New York Academy of Sciences, 1,* 1–26.

Gabler, J., & Frank, D. J. (2005). The natural sciences in the university: Change and variation over the 20th century. *Sociology of Education, 78,* 183–206.

Gaillard, J. (1991). *Scientists in the Third World.* Lexington, KY: University Press of Kentucky.

Gaillard, J. (1992a). Science policies and cooperation in Africa: Trends in the production and utilization of knowledge. *Science Communication, 14,* 212–33.

Gaillard, J. (1992b). Use of publication lists to study scientific production and strategies of scientists in developing countries. In R. Arvanitis & J. Gaillard (eds.), *Science Indicators for Developing Countries* (pp. 439–56). Paris: ORSTOM.

Gaillard, J. (1994). North-South research partnership: Is collaboration possible between unequal partners? *Knowledge and Policy, 2,* 195–228.

Gaillard, J. (1997). The Senegalese scientific community: Africanization, dependence and crisis. In J. Gaillard, V. V. Krishna & R. Waast (eds.), *Scientific Communities in the Developing World* (pp. 155–82). New Delhi: Sage Publications.

Gaillard, J. (2000). *Science in Africa at the Dawn of the 21st Century: Country Report Tanzania.* Paris: IRD (Institut de Recherches pour le Développement).

Gaillard, J. (2003a). Overcoming the scientific generation gap in Africa: an urgent priority. *Interdisciplinary Science Reviews, 28,* 15–25.

Gaillard, J. (2003b). Tanzania: a case of 'dependent science'. *Science, Technology & Society, 8,* 317–43.

Gaillard, J., Hassan, M., & Waast, R., with Schaffer, D. (2005). Africa. In *UNESCO Science Report 2005* (pp. 177–201). Paris: UNESCO.

Gaillard, J., & Khelfaoui, H., with Ngatchou, N. (n.d.). *Mapping Research Systems in Developing Countries: Country Report – the Science and Technology System of Cameroon.* Paris: UNESCO.

Gaillard, J., Krishna, V. V., & Waast, R. (1997). Introduction: Scientific communities in the developing world. In J. Gaillard, V. V. Krishna & R. Waast (eds.), *Scientific Communities in the Developing World* (pp. 11–49). New Delhi: Sage Publications.

Gaillard, J., & Tullberg, F. A. (2001). *Questionnaire survey of African scientists, IFS grantees and INCO beneficiaries* (Report No. 2). Stockholm: International Foundation for Science.

Gaillard, J., & Waast, R. (1992). The uphill emergence of scientific communities in Africa. *Journal of Asian and African Studies, 27,* 41–67.

Garfield, E. (1990). Michael J. Moravcsik: Multidimensionals scholar and hero of third world science. *Journalology, 13,* 9–12.

Gathiram, P., & Hänninen, O. (2014). Medicine and medical sciences in Africa. *Pathophysiology, 21,* 129–33.

Gaudin, T. (2000). The feasibility of science foresight: What are the priorities? In *UNESCO, World Conference on Science: Science for the Twenty-First Century – a New Commitment* (pp. 316–19). Paris: UNESCO.

Gazni, A., & Ghaseminik, Z. (2019). The increasing dominance of science in the economy: Which nations are successful? *Scientometrics, 120,* 1411–26.

Gazni, A., Lariviére, V., & Didegah, F. (2016). The effect of collaborators on institutions' scientific impact. *Scientometrics, 109,* 1209–30.

Gazni, A., Sugimoto, C. R., & Didegah, F. (2012). Mapping world scientific collaboration: Authors, institutions, and countries. *Journal of the American Society for Information Science and Technology, 63,* 323–35.

GCIS (Government Communications). (2015). Science and technology. In E. Tibane & M. Honwane (eds.), *South Africa Yearbook 2014/15* (pp. 339–54). Pretoria: Government Communications, the Government of the Republic of South Africa.

Gemedo-Dalle, J. I., & Maass, B. L. (2006). Indigenous ecological knowledge of Borana pastoralists in southern Ethiopia and current challenges. *The International Journal of Sustainable Development and World Ecology*, 13, 113–30.

Gerdin, A. (2002). Productivity and economic growth in Kenyan agriculture, 1964–1996. *Agricultural Economics*, 27, 7–13.

Gibbons, M. (1999). Science's new social contract with society. *Nature*, 402, C81–C84.

Gibbons, M., Limoges, C., Nowotny, H., Schwartzman, S., Scott, P., & Trow, M. (1994). *The New Production of Knowledge: the Dynamics of Science and Research in Contemporary Societies*. London: Sage Publications.

Gladwell, M. (2009). *Outliers: the Story of Success*. London: Penguin Books.

Glatz, F. (2000). Opening address: Science in the 21st century. In *UNESCO, World Conference on Science: Science for the Twenty-First Century – a New Commitment* (pp. 18–21). Paris: UNESCO.

GOBF (Government of Burkina Faso). (2016). *National Plan for Economic and Social Development (PNDES), 2016–2020*. Ouagadougou: Government of Burkina Faso.

Godin, B. (2005). *Measurement and Statistics on Science and Technology: 1930s to the Present*. London: Routledge.

Godin, B., & Gingras, Y. (2000). The place of universities in the system of knowledge production. *Research Policy*, 29, 273–8.

Goldemberg, J. (1998). What is the role of science in developing countries? *Science*, 279, 1140–1.

GOM (Government of Malawi). (n.d.). *Malawi Growth and Development Strategy II, 2011–2016*. Lilongwe: Department of Development Planning, Government of Malawi.

GOROU (Government of the Republic of Uganda). (2009). *National Science, Technology and Innovation Policy*. Kampala: Ministry of Finance, Planning and Development, Government of the Republic of Uganda.

GOROU (Government of the Republic of Uganda). (2013). *Uganda Vision 2040*. Kampala: Government of the Republic of Uganda.

Gow, J., & Parton, K. A. (1992). *The evolution of Kenyan agricultural policy*. Paper presented at the Australian Agricultural Economics Society Conference, Canberra.

GOZ (Government of Zimbabwe). (2012). *Second Science, Technology and Innovation Policy of Zimbabwe*. Harare: Government of Zimbabwe.

Grange, L. L. (2007). Integrating western and indigenous knowledge Systems: the basis for effective science education in South Africa? *International Review of Education*, 53, 577–91.

Green, L. J. F. (2012). Beyond South Africa's 'indigenous knowledge–science' wars. *South African Journal of Science, 108.* http://dx.doi.org/10.4102/sajs.v4108i4107/4108.4631.

Gruhn, I. V. (1984). Towards scientific and technological independence? *The Journal of Modern African Studies, 22,* 1–17.

Gulland, A. (2012). Uganda launches vaccine programme to fight its commonest cancer. *BMJ: British Medical Journal, 345,* 4.

Gupta, B. M., & Karisiddippa, C. R. (1999). Collaboration and author productivity: a study with a new variable in Lotka's Law. *Scientometrics, 44,* 129–34.

Hackmann, H., & Boulton, G. (2015). Science for a sustainable and just world: a new framework for global science policy? In *UNESCO Science Report: Towards 2030* (pp. 12–14). Paris: UNESCO.

Hagopian, A., Thompson, M. J., Fordyce, M., Johnson, K. E., & Hart, L. G. (2004). The migration of physicians from sub-Saharan Africa to the United States of America: Measures of the African brain drain. *Human Resources for Health, 2*(17). www.human-resources-health.com/content/2/1/1.

Hans, B. (2016). Free varsity access the key. *The Mercury,* 20 December, p. 2.

Harding, S. (2011). Beyond postcolonial theory: Two undertheorized perspectives on science and technology. In S. Harding (ed.), *The Postcolonial Science and Technology Studies Reader* (pp. 1–31). London: Duke University Press.

Hassan, M. H. A. (2001). Can science save Africa? *Science, 292,* 1609.

Hassan, M. H. A. (2007). Editorial: a new dawn for science in Africa. *Science,* New Series, *316,* 1813.

He, Z.-L., Geng, X.-S., & Campbell-Hunt, C. (2009). Research collaboration and research output: a longitudinal study of 65 biomedical scientists in a New Zealand university. *Research Policy, 38,* 306–17.

Hetman, F. (1979). Planning: Perspective analysis and science and technology policy. In V. L. Urquidi (ed.), *Science and Technology in Development Planning: Science, Technology and Global Problems* (pp. 21–32). Oxford: Pergamon Press.

Hewitt, T., & Albu, M. (1998). Structural adjustment, industrialisation and technological capabilities in Africa. *Science, Technology & Society, 3,* 335–64.

Hicks, D. M., & Katz, J. S. (1996). Where is science going? *Science, Technology and Human Values, 21,* 379–406.

Hofmeyr, J. H. (1929a). Africa and science. *Science,* New Series, *70,* 269–74.

Hofmeyr, J. H. (1929b). Africa and science II. *Science,* New Series, *70,* 294–9.

Hooli, L. J., & Jauhiainen, J. S. (2018). Building an innovation system and indigenous knowledge in Namibia. *African Journal of Science, Technology, Innovation and Development, 10*, 183–96.

Hoyningen-Huene, P. (2000). The nature of science. In *UNESCO, World Conference on Science: Science for the Twenty-First Century – a New Commitment* (pp. 52–6). Paris: UNESCO.

Hu, Z., Chen, C., & Liu, Z. (2013). *How are collaboration and productivity correlated at various career stages of scientists?* Paper presented at the 14th International Society of Scientometrics and Informetrics Conference, 15–19 July, Vienna.

Hubbard, G. G. (1889). Africa, its past and future. *Science, 13*, 42–50.

Hudu, A. (2015). Working and living conditions of academic staff in Nigeria: Strategies for survival at Ahmadu Bello university. In Y. Lebeau & M. Ogunsanya (eds.), *The Dilemma of Post-Colonial Universities* (pp. 209–40). Ibadan: IFRA/ABB.

Hung, W.-C., Lee, L.-C., & Tsai, M.-M. (2009). An international comparison of relative contributions to academic productivity. *Scientometrics, 81*, 703–18.

Hunter, J. (2005). The role of information technologies in indigenous knowledge management. *Australian Academic & Research Libraries, 36*, 109–24.

IAC (Inter Academy Council). (2004). *Inventing a Better Future: a Strategy for Building Worldwide Capacities in Science and Technology.* Amsterdam: Inter Academy Council.

IAP (Inter Academy Partnership). (2016). *Doing Global Science: a Guide to Responsible Conduct in the Global Research Enterprise.* Princeton, NJ: Princeton University Press.

ICIPE (1988). *The new challenge of science and technology for development Africa.* Paper presented at the Symposium on Scientific Institution Building in Africa, March 14–18, Bellagio, Italy.

ICSU (International Council for Science). (2005). *Science and Society: Rights and Responsibilities.* Paris: International Council for Science.

IEP (Institute for Economics and Peace). (2016). *Global Peace Index 2016.* New York, NY: Institute for Economics and Peace.

Iganiga, B. O., & Unemhilin, D. O. (2017). The impact of federal government agricultural expenditure on agricultural output in Nigeria. *Journal of Economics, 2*, 81–8.

Ikram, K. (2006). *The Egyptian Economy, 1952–2000: Performance, Policies and Issues.* London: Routledge.

Imam, H. (2012). Educational policy in Nigeria from the colonial era to the post-independence period. *Italian Journal of Sociology of Education, 1*, 181–204.

IMF (International Monetary Fund). (2012). *Burkina Faso: Strategy for Accelerated Growth and Sustainable Development 2011–2015*. Washington, DC: International Monetary Fund Publication Services.

Ingwersen, P., & Jacobs, D. (2004). South African research in selected scientific areas: Status 1981–2000. *Scientometrics, 59*, 405–23.

Inkster, I. (1985). Scientific enterprise and the colonial 'Model': Observations on Australian experience in historical context. *Social Studies of Science, 15*, 677–704.

Inönü, E. (2003). The influence of cultural factors on scientific production. *Scientometrics, 56*, 137–46.

Irikefe, V., Vaidyanathan, G., Nordling, L., Twahirwa, A., Nakkazi, E., & Monastersky, R. (2011). Science in Africa: View from the front line. *Nature, 474*, 556–9.

Ivowi, U. M. O. (1998). Curriculum and content of education. In UNESCO (ed.), *The State of Education in Nigeria* (pp. 22–33). Lagos: UNESCO.

Jacobs, D., & Ingwersen, P. (2000). A bibliometric study of the publication patterns in the sciences of South African scholars, 1981–96. *Scientometrics, 47*, 75–93.

Jalali, A. (2000). Science, development and globalization. In *UNESCO, World Conference on Science: Science for the twenty-First century – a new commitment* (pp. 303–6). Paris: UNESCO.

Jauhiainen, J. S., & Hooli, L. (2017). Indigenous knowledge and developing countries' innovation Systems: the case of Namibia. *International Journal of Innovation Studies, 1*, 89–106.

Jeenah, M., & Pouris, A. (2008). South African research in the context of Africa and globally. *South African Journal of Science, 104*, 351–4.

Jomo, K. S. (2004). *The New Economic Policy and Interethnic Relations in Malaysia*. Geneva: United Nations Research Institute for Social Development.

Jonathan, A., Christopher, K., & Daniel, H. (2010). *Global Research Report Africa*. Leeds: Evidence, Thomson Reuters.

Jones, N., Bailey, M., & Lyytikäinen, M. (2007). *Research Capacity Strengthening in Africa: Trends, Gaps and Opportunities*. London: Overseas Development Institute.

Juma, C., & Clark, N. (1995). Policy research in Sub-Saharan Africa: an exploration. *Public Administration and Development, 15*, 121–37.

Juma, C., & Yee-Cheong, L. (2005). Reinventing global health: the role of science, technology, and innovation. *The Lancet, 365*, 1105–7.

Kaba, A. J. (2011). The status of Africa's emigration brain drain in the 21st century. *The Western Journal of Black Studies, 35*, 187–95.

Kagame, P. (2009). Challenges and prospects of advancing science and technology in Africa: the case of Rwanda. *Science, 322,* 545–51.

Kalipeni, E., Semu, L. L., & Mbilizi, M. A. (2012). The brain drain of health care professionals from sub-Saharan Africa: a geographic perspective. *Progress in Development Studies, 12,* 153–71.

Kaplan, D. (2004). South Africa's national research and development strategy: a review. *Science, Technology and Society, 9,* 273–94.

Kaplan, D. (2008). Science and technology policy in South Africa: Past performance and proposals for the future. *Science, Technology & Society, 13,* 95–122.

Kargbo, J. A. (2005). Managing indigenous knowledge: What role for public librarians in Sierra Leone? *The International Information & Library Review, 37,* 199–207.

Karlovčec, M., Lužar, B., & Mladenić, D. (2016). Core-periphery dynamics in collaboration networks: the case study of Slovenia. *Scientometrics, 109,* 1561–78.

Kashim, I. B., & Adelabu, O. S. (2010). The current emphasis on science and technology in Nigeria: Dilemmas for Art education. *Leonardo, 43,* 269–73.

Keay, R. (1976). Scientific cooperation in Africa. *African Affairs, 75,* 86–97.

Kennedy, D. (2001). Editorial: Science and development. *Science,* New Series, *294,* 2053.

Kenz, A. E., & Waast, R. (1997). Sisyphus or the scientific communities of Algeria. In J. Gaillard, V. V. Krishna & R. Waast (eds.), *Scientific Communities in the Developing World* (pp. 53–80). New Delhi: Sage Publications.

Khelfaoui, H. (2004). Scientific research in Algeria institutionalisation versus professionalisation. *Science, Technology & Society, 9,* 75–101.

Khelfaoui, H. (n.d.). *Mapping Research Systems in Developing Countries: Country Report – The Science and Technology System of Burkina Faso.* Paris: UNESCO.

Khor, K. A., & Yu, L.-G. (2016). Influence of international co-authorship on the research citation impact of young universities. *Scientometrics, 107,* 1095–110.

Kibuka-Sebitosi, E. (2006). Lessons from Ugandan indigenous knowledge systems regarding the management of HIV and AIDS. *International Journal of African Renaissance Studies in Higher Education, 1,* 111–28.

Kigotho, W. (2013). Migration and brain drain from Africa acute. *University World News* (291). www.universityworldnews.com/article.php?story=20131011121316706.

King, D. A. (2004). The scientific impact of nations. *Nature, 430,* 311–16.

Kirkland, J., & Ajai-Ajagbe, P. (2013). *Research Management in African Universities: From Awareness Raising to Developing Structures*. London: The Association of Commonwealth Universities.

Klein, J. (2011). Indigenous knowledge and education: the case of the Nama people in Namibia. *Education as Change, 15*, 81–94.

Koenig, R. (2007). Egypt plans a shakeup of research programs. *Science, New Series, 317*, 30.

Kolinsky, M. (1985). The growth of Nigerian universities 1948–1980: the British share. *Minerva, 23*, 29–61.

Kowalski, P., Lattimore, R., & Bottini, N. (2009). South Africa's Trade and Growth. OECD Trade Policy Working Paper No. 91, Paris: OECD.

Kraemer–Mbula, E., & Scerri, M. (2015). Southern Africa. In *UNESCO Science Report: Towards 2030* (pp. 535–65). Paris: UNESCO.

Krieger, E. M. (2000). Scientific capabilities in the research on basic needs for development. In *UNESCO, World Conference on Science: Science for the Twenty-First Century – a New Commitment* (pp. 104–6). Paris: UNESCO.

Krishna, V. V. (2014). Changing social relations between science and society: Contemporary challenges. *Science, Technology & Society, 19*, 133–59.

Krishna, V. V., Waast, R., & Gaillard, J. (1998). Globalization and scientific communities in developing countries. In *UNESCO, World Science Report 1998* (pp. 273–87). Paris: UNESCO.

Krishna, V. V., Waast, R., & Gaillard, J. (2000). The changing structure of science in developing countries. *Science, Technology & Society, 5*, 209–24.

Kwapong, A. A. (1988). *African scientific and technological institution building and the role of international co-operation*. Paper presented at the Symposium on Scientific Institution Building in Africa, March 14–18. Bellagio, Italy.

Lakitan, B., Hidayat, D., & Herlinda, S. (2012). Scientific productivity and the collaboration intensity of Indonesian universities and public R&D institutions: Are there dependencies on collaborative R&D with foreign institutions? *Technology in Society, 34*, 227–38.

Lancaster, F. W., & Abdullah, S. B. (1992). Science and politics: Some bibliometric analysis. In R. Arvanitis & J. Gaillard (eds.), *Science Indicators for Developing Countries* (pp. 319–31). Paris: ORSTOM.

Lancet, The (2009). Strengthening research capacity in Africa. Editorial. *Lancet, 374*, 1.

Landini, F., Malerba, F., & Mavilia, R. (2015). The structure and dynamics of networks of scientific collaborations in Northern Africa. *Scientometrics, 105*, 1787–807.

Lane, N. (2000). The scientist as global citizen. In *UNESCO, World Conference on Science: Science for the Twenty-First Century – a New Commitment* (pp. 41–4). Paris: UNESCO.

Lawler, A. (2011). A new day for Egyptian science? *Science*, New Series, *333*, 278–84.

Lebeau, Y. (2003). Extraversion strategies within a peripheral research community: Nigerian Scientists' responses to the state and changing patterns of international science and development cooperation. *Science, Technology & Society*, *8*, 183–213.

Lebeau, Y., Onyeonoru, I., & Ukah, F. K. (2000). *Science in Africa at the Dawn of the 21st Century: Country Report Nigeria, Rapport pays*. Paris: IRD (Institut de Recherches pour le Développement).

Lee, S., & Bozeman, B. (2005). The impact of research collaboration on scientific productivity. *Social Studies of Science*, *35*, 673–702.

Lemma, A. (1988). *Science and Technology in Africa: Some Reflections on Lessons learned and Prospects and Challenges for the Future*. Paper presented at the Symposium on Scientific Institution Building in Africa. March 14–18, Bellagio, Italy.

Lewin, K. M. (2009). Access to education in sub-Saharan Africa: Patterns, problems and possibilities. *Comparative Education*, *45*, 151–74.

Lewison, G., Kumar, S., Wong, C.-Y., Roe, P., & Webber, R. (2016). The contribution of ethnic groups to Malaysian scientific output, 1982–2014, and the effects of the new economic policy. *Scientometrics*, *109*, 1877–93.

Liao, C., Ruelle, M. L., & Kassam, K.-A. S. (2016). Indigenous ecological knowledge as the basis for adaptive environmental management: Evidence from pastoralist communities in the Horn of Africa. *Journal of Environmental Management*, *182*, 70–9.

Licari, J. (1997). Economic Reform in Egypt in a Changing Global Economy. Working Paper No. 129. Paris: OECD Development Centre.

Liebenberg, F., & Pardey, P. G. (2011). South African agricultural R&D: Policies and public institutions, 1880–2007. *Agrekon*, *50*, 1–15.

Liebenberg, F., Pardey, P. G., & Kahn, M. (2011). South African agricultural R&D investments: Sources, structure, and trends, 1910–2007. *Agrekon*, *50*, 1–26.

Lipsey, R. G. (2001). *Understanding technological change: East West Center Working Papers* (Vol. 13). Honolulu, HA: East West Center.

Lipsey, R. G. (2009). Economic growth related to mutually interdependent institutions and technology. *Journal of Institutional Economics*, *5*, 259–88.

Lipsey, R. G., Carlaw, K. I., & Bekar, C. T. (2005). *Economic Transformations: General Purpose Technologies and Longer Term Economic Growth*. New York, NY: Oxford University Press.

Lomnitz, L., & Salord, S. G. (1992). Ambiguities and discrepancies in the criteria for evaluating technological research in Mexico. In R. Arvanitis &

J. Gaillard (eds.), *Science Indicators for Developing Countries* (pp. 115–24). Paris: ORSTOM.

Long, F. A. (1981). Role of scientists in the development of science policy. *Bulletin of Science, Technology & Society, 1*, 225–33.

Lotka, A. J. (1926). The frequency of distribution of scientific productivity. *Journal of the Washington Academy of Science, 16*(12), 317–23.

Loyson, P. (2011). Chemistry in the time of the Pharaohs. *Journal of Chemical Education, 88*, 146–50.

Lubchenco, J. (2000). A new social contract for science. In *UNESCO, World Conference on Science: Science for the Twenty-First Century – a New Commitment* (pp. 278–80). Paris: UNESCO.

Lubowa, M. W. (1992). Access to national and international scientific information as revealed by scientific activities in three peripheral countries. In R. Arvanitis & J. Gaillard (eds.), *Science Indicators for Developing Countries* (pp. 487–95). Paris: ORSTOM.

Lwoga, E. T., Ngulube, P., & Stilwell, C. (2010). Managing indigenous knowledge for sustainable agricultural development in developing countries: Knowledge management approaches in the social context. *The International Information & Library Review, 42*, 174–85.

Maassen, P. (2012). Universities and the effects of external funding: Sub-Saharan Africa and the Nordic countries. In A. R. Nelson & I. P. Wei (eds.), *The Global University: Past, Present and Future Perspectives* (pp. 231–54). New York, NY: Palgrave Macmillan.

MacGregor, K. (2009). Africa: Higher education and development. *University World News, 96*. www.universityworldnews.com/article.php?story=2009101110573812. 6 December 2016.

Macías-Chapula, C. A., & Mijangos-Nolasco, A. (2002). Bibliometric analysis of AIDS literature in Central Africa. *Scientometrics, 54*, 309–17.

Macleod, R. (1996). Reading the discourse of colonial science. In P. Petitjean (ed.), *Les Sciences Coloniales: Figures et Institutions* (Vol. 2, pp. 87–96). Paris: L'institfurta Nçaidse Recherchsec Ientifipquoeu Rle Développement En Cooperation.

Macnaghten, P., & Chilvers, J. (2014). The future of science governance: Publics, policies, practices. *Environment and Planning C: Government and Policy, 32*, 530–48.

Madikizela, M. (n.d.). *Mapping Research Systems in Developing Countries: Country Report – The Science and Technology System of the Republic of Tunisia*. Paris: UNESCO.

Madox, J. (1968). Choice and the scientific community. In E. Shils (ed.), *Criteria for Scientific Development: Public Policy and National Goals* (pp. 44–62). Cambridge, MA: The MIT Press.

Makgetlaneng, S. (2003). South African economic and trade policy in Africa: a critical analysis. *African Journal of Political Science, 8,* 87–107.

Malcom, S., Cetto, A. M., Dickson, D., Gaillard, J., Schaeffer, D., & Quere, Y. (2002). *Science Education and Capacity Building for Sustainable Development.* Paris: International Council for Science.

Mamo, A., Mekuriaw, A., & Woldehanna, F. (2014). *IFS-AAS Project on Developing an Enabling Scientific Equipment Policy in Africa: Ethiopia Country Study.* Addis Ababa: International Foundation for Science and MacArthur Foundation.

Manyaka, J. (2006). Tracing a sound knowledge base from indigenous knowledge: the integration of indigenous and Western medical systems. *South African Journal of African Languages, 26,* 69–76.

Martin, B. R. (2012). The evolution of science policy and innovation studies. *Research Policy, 41,* 1219–39.

Marton-Lefèvre, J. (2000). International cooperation in science. In *UNESCO, World Conference on Science: Science for the Twenty-First Century – a New Commitment* (pp. 66–72). Paris: UNESCO.

Matemba, Y. H., & Lilemba, J. M. (2015). Challenging the status quo: Reclaiming indigenous knowledge through Namibia's postcolonial education system. *Diaspora, Indigenous, and Minority Education, 9,* 159–74.

Matsuura, K. (2000). S&T cooperation and the role of Asia. In *UNESCO, World Conference on Science: Science for the Twenty-First Century – a New Commitment* (pp. 378–81). Paris: UNESCO.

Maunganidze, L. (2016). A moral compass that slipped: Indigenous knowledge systems and rural development in Zimbabwe. *Cogent Social Sciences, 2.* http://dx.doi.org/10.1080/23311886.23312016.21266749.

Mazrui, A. A. (2002). Brain drain between counterterrorism and globalization. *African Issues, 30,* 86–89.

Mbarga, G. (2000). L'Afrique refuse-t-elle la science? In *UNESCO, World Conference on Science: Science for the Twenty-First Century – a New Commitment* (pp. 298–300). Paris: UNESCO.

McConney, A., Oliver, M. C., Woods-McConney, A., Renato Schibeci, & Maor, D. (2014). Inquiry, engagement, and literacy in science: a retrospective, cross-national analysis using PISA 2006. *Science Education, 98,* 963–80.

Mêgnigbêto, E. (2013a). International collaboration in scientific publishing: the case of West Africa (2001–2010). *Scientometrics, 96,* 761–83.

Mêgnigbêto, E. (2013b). Scientific publishing in West Africa: Comparing Benin with Ghana and Senegal. *Scientometrics, 95,* 1113–39.

Meneghini, R. (1992). Brazilian production in biochemistry: International versus domestic publication. In R. Arvanitis & J. Gaillard (eds.), *Science Indicators for Developing Countries* (pp. 457–85). Paris: ORSTOM.

Merton, R. K. (1937). Science, population and society. *The Scientific Monthly, 44,* 165–71.

Merton, R. K. (1938). Science and the social order. *Philosophy of Science, 5,* 321–37.

Merton, R. K. (1939). Science and the economy of seventeenth century England. *Science and Society, 3,* 3–27.

Merton, R. K. (1942). Science and technology in a democratic order. *Journal of Legal and Political Sociology, 1,* 115–26.

Merton, R. K. (1968). The Matthew Effect in science: the reward and communication systems of science are considered. *Science,* New Series, *159,* 56–63.

Merton, R. K. (1969). Behaviour patterns of scientists. *American Scientist, 57,* 1–23.

Merton, R. K. (1973). *The Sociology of Science: Theoretical and Empirical Investigations* (edited and with an introduction by Norman W. Storer). Chicago, IL: The University of Chicago Press.

Merton, R. K. (1978). *Science, Technology and Society in Seventeenth Century England.* New Jersey: Humanities Press.

Merton, R. K. (1988). The Matthew Effect in science, II: Cumulative advantage and the symbolism of intellectual property. *Isis, 79,* 606–3.

Merton, R. K. (1995). The Thomas Theorem and the Matthew Effect. *Social Forces, 74,* 379–424.

Merton, R. K., with Barber, B. (1973). Sorokin's formulations in the sociology of science. In R. K. Merton, *The Sociology of Science: Theoretical and Empirical Investigations* (pp. 142–72). (edited and with an introduction by Norman W. Storer). Chicago, IL: The University of Chicago Press.

Meyer, J.-B. (1997). Science and technology in South Africa: a new society in the making. In J. Gaillard, V. V. Krishna & R. Waast (eds.), *Scientific Communities in the Developing World* (pp. 183–204). New Delhi: Sage Publications.

Michelson, E. S. (2006). The transformation of African academies of science: the evolution of new institutions. *Bulletin of Science, Technology & Society, 26,* 419–29.

Midgley, M. (2005). Mapping science: in memory of John Ziman. *Interdisciplinary Science Reviews, 30,* 195–7.

Mills, G. (2016). *Why Africa Is Poor.* Cape Town: Penguin Books.

Mkandawire, T. (1994). Social sciences in Africa: Some lessons for South Africa. *South African Sociological Review, 6,* 1–13.

Moahi, K. H. (2012). Promoting African indigenous knowledge in the knowledge economy: Exploring the role of higher education and libraries. *Aslib Proceedings, 64*, 540–54.

Moravcsik, M. J. (1966). Some practical suggestions for the improvement of science in developing countries. *Minerva, 4*, 381–90.

Moravcsik, M. J. (1973). The transmission of a scientific civilization. *Bulletin of the Atomic Scientists, 29*, 25–8.

Moravcsik, M. J. (1975). *Science Development: the Building of Science in Less Developed Countries*. Bloomington, IN: Program for Advanced Studies in Institution Building and Technical Assistance Methodology, Indiana University.

Moravcsik, M. J. (1982). The effectiveness of research in developing countries. *Social Studies of Science, 12*, 114–47.

Moravcsik, M. J. (1983). The role of science in technology transfer. *Research Policy, 12*, 287–96.

Moravcsik, M. J. (1984). Can we plan science? (semantics and pitfalls). *Bulletin of Science, Technology & Society, 4*, 361–78.

Moravcsik, M. J. (1985). Science in the developing countries: an unexplored and fruitful area for research in science studies. *4S Review, 3*, 2–13.

Moravcsik, M. J. (1986a). The prospect of scientific growth in poor countries. *Minerva, 24*, 137–42.

Moravcsik, M. J. (1986b). Two perceptions of science development. *Research Policy, 15*, 1–11.

Moravcsik, M. J. (1987). Science policy and development in the third world. *Bulletin of Science, Technology & Society, 7*, 598–604.

Moravcsik, M. J. (1989). Dependence and science scenarios for the Third World. *Social Science Information, 28*, 445–52.

Moravcsik, M. J., & Ziman, J. M. (1975). Paradisia and dominatia: Science and the developing world. *Foreign Affairs, 53*, 699–724.

Moravcsik, M. J., Ziman, J., & Szmant, H. H. (1975). Third world science and technology. *Science*, New Series, *190*, 938

Morley, B., & Perdikis, N. (2000). Trade liberalisation, government expenditure, and economic growth in Egypt. *Journal of development Studies, 36*, 38–54.

Mouton, J. (2008). Africa's science decline: the challenge of building scientific institutions. *Harvard International Review, 30*, 46–51.

Mouton, J., & Boshoff, N. (n.d.). *Mapping Research Systems in Developing Countries: Country Report – the Science and Technology System of Ethiopia*. Paris: UNESCO.

Mouton, J., & Gevers, W. (2010). Introduction. In R. Diab & W. Gevers (eds.), *The State of Science in South Africa* (pp. 39–67). Pretoria: ASSAf.

Mouton, J., & Hackmann, H. (1997). *Survey on Scholarship, Research and Development*. Stellenbosch: University of Stellenbosch, Centre for Interdisciplinary Studies for the Department of Arts, Culture, Science and Technology, the Government of the Republic of South Africa.

Mouton, J., & Waast, R. (2009). *Mapping Research Systems in Developing Countries: Synthesis Report*. Paris: UNESCO.

Mouton, J., Gaillard, J., & Lill, M. v. (2015). Functions of science granting councils in sub-Saharan Africa. In N. Cloete, P. Maassen & T. Bailey (eds.), *Knowledge Production and Contradictory Functions in African Higher Education* (pp. 148–71). Cape Town: African Minds.

MPCDU & UNDP (Ministry of Planning and Communal Development/ Forecasting Unit and United Nations Development Programme). (2011). *Complete Vision Burundi 2025*. Burundi: Ministry of Planning and Communal Development/Forecasting Unit and United Nations Development Programme.

Mubangizi, J., & Kaya, H. (2015). African indigenous knowledge systems and human rights: Implications for higher education. *International Journal of African Renaissance Studies – Multi-, Inter- and Transdisciplinarity, 10*, 125–42.

Muriithi, P., Horner, D., Pemberton, L., & Wao, H. (2018). Factors influencing research collaborations in Kenyan universities. *Research Policy, 47*, 88–97.

Musiige, G., & Maassen, P. (2015). Faculty perceptions of the factors that influence research productivity at Makerere University. In N. Cloete, P. Maassen & T. Bailey (eds.), *Knowledge Production and Contradictory Functions in African Higher Education* (pp. 109–27). Cape Town: African Minds.

Mutapi, F. (2012). Advances in parasite control in Africa: From basic science to translation. *Journal of Parasitology Research, 2012*, 1–2.

Nagtegaal, L. W., & de Brun, R. E. (1994). The French connection and other neo-colonial patterns in the global network of science. *Research Evaluation, 4*, 119–27.

Narváez-Berthelemot, N., Russel, J. M., Arvanitis, R., Waast, R., & Gaillard, J. (2002). Science in Africa: an overview of mainstream scientific output. *Scientometrics, 54*, 229–41.

Nasir, A., Ali, T. M., Shahdin, S., & Rahman, T. U. (2011). Technology Achievement Index 2009: Ranking and comparative study of nations. *Scientometrics, 87*, 41–62.

Nature Index. (2014). Africa. *Nature Index Global*, S92–93.

Ndemo, B. (2015). Effective innovation policies for development: the case of Kenya. In S. Dutta, B. Lanvin & S. Wunsch-Vincent (eds.), *The Global*

Innovation Index 2015: Effective Innovation Policies for Development (pp. 131–8). Geneva: The World Intellectual Property Organization.

NEPAD (New Partnership for Africa's Development). (2005). *Africa's Science and Technology Consolidated Plan of Action.* Johannesburg: NEPAD.

NEPAD (New Partnership for Africa's Development). (2011). *NEPAD: a Continental Thrust: Advancing Africa's Development.* Johannesburg: NEPAD.

Ngwainmbi, E. K. (2000). Africa in the global infosupermarket: Perspectives and prospects. *Journal of Black Studies, 30,* 534–52.

Nicholls, P. T. (1989). Bibliometric modeling processes and the empirical validity of Lotka's Law. *Journal of the American Society for Information Science, 40,* 379–85.

Nordling, L. (2010a). African nations vow to support science. *Nature, 465,* 994–5.

Nordling, L. (2010b). Ethiopia launches first science academy. *Nature,* doi:10.1038/news.2010.173.

Nordling, L. (2014). Africa science plan attacked. *Nature, 510,* 454–3.

Nour, S. S. O. M. (2005). Science and technology development indicators in the Arab region: a comparative study of Arab gulf and Mediterranean countries. *Science, Technology & Society, 10,* 249–74.

Nour, S. S. O. M. (2011). National, regional and global perspectives of higher education and science policies in the Arab region. *Minerva, 49,* 387–423.

Nour, S. S. O. M. (2012). Assessment of science and technology indicators in Sudan. *Science, Technology & Society, 17,* 323–54.

Nowotny, H., Scott, P., & Gibbons, M. (2003). 'Mode 2' revisited: the new production of knowledge. *Minerva, 41,* 179–94.

NPCA (NEPAD Planning and Coordinating Agency). (2014). *African Innovation Outlook 2014.* Pretoria: NEPAD Planning and Coordinating Agency.

Ntiri, D. W. (1993). Africa's educational dilemma: Roadblocks to universal literacy for social integration and change. *International Review of Education, 39,* 357–72.

Nwagwu, N. A. (1998). Management, structure and financing of education. In UNESCO (ed.), *The State of Education in Nigeria* (pp. 10–21). Lagos: UNESCO.

Nwagwu, W. E., & Iheanetu, O. (2011). Use of scientific information sources by policymakers in the science and technology sector of Nigeria. *African Journal of Library, Archives and Information Science, 21,* 59–71.

Nyiira, Z. M. (2005). *New Directions for Namibia's Science and Technology Sector: Towards a Science and Technology Plan.* Report submitted to UNESCO and the Government of the Republic of Namibia.

Oanda, I., & Sall, E. (2016). From peril to promise: Repositioning higher education for the reconstruction of Africa's future. *International Journal of African Higher Education, 3*, 51–78.

OAU (Organization of African Unity). (1981). *Lagos Plan of Action for the Economic Development of Africa, 1980–2000.* Addis Ababa: Organization of African Unity.

Odhiambo, T. R. (1967). East Africa: Science for development. *Science*, New Series, *158*, 876–81.

Odhiambo, T. R. (1993). Africa. In *UNESCO, World Science Report, 1993* (pp. 86–95). Paris: UNESCO.

Odhiambo, T. R. (2000). The responsibility of science in the alleviation of poverty in the world. In *UNESCO, World Conference on Science: Science for the Twenty-First Century – a New Commitment* (pp. 118–20). Paris: UNESCO.

OECD (Organisation of Economic Cooperation and Development). (2000). Science, technology and innovation in the new economy. *OECD Observer*, www.oecd.org/science/sci-tech/1918259.pdf. Accessed 15 April 2017.

OECD (Organisation of Economic Cooperation and Development). (2006). *OECD Review of Agricultural Policies: South Africa.* Paris: OECD.

OECD (Organisation of Economic Cooperation and Development). (2015). *OECD Investment Policy Reviews: Nigeria 2015.* Paris: OECD.

OECD (Organisation of Economic Cooperation and Development). (2016). *Main Science and Technology Indicators.* Paris: OECD.

OECD-UNDESA (Organisation of Economic Cooperation and Development-United Nations Department of Economic and Social Affairs Population Division). (2013). *World Migration in Figures.* Paris: OECD-UNDESA.

OECD-World Bank. (2010). *Reviews of National Policies for Education: Higher Education in Egypt 2010.* Paris: OECD/The International Bank Reconstruction and Development/The World Bank.

OECD (Organisation of Economic Cooperation and Development). (2008). *Reviews of National Policies for Education: South Africa.* Paris: OECD.

Okeke, I. N. (2010). African researchers underrepresented. *Science*, New Series, *328*(5982), 1103.

Okoti, M., Keya1, G. A., Esilaba, A. O., & Cheruiyot, H. (2006). Indigenous technical knowledge for resource monitoring in Northern Kenya. *Journal of Human Ecology, 20*, 183–9.

Olaore, A. Y., & Drolet, J. (2017). Indigenous knowledge, beliefs, and cultural practices for children and families in Nigeria. *Journal of Ethnic & Cultural Diversity in Social Work, 26*, 254–70.

Oluwatoyese, O. P., Applanaidu, S. D., & Razak, N. A. A. (2016). Macroeconomic factors and agricultural sector in Nigeria. *Procedia Social and Behavioral Sciences, 219*, 562–70.

Omolewa, M. (2008). Adult literacy in Africa: the push and pull factors. *International Review of Education, 54*, 697–711.

Ondari-Okemwa, E. (2007). Scholarly publishing in sub-Saharan Africa in the twenty-first century: Challenges and opportunities. *First Monday, 12*, http://firstmonday.org/article/view/1966/1842.

Onwu, G., & Mosimege, M. (2004). Indigenous knowledge systems and science and technology education: a dialogue. *African Journal of Research in Mathematics, Science and Technology Education, 8*, 1–12.

Onyancha, O. B. (2011). Research collaborations between South Africa and other countries, 1986–2005: an informetric analysis. *African Journal of Library & Information Science, 21*, 99–112.

Onyancha, O. B., & Maluleka, J. R. (2011). Knowledge production through collaborative research in sub-Saharan Africa: How much do countries contribute to each other's knowledge output and citation impact? *Scientometrics, 87*, 315–36.

Osabutey, E. L. C., & Jin, Z. (2016). Factors influencing technology and knowledge transfer: Configurational recipes for Sub-Saharan Africa. *Journal of Business Research, 69*, 5390–5.

Osabutey, E. L., & Debrah, Y. A. (2012). Foreign direct investment and technology transfer policies in Africa: a review of the Ghanaian experience. *Thunderbird International Business Review, 54*, 441–56.

Osborne, D. (1971). The use and promotion of science in developing countries. *Minerva, 9*, 45–55.

Osborne, M. A. (1999). Introduction: the social history of science, technoscience and imperialism. *Science, Technology & Society, 4*, 161–70.

Ouédraogo, I., Nacoulma, B. M. I., Hahn, K., & Thiombiano, A. (2014). Assessing ecosystem services based on indigenous knowledge in southeastern Burkina Faso (West Africa). *International Journal of Biodiversity Science, Ecosystem Services & Management, 10*, 313–21.

Oukem-Boyer, O., Djikeng, A., Cappelli, G., & Fouda, P. (2009). Tackling human resources in Africa: How one institute leverages overseas talent to develop its research strategy. *The Scientist, 23*, 24.

Owusu-Ansah, F. E., & Mji, G. (2013). African indigenous knowledge and research. *African Journal of Disability, 2*. http://dx.doi.org/10.4102/ajod.v4102i4101.4130.

Owusu-Nimo, F., & Boshoff, N. (2017). Research collaboration in Ghana: Patterns, motives and roles. *Scientometrics, 110*, 1099–121.

Özden, Ç. g., & Schiff, M. (2006). Overview. In Ç. Özden & M. Schiff (eds.), *International Migration, Remittances, and the Brain Drain* (pp. 1–18). Washington, DC: World Bank and Palgrave Macmillan.

Ozor, F. U. (2014). Research governance and scientific knowledge production in The Gambia. *South African Journal of Science, 110.* http://dx.doi.org/10.1590/.

Padilla-Pérez, R., & Gaudin, Y. (2014). Science, technology and innovation policies in small and developing economies: the case of Central America. *Research Policy, 43,* 749–59.

Pan, R. K., Kaski, K., & Fortunato, S. (2012). World citation and collaboration networks: Uncovering the role of geography in science. *Scientific Reports, 2,* 1–7.

Patel, I. G. (1993). Keynote address: Higher education and economic development. In A. Ransom, S.-M. Khoo & V. Selvaratnam (eds.), *Improving Higher Education in Developing Countries* (pp. 45–56). Washington, DC: The International Bank for Reconstruction and Development / The World Bank.

Peimbert, M. (2000). Fundamental science: a view from the South. In UNESCO, *World Conference on Science: Science for the Twenty-First Century – a New Commitment* (pp. 98–100). Paris: UNESCO.

Philander, S. G. (2009). How many scientists does South Africa need? *South African Journal of Science, 105,* 172–3.

Pillay, K. (2017). Durban joins the march of science. The Mercury, 24 April, p. 3.

Polanyi, M. (1962). The republic of science: Its political and economic theory. *Minerva, 1,* 54–73.

Polanyi, M. (1968). The growth of science in society. In E. Shils (ed.), *Criteria for Scientific Development: Public Policy and National Goals* (pp. 187–99). Cambridge, MA: The MIT Press.

Pouris, A. (1989). A scientometric assessment of agricultural research in South Africa. *Scientometrics, 17,* 401–13.

Pouris, A. (2003). South Africa's research publication record: the last ten years. *South African Journal of Science, 99,* 425–28.

Pouris, A. (2010). A scientometric assessment of the Southern Africa Development Community: Science in the tip of Africa. *Scientometrics, 85,* 145–54.

Pouris, A., & Ho, Y.-S. (2014). Research emphasis and collaboration in Africa. *Scientometrics, 98,* 2169–84.

Pouris, A., & Pouris, A. (2009). The state of science and technology in Africa (2000–2004): a scientometric assessment. *Scientometrics, 79,* 297–309.

Pouris, A., & Pouris, A. (2011). Scientometrics of a pandemic: HIV/AIDS research in South Africa and the world. *Scientometrics, 86,* 541–52.

Power, C. N. (2000). Science education in schools. In *UNESCO, World Conference on Science: Science for the Twenty-First Century – A New Commitment* (pp. 156–9). Paris: UNESCO.

Price, D. J. d. S. (1965). The science of science. *Bulletin of the Atomic Scientists, 21*, 2–8.

Prpić, K. (2011). Science, the public, and social elites: How the general public, scientists, top politicians and managers perceive science. *Public Understanding of Science, 20*, 733–50.

Puuska, H.-M., Muhonen, R., & Leino, Y. (2014). International and domestic co-publishing and their citation impact in different disciplines. *Scientometrics, 98*, 823–39.

Quartey, J. A. K. (1971). Science in developing countries. *Minerva, 9*, 548–50.

Rabkin, Y. M., Eisemon, T. O., Lafitte-Houssat, J.-J., & Rathgeber, E. M. (1979). Citation visibility of Africa's science. *Social Studies of Science, 9*, 499–506.

Radnitzky, G. (1983). Science, technology, and political responsibility. *Minerva, 21*, 234–64.

Radosevic, S., & Yoruk, E. (2014). Are there global shifts in the world science base? Analysing the catching up and falling behind of world regions. *Scientometrics, 101*, 1897–924.

Raina, D. (1999). From west to non-west? Basalla's three-stage model revisited. *Science as Culture, 8*, 497–516.

Ransom, A., Khoo, S.-M., & Selvaratnam, V. (1993). *Improving higher education in developing countries*. Washington, DC: The International Bank for Reconstruction and Development/The World Bank.

Raseroka, K. (2008). Information transformation Africa: Indigenous knowledge – Securing space in the knowledge society. *The International Information & Library Review, 40*(4), 243–50.

Rath, A. (1990). Science, technology, and policy in the periphery: a perspective from the centre. *World Development, 18*, 1429–43.

Rivlin, B. (1969). Research in North Africa: a report to the Research Liaison Committee of the African Studies Association. *African Studies Bulletin, 12*, 343–5.

ROG (Republic of Ghana). (2010). *National Science, Technology and Innovation Policy*. Accra: Ministry of Environment, Science and Technology: Republic of Ghana.

Rogers, C. L. (2005). Report – The nexus: Where science meets society. *Science Communication, 27*, 146–9.

ROK (Republic of Kenya). (2008). *Science, Technology and Innovation Policy and Strategy*. Nairobi: Ministry of Science and Technology, Republic of Kenya.

ROK (Republic of Kenya). (2012). *A Policy Framework for Science, Technology and Innovation*. Nairobi: Ministry of Higher Education, Science and Technology, Republic of Kenya.

ROK (Republic of Kenya). (n.d.). *Sector Plan for Science, Technology and Innovation, 2013–2017*. Nairobi: Ministry of Education, Science and Technology, Republic of Kenya.

ROM (Republic of Mauritius). (n.d.). *National Report of the Republic of Mauritius*. Geneva: UNDESA and UNDP.

ROM (Republic of Mozambique). (2006). *Mozambique Science Technology and Innovation Strategy (MOSTIS)*. Maputo: Republic of Mozambique.

RON (Republic of Namibia). (2014). *The National Programme on Research, Science, Technology and Innovation*. Windhoek: Republic of Namibia.

Rooyen, J. v., & Machethe, C. (2010). Determining the agricultural sector's role in regional development in South Africa. *Agrekon, 30*, 175–81.

ROR (The Republic of Rwanda). (2006). *The Republic of Rwanda Policy on Science, Technology and Innovation*. Kigali: Ministry in the President's Office in Charge of Science, Technology and Scientific Research, The Government of the Republic of Rwanda.

ROS (Republic of Seychelles). (2013). *National report. prepared for the 3rd international conference on small island development states to be held in Apia Samoa, 2014*. Victoria: The Government of the Republic of Seychelles.

ROZ (Republic of Zambia). (1996). *National Policy on Science and Technology*. Lusaka: The Government of the Republic of Zambia.

Safonova, M., & Sokolov, M. (2013). *The construction of the world-system: Regression and social network approaches to analysis of international academic ties*. Paper presented at the 14th International Society of Scientometrics and Informetrics Conference, 15–19 July, Vienna.

Sagasti, F. R. (1973). Underdevelopment, science and technology: the point of view of the underdeveloped countries. *Science Studies, 3*, 47–59.

Sagasti, F. R. (1979). Notes on science, technology and development planning. In V. L. Urquidi (ed.), *Science and Technology in Development Planning: Science, Technology and Global Problems* (pp. 117–33). Oxford: Pergamon Press.

Salam, A. (1968). The isolation of the scientist in developing countries. In E. Shils (ed.), *Criteria for Scientific Development: Public Policy and National Goals* (pp. 200–4). Cambridge, MA: The MIT Press.

Saldaña, J. J. (2000). Western and non-Western science: History and perspectives. In UNESCO, *World Conference on Science: Science for the Twenty-First Century – a New Commitment* (pp. 86–7). Paris: UNESCO.

Sancho, R. (1992). Misjudgements and shortcomings in the measurements of scientific activities in less developed countries. In R. Arvanitis & J. Gaillard (eds.), *Science Indicators for Developing Countries* (pp. 411–23). Paris: ORSTOM.

Sanderson, G. N. (1985). The European partition of Africa: Origins and dynamics. In R. Oliver & G. N. Sanderson (eds.), *The Cambridge History of Africa* (Vol. 6, pp. 96–158). London: Cambridge University Press.

Sandrey, R., Punt, C., Jensen, H. G., & Vink, N. (2011). Agricultural Trade and Employment in South Africa. OECD Trade Policy Working Papers, No. 130. Paris: OECD.

Sanyal, B. C., & Varghese, N. V. (2006). *Research Capacity of the Higher Education Sector in Developing Countries.* Paris: UNESCO.

Sarton, G. (1918). The teaching of the history of science. *The Scientific Monthly, 7,* 193–211.

Sarton, G. (1936). Remarks concerning the history of twentieth century science. *Isis, 26,* 53–62.

Savanur, K., & Srikanth, R. (2010). Modified collaborative coefficient: a new measure for quantifying the degree of research collaboration. *Scientometrics, 84,* 365–71.

Sawahel, W. (2009). IRAN: 20-year plan for knowledge based economy. *University News, 90,* 1–3.

Sawyerr, A. (2004). Challenges facing African universities: Selected issues. *African Studies Review, 47,* 1–59.

Schafer, M. J., Shrum, W. M., Paige Miller, B., Mbatia, P. N., Palackal, A., & Dzorgbo, D.-B. S. (2016). Access to ICT and research output of agriculture researchers in Kenya. *Science, Technology & Society, 21,* 250–70.

Schemm, Y. (2013). Africa doubles research output over past decade, moves towards a knowledge-based economy. *Research Trends, 35.* www .researchtrends.com/issue-35-december-2013/africa-doubles-research-out put/.

Schettkat, R. (2008). Introductory summary: Prosperity for Germany and Europe. In R. Schettkat & J. Langkau (eds.), *Economic Policy Proposals for Germany and Europe* (pp. 1–19). London: Routledge.

Schoepf, B. G. (2003). Lessons for AIDS control in Africa. *Review of African Political Economy, 30,* 553–72.

Scholes, R. J., Anderson, F., Kenyon, C., Napier, J., Ngoepe, P., Wilgen, B. v., et al. (2008). Science councils in South Africa. *South African Journal of Science, 104,* 435–38.

Schoole, C. T. (2012). The unequal playing field: Academic remuneration in South Africa. In P. G. Altbach, L. Reisberg, M. Yudkevich, G. Androushchak & I. F. Pacheco (eds.), *Paying the Professoriate:*

a Global Comparison of Compensation and Contracts (pp. 288–96). New York, NY: Routledge.

Schott, T. (1991). The world scientific community: Globality and globalisation. *Minerva, 29,* 440–62.

Schott, T. (1993). World science: Globalization of institutions and participation. *Science, Technology and Human Values, 18,* 196–208.

Schott, T. (1998). Ties between center and periphery in the scientific world system: Accumulation of rewards, dominance and self-reliance in the center. *Journal of World Systems Research, 4,* 112–44.

Schubert, A., & Braun, T. (1992). Three scientometric studies on developing countries as a tribute to Michael Moravcsik. In R. Arvanitis & J. Gaillard (eds.), *Science Indicators for Developing Countries* (pp. 49–64). Paris: ORSTOM.

Schubert, T., & Sooryamoorthy, R. (2010). Can the centre–periphery model explain patterns of international scientific collaboration among threshold and industrialised countries? The case of South Africa and Germany. *Scientometrics, 83,* 181–203.

Seaborg, G. T. (1970). A scientific safari to Africa. *Science,* New Series, *169,* 554–61.

Sebastian, P. (2013). Trade policy is science policy. *Issues in Science & Technology, 30,* 85–88.

Shils, E. (1968). Introduction. In E. Shils (ed.), *Criteria for Scientific Development: Public Policy and National Goals* (pp. v–xiv). Cambridge, MA: The MIT Press.

Shrum, W. (1997). View from afar: 'Visible' productivity of scientists in the developing world. *Scientometrics, 40,* 215–35.

Siino, F. (2003). Tunisian science in search of legitimacy. *Science, Technology & Society, 8,* 261–81.

Sin, S.-C. J. (2011). International coauthorship and citation impact: a bibliometric study of six LIS journals, 1980–2008. *Journal of the American Society for Information Science and Technology, 62,* 1770–83.

Singh-Pillay, A., Alant, B. P., & Nwokocha, G. (2017). Tapping into Basic 7–9 science and technology teachers' conceptions of indigenous knowledge in Imo State, Nigeria. *African Journal of Research in Mathematics, Science and Technology Education, 21,* 125–35.

Singh, K., Granville, M., & Dika, S. (2002). Mathematics and science achievement: Effects of motivation, interest, and academic engagement. *Journal of Educational Research, 95,* 323–32.

Siyanbola, W., Adeyeye, A., Olaopa, O., & Hassan, O. (2016). Science, technology and innovation indicators in policy-making: the Nigerian experience. *Palgrave Communications, 2,* DOI:10.1057/palcomms.2016.15,1–9.

Smith, D. (1967). Scientific research centres in Africa. *African Studies Bulletin*, *10*, 20–47.

Smith, E. (2010). Do we need more scientists? A long-term view of patterns of participation in UK undergraduate science programmes. *Cambridge Journal of Education*, *40*, 281–98.

Soete, L., Schneegans, S., Eröcal, D., Angathevar, B., & Rasiah, R. (2015). A world in search of an effective growth strategy. In *UNESCO Science Report: Towards 2030* (pp. 21–55). Paris: UNESCO.

Sofolahan., J. O. (1998). National policy review issues. In UNESCO (ed.), *The State of Education in Nigeria* (pp. 3–9). Lagos: UNESCO.

Sooryamoorthy, R. (2009a). Collaboration and publication: How collaborative are scientists in South Africa? *Scientometrics*, *80*, 419–39.

Sooryamoorthy, R. (2009b). Do types of collaboration change citation? Collaboration and citation patterns of South African science publications. *Scientometrics*, *81*, 171–93.

Sooryamoorthy, R. (2011). Collaboration in South African engineering research. *South African Journal of Industrial Engineering*, *22*, 18–36.

Sooryamoorthy, R. (2013). Scientific Collaboration in South Africa. *South African Journal of Science*, *109*, 1–5.

Sooryamoorthy, R. (2014). Publication productivity and collaboration of researchers in South Africa: New empirical evidence. *Scientometrics*, *98*, 531–45.

Sooryamoorthy, R. (2015). *Transforming Science in South Africa: Development, Collaboration and Productivity*. Hampshire and New York, NY: Palgrave Macmillan.

Sooryamoorthy, R. (2016). Producing information: Communication and collaboration in the South African scientific community. *Information, Communication & Society*, *19*, 141–59.

Sooryamoorthy, R. (2017a). Do types of collaboration change citation? A scientometric analysis of social science publications in South Africa. *Scientometrics*, *111*, 379–400.

Sooryamoorthy, R. (2017b). *Networks of Communication in South Africa: New Media, New Technologies*. Cambridge: Cambridge University Press.

Sooryamoorthy, R. (2018). The production of science in Africa: an analysis of publications in the science disciplines, 2000–2015. *Scientometrics*, *115*, 317–49.

Sooryamoorthy, R., & Shrum, W. (2007). Does the Internet promote collaboration and productivity? Evidence from the scientific community in South Africa. *Journal of Computer-Mediated Communication*, *12*, 733–51.

Strohecker, K. (2016). African nations register declines in governance: Continent's progress being held back. *Business Report*, October 4, p. 19.

Sun, X., Kaur, J., Milojević, S., Flammini, A., & Menczer, F. (2013). Social dynamics of science. *Scientific Reports*, *3*, DOI:10.1038/srep01069.

Sutton, F. X. (1988). *A foundation perspective on African science.* Paper presented at the Symposium on Scientific Institution Building in Africa, March 14–18, Bellagio, Italy.

Taylor, D. L., & Cameron, A. (2016). Valuing IKS in successive South African physical sciences curricula. *African Journal of Research in Mathematics, Science and Technology Education As Change*, *20*, 35–44.

Teferra, D. (2013). Funding higher education in Africa: State, trends and perspectives. *Journal of Higher Education in Africa*, *11*, 19–51.

Teng-Zeng, F. (n.d.). *Mapping Research Systems in Developing Countries: Country Report – The Science and Technology System of Ghana.* Paris: UNESCO.

Tharakan, J. (2015). Integrating indigenous knowledge into appropriate technology development and implementation. *African Journal of Science, Technology, Innovation and Development*, *7*, 364–70.

THDR (Tanzania Human Development Report 2014). (2015). Dodoma, Tanzania: Economic and Social Research Foundation, United Nations Development Programme, and the Government of the United Republic of Tanzania.

The Royal Society. (2011). *Knowledge, Networks and Nations: Global Scientific Collaboration in the 21st Century.* London: The Royal Society.

Thisen, J. K. (1993). The development and utilization of science and technology in productive sectors: Case of developing Africa. *Africa Development*, *18*, 5–35.

Thomas, S. M. (1992). The evaluation of plant biomass research: a case study of the problems inherent in bibliometric methods. In R. Arvanitis & J. Gaillard (eds.), *Science Indicators for Developing Countries* (pp. 149–64). Paris: ORSTOM.

Thulstrup, E. W., Fekadu, M., & Negewo, A. (1996). *Building Research Capacity in Ethiopia.* Stockholm: Swedish International Development Cooperation Agency.

Tijani, A. A., Oluwasola, O., & B, O. I. (2015). Public sector expenditure in agriculture and economic growth in Nigeria: an empirical investigation. *Agrekon*, *54*, 76–92.

Tijssen, R. (2015). Research output and international research cooperation in African flagship universities. In N. Cloete, P. Maassen & T. Bailey (eds.), *Knowledge Production and Contradictory Functions in African Higher Education* (pp. 61–74). Cape Town: African Minds.

Tijssen, R. J. W. (2007). Africa's contribution to the worldwide research literature: New analytical perspectives, trends, and performance indicators. *Scientometrics, 71*, 303–27.

Tilley, H. (2011). *Africa as a Living Laboratory: Empire, Development and the Problem of Scientific Knowledge, 1870–1950.* Chicago, IL: The University of Chicago Press.

Tindemans, P. (2005). Producing knowledge and benefiting from it: The new rules of the game. In *UNESCO Science Report 2005* (pp. 1–24). Paris: UNESCO.

Toivanen, H., & Ponomariov, B. (2011). African regional innovation systems: Bibliometric analysis of research collaboration patterns, 2005–2009. *Scientometrics, 88*, 471–93.

Traoré, N., & Landry, R. (1997). On the determinants of scientists' collaboration. *Science Communication, 19*, 124–40.

Turpin, T., & Martinez-Fernandez, C. (2003). Bridging knowledge boundaries: a challenge for S& T policy in Mozambique. *Science, Technology & Society, 8*, 215–34.

Ubogu, F. N., & Van den Heever, M. (2013). Collaboration on academic research support among five African universities *Qualitative and Quantitative Methods in Libraries, 2*, 207–19.

UIS (UNESCO Institute for Statistics). (2016a). *Dataset: Science, Technology and Innovation – Researchers FTE.* Paris: UNESCO Institute for Statistics.

UIS (UNESCO Institute for Statistics). (2016b). *Dataset: Education – Adult Literacy Rate, Population15+ years, both sexes.* Paris: UNESCO Institute for Statistics.

UIS (UNESCO Institute for Statistics). (2016c). *Science, technology and innovation: Gross domestic expenditure on R&D (GERD), GERD as a percentage of GDP, GERD per capita and GERD per researcher.* Paris: UNESCO Institute for Statistics.

UIS (UNESCO Institute for Statistics). (2017). *Global investments in R&D.* Fact Sheet No. 42. Paris: UNESCO Institute for Statistics.

UN (United Nations). (1980). *Economic Commission for Africa: Strategy for the African Region in the International Development Strategy or the United Nations Third Development Decade.* Addis Ababa: United Nations.

UN (United Nations). (2015). *Transforming Our World: the 2030 Agenda for Sustainable Development.* New York, NY: United Nations.

UNCTAD (United Nations Conference on Trade and Development). (2008). *Science, Technology and Innovation Policy (STIP) Review of Angola.* Geneva: United Nations.

UNCTAD (United Nations Conference on Trade and Development). (2010a). *Science, Technology & Innovation Policy Review: Lesotho.* Geneva: United Nations.

UNCTAD (United Nations Conference on Trade and Development). (2010b). *Science, Technology & Innovation Policy Review: Mauritania.* Geneva: United Nations.

UNCTAD (United Nations Conference on Trade and Development). (2011). *A Framework for Science, Technology and Innovation Policy Reviews: Helping Countries Leverage Knowledge and Innovation for Development.* Geneva: United Nations.

UNCTAD (United Nations Conference on Trade and Development). (2015). *Technology and Innovation Report 2015: Fostering Innovation Policies for Industrial Development.* Geneva: UNCTAD.

UNDP (United Nations Development Programme). (2003). *The Arab Human Development Report 2003: Building a Knowledge Society.* New York, NY: United Nations Development Programme.

UNECA (United Nations Economic Commission for Africa). (2013). *National Experiences in the Transfer of Publicly Funded Technologies in Africa: Ghana, Kenya and Zambia.* Addis Ababa: Economic Commission for Africa.

UNECA (United Nations Economic Commission for Africa). (2014). *Dynamic Industrial Policy in Africa: Innovative Institutions, Effective Processes and Flexible Mechanisms.* Addis Ababa: Economic Commission for Africa.

UNECA (United Nations Economic Commission for Africa). (2016). Africa's science, technology and innovation policies – National, regional and continental. In UNECA (ed.), *Assessing Regional Integration in Africa VII: Innovation, Competitiveness and Regional Integration* (pp. 83–104). New York, NY: United Nations.

UNESCO (United Nations Educational, Scientific and Cultural Organization). (1974a). *National Science Policies in Africa: Situation and Future Outlook.* Paris: UNESCO.

UNESCO (United Nations Educational, Scientific and Cultural Organization). (1974b). *Science and Technology in African Development.* Paris: UNESCO.

UNESCO (United Nations Educational, Scientific and Cultural Organization). (1986). *Comparative Study on the National Science and Technology Policy-Making Bodies in the Countries of West Africa.* Paris: UNESCO.

UNESCO (United Nations Educational, Scientific and Cultural Organization). (1987). *Intra-African and inter-regional scientific and technological co-operation.* CASTAFRICA II: Second conference on ministers responsible for the application of science and technology to development in Africa, 6–15 July, Arusha, Tanzania.

UNESCO (United Nations Educational, Scientific and Cultural Organization). (1998). *The State of Education in Nigeria.* Lagos: UNESCO.

UNESCO (United Nations Educational Scientific and Cultural Organization). (2000). *World Conference on Science: Science for the Twenty-First Century – a New Commitment*. Paris: UNESCO.

UNESCO (United Nations Educational, Scientific and Cultural Organization). (2006). From brain drain to brain gain. *Education Today*, *18*, 4–6.

UNESCO (United Nations Educational, Scientific and Cultural Organization). (2007a). *Review of Science and Technology Meetings at Ministerial Level 1996–2006*. Paris: UNESCO.

UNESCO (United Nations Educational, Scientific and Cultural Organization). (2007b). *Science in Africa: UNESCO's Contribution to Africa's Plan for Science and Technology to 2010*. Paris: UNESCO.

UNESCO (United Nations Educational, Scientific and Cultural Organization) (2009). *Science, Technology & Innovation Policy Initiative: Responding to the Needs of Africa*. Paris: UNESCO.

UNESCO (United Nations Educational, Scientific and Cultural Organization). (2015a). *UNESCO Science Report: Towards 2030*. Paris: UNESCO.

UNESCO (United Nations Educational, Scientific and Cultural Organization). (2015b). *Global Investments in R&D: UIS Fact Sheet*. Paris: UNESCO.

UNESCO (United Nations Educational, Scientific and Cultural Organisation). (2019). *Global Investments in R&D*. Paris: UNESCO.

Urama, K., Muchie, M., & Twiringiyimana, R. (2015). East and Central Africa. In *UNESCO Science Report: Towards 2030* (pp. 499–533). Paris: UNESCO.

van den Brink, J., & Snyman, I. (2007). Advancing science in Africa. *Nature Materials*, *6*, 792–93.

Vavakova, B. (1998). The new social contract between governments, universities and society: Has the old one failed? *Minerva*, *36*, 209–28.

Vellho, L. (2004). Research capacity building for development: From old to new assumptions. *Science, Technology & Society*, *9*, 171–207.

Verran, H. (2001). *Science and an African Logic*. Chicago, IL: Chicago University Press.

Viljoen, M. F. (2005). South African Agricultural Policy, 1994 to 2004: Some reflections. *Agrekon*, *44*, 1–16.

Virasoro, M. A. (2000). The universal value of fundamental science. In *UNESCO, World Conference on Science: Science for the Twenty-First Century – a New Commitment* (pp. 57–62). Paris: UNESCO.

Vitta, P. B. (1990). Technology policy in sub-Saharan Africa: Why the dream remains unfulfilled. *World Development*, *18*, 1471–80.

Waast, R. (2002). *The State of Science in Africa: an Overview*. Paris: L'Institut de Recherches pour le Développement.

Waast, R. (2009). Science and technology policy in Africa. In R. Arvanitis (ed.), *Science and Technology Policy: Encyclopaedia of Life Support Systems* (Vol. 2, pp. 65–81). Oxford: Eolss Publishers.

Waast, R. (2010). Science in West Asia and North Africa: an introduction. *Science, Technology & Society, 15*, 181–6.

Waast, R., & Krishna, V. V. (2003a). The status of science in Africa. *Science, Technology & Society, 8*, 145–52.

Waast, R., & Krishna, V. V. (2003b). Science in Africa: From institutionalisation to scientific free market – What options for development? *Science, Technology & Society, 8*, 153–81.

Waast, R., & Rossi, P.-L. (2010). Scientific production in Arab countries: a bibliometric perspective. *Science, Technology & Society, 15*, 339–70.

Wachira, B., & Martin, I. B. K. (2011). The state of emergency care in the Republic of Kenya. *African Journal of Emergency Medicine, 1*, 160–5.

Wad, A. (1984). Science, Technology and industrialisation in Africa. *Third World Quarterly, 6*, 327–50.

Wagner, C. S., Brahmakulam, I., Jackson, B., Wong, A., & Yoda, T. (2001). *Science and Technology Collaboration: Building Capacity in Developing Countries?* Santa Monica, CA: RAND.

Wakhungu, J. W. (2001). Science, technology, and public policy in Africa: a framework for action. *Bulletin of Science, Technology & Society, 21*, 246–52.

Walker, C., & Chinigò, D. (2018). Disassembling the Square Kilometre Array: astronomy and development in South Africa. *Third World Quarterly, 39*, 1979–97.

Wandiga, S. O. (2000). Science for development. In *UNESCO, World Conference on Science: Science for the Twenty-First Century – a New Commitment* (pp. 260–63). Paris: UNESCO.

Wane, N. N. (2014). *Indigenous African Knowledge Production: Food Processing Practices among Kenyan Rural Women.* Toronto: University of Toronto Press.

Wangenge-Ouma, G., Lutomiah, A., & Langa, P. (2015). Academic incentives for knowledge production in Africa: Case studies of Mozambique and Kenya. In N. Cloete, P. Maassen & T. Bailey (eds.), *Knowledge Production and Contradictory Functions in African Higher Education* (pp. 128–47). Cape Town: African Minds.

Webersik, C., & Wilson, C. (2009). Achieving environmental sustainability and growth in Africa: the role of science, technology and innovation. *Sustainable Development, 17*, 400–413.

Weinberg, A. M. (1968a). Criteria for scientific choice. In E. Shils (ed.), *Criteria for Scientific Development: Public Policy and National Goals* (pp. 21–33). Cambridge, MA: The MIT Press.

Weinberg, A. M. (1968b). Criteria for scientific choice II: the two cultures. In E. Shils (ed.), *Criteria for Scientific Development: Public Policy and National Goals* (pp. 80–91). Cambridge, MA: The MIT Press.

Weinberg, B. A. (2011). Developing science: scientific performance and brain drains in the developing world. *Journal of Development Economics, 95*, 95–104.

Wendo, C. (2001). Uganda launches HIV/AIDS treatment and training centre for Africa. *The Lancet, 357,* 1957.

Wendo, C. (2003). Uganda agrees to increase health spending using Global Fund's grant. *The Lancet, 361,* 319.

Whitney, G. (1992). Access to third world science in international scientific and technical bibliographic databases. In R. Arvanitis & J. Gaillard (eds.), *Science Indicators for Developing Countries* (pp. 391–409). Paris: ORSTOM.

Widstrand, C. (1992). *Tanzania: Development of Scientific Research and SAREC's Support 1977–1991.* Stockholm: SIDA-SAREC.

Williams, B. R. (1968). Research and economic growth – What should we expect? In E. Shils (ed.), *Criteria for Scientific Development: Public Policy and National Goals* (pp. 92–106). Cambridge, MA: The MIT Press.

Williams, D. (1984). English speaking West Africa. In M. Crowder (ed.), *The Cambridge History of Africa* (Vol. 8, pp. 331–82). Cambridge: Cambridge University Press.

Wionczek, M. S. (1979). The planning of science and technology in the development process. In V. L. Urquidi (ed.), *Science and Technology in Development Planning: Science, Technology and Global Problems* (pp. 109–16). Oxford: Pergamon Press.

Wiseman, A. W. (2012). The impact of student poverty on science teaching and learning: a cross-national comparison of the South African case. *American Behavioral Scientist, 56,* 941–60.

Wolhuter, B. (2016a). Big problems with Maths: Minister's frank assessment. *The Mercury,* 13 December, p. 3.

Wolhuter, B. (2016b). SA Maths improves to 'low'. *The Mercury,* 30 November, p. 2.

Worboys, M. (1996). British colonial science policy, 1918–1939. In P. Petitjean (ed.), *Les sciences Coloniales: Figures et Institutions* (Vol. 2, pp. 99–111). Paris: L'institfurta Nçaidse Recherchsec Ientifipquoeu Rle Développement En Cooperation.

World Bank. (1981). *Accelerated Development in sub-Saharan Africa: an Agenda for Action.* Washington, DC: World Bank.

World Bank. (2006). *World Development Report 2017. Development and the Next Generation.* Washington, DC: World Bank.

World Bank. (2007). *Knowledge Economy Index (KEI) 2007 Rankings.* Washington, DC: World Bank.

World Bank. (2016). *Ethiopia Public Expenditure Review.* Washington, DC: The International Bank for Reconstruction and Development / The World Bank.

Worthington, E. B. (1938). *Science in Africa: a Review of Scientific Research Relating to Tropical and Southern Africa.* London: Oxford University Press.

Worthington, E. B. (1952). Organization of research in Africa. *The Scientific Monthly, 74,* 39–44.

Xu, J., Coats, L. T., & Davidson, M. L. (2012). Promoting student interest in science: the perspectives of exemplary African American teachers. *American Educational Research Journal, 49,* 124–54.

Xun, Z. K. (2000). Public understanding of science: Essentials and its practice. In *UNESCO, World Conference on Science: Science for the Twenty-First Century – a New Commitment* (pp. 290–2). Paris: UNESCO.

Ynalvez, M., & Shrum, W. (2011). Professional networks, scientific collaboration, and publication productivity in resource-constrained research institutions in a developing country. *Research Policy, 40,* 204–16.

Yongxiang, L. (2000). Science for future generations. In *UNESCO, World Conference on Science: Science for the Twenty-First Century – a New Commitment* (pp. 281–6). Paris: UNESCO.

Yriart, M. F. (2000). Science in the Third World: a paradox of prestige and neglect. In *UNESCO, World Conference on Science: Science for the Twenty-First Century – a New Commitment* (pp. 295–7). Paris: UNESCO.

Zachariah, M., & Sooryamoorthy, R. (1994). *Science for Social Revolution? Achievements and Dilemmas of a Development Movement-the Kerala Sastra Sahitya Parishad.* London: Zed Books.

Zahlan, A. B. (1997). Scientific communities in Egypt: Emergence and effectiveness. In J. Gaillard, V. V. Krishna & R. Waast (eds.), *Scientific Communities in the Developing World* (pp. 81–104). New Delhi: Sage Publications.

Zegeye, A., & Vambe, M. (2006). Knowledge production and publishing in Africa. *Development Southern Africa, 23,* 333–49.

Ziman, J. (1971). Three patterns of research in developing countries. *Minerva, 9,* 32–7.

Ziman, J. (1978). *Reliable Knowledge: an Exploration of the Grounds for Belief in Science.* Cambridge: Cambridge University Press.

Ziman, J. (1984). *An Introduction to Science Studies: the Philosophical and Social Aspects of Science and Technology.* Cambridge: Cambridge University Press.

Ziman, J. (2000a). *Real Science: What It Is, and What It Means.* Cambridge: Cambridge University Press.

Ziman, J. (2000b). The Republic of Science: Its political and economic theory. *Minerva: a Review of Science, Learning & Policy, 38,* 21–5.

Zoller, U. (2013). Science, Technology, Environment, Society (STES) literacy for sustainability: What should it take in Chem/Science education? *Educación Química, 24,* 207–14.

Zymelman, M. (1990). *Science, Education, and Development in Sub-Saharan Africa.* Washington, DC: The International Bank for Reconstruction and Development/The World Bank.

Index